Estates Gazette Law Reports

Cumulative Index 1995-2004

**Supplement to
Estates Gazette Law Reports 2004, Vol 3**

EG BOOKS

Routledge
Taylor & Francis Group
LONDON AND NEW YORK

EG Books is an imprint of Routledge
2 Park Square, Milton Park, Abingdon, Oxon OX14 4RN
711 Third Avenue, New York, NY 10017, USA
Routledge is an imprint of the Taylor & Francis Group, an informa business

ISBN for this volume 0 7282 0437 1

For information on all EG Books publications
visit our website www.routledge.com

TABLE OF CASES

Titles of cases shown in bold type
Names of parties reversed shown in ordinary type

References are to year, volume number and page,
eg **[1998]** 1 **205** refers to vol 1 of 1998 at p205

A, X v, [2000] 1 19
Abbahall Ltd v Smee, (CA), [2003] 2 66
Abbathure Flat Management Co Ltd, Maryland Estates Ltd v, [1999] 1 100
Abbey National Mortgages plc v Key Surveyors Nationwide Ltd, (Off Ref), [1995] 2 134; (CA), [1996] 2 99
Aberdeen Steak Houses Group Ltd v Crown Estate Commissioners, (Ch), [1997] 2 107
Able UK Ltd, *ex parte*. R v Northumbrian Water Ltd, *ex parte* Able UK Ltd (QB), [1996] 2 15
Academic Agency Ltd, Bacchiocchi v, [1998] 3 157
Active Estates Ltd v Parness, (Ch), [2002] 3 13
Adagio Properties Ltd v Ansari, (CA), [1998] 2 69
Adami v Lincoln Grange Management Ltd, (CA), [1998] 1 58
Adams, Batt v, [2001] 2 92
Adams v Rhymney Valley District Council, (CA), [2000] 3 25
Adams, Sevington Properties Ltd v, [1998] 2 57
Addin v Secretary of State for the Environment, (Ch), [1997] 1 99
Addison, Northern Electric plc v, [1997] 2 111
Adur District Council, International Traders Ferry Ltd v, [2004] 2 89
Agricultural Mortgage Corporation plc v Woodward, (CA), [1995] 1 1
Ahmed, National Provincial Building Society v, [1995] 2 127
Ainsdale Investments Ltd v First Secretary of State, (Admin), [2004] 2 9
Akhtar, Hazel v, [2002] 1 45
Al Tamimi, Channel Hotels & Properties Ltd v, [2004] 1 16
Alan Wibberley Building Ltd v Insley, (CA), [1998] 2 107; (HL), [1999] 2 89
Aldavon Co Ltd v Deverill, [1999] 2 69
Alexander v Royal Hotel (Caithness) Ltd, (Ct Sess: Inner House), [2001] 1 6
Ali, Taj v, [2000] 3 35
Allied Domecq Retailing Ltd v Williams, (VO), (LT), [2000] 2 171
Allied Dunbar Assurance plc v Homebase, (CA), [2002] 2 23
Allied Irish Bank plc, Ropaigealach v, [2002] 1 83
Allotey, Thamesmead Town Ltd v, [1998] 3 97
Alltime Securities Ltd, Doorbar v, [1996] 2 33
Amber Pass Ltd, John Laing Construction Ltd v, [2004] 2 128
Ambrose v Kay, (CA) [2002] 1 49
Amec Developments Ltd v Jury's Hotel Management (UK) Ltd, (Ch), [2001] 1 81
AMEC Properties Ltd, Lotteryking Ltd v, [1995] 2 13
American Express Ltd, Reed Personnel Services plc v, [1997] 1 229
Amplight Properties Ltd, Footwear Corporation Ltd v, [1998] 2 38
Amsprop Trading Ltd v Harris Distribution Ltd, (Ch), [1997] 2 78
Anglian Water Services Ltd, Kettering Borough Council v, [2001] 2 157
Anker, Childers (JW) Trustees v, [1996] 1 1
Ansari, Adagio Properties Ltd v, [1998] 2 69
Anston Properties Ltd, Benjamin (VO) v, [1998] 2 147
Antonelli v Secretary of State for Trade and Industry, (QB), [1996] 1 229; (CA), [1998] 1 9
Apostole, Luminar Leisure Ltd v, [2001] 3 23
ARC Ltd, Hillingdon London Borough Council v, [1997] 2 21; [1998] 3 18
ARC Ltd (No 2), Hillingdon London Borough Council v, [2000] 3 97
ARC Ltd, Hillingdon London Borough Council v, (No 2), [1999] 3 125
Aran Catering Ltd v Stepien Lake Gilbert & Paling, (Ch), [2002] 1 69
Argyll Stores (Holdings) Ltd, Co-operative Insurance Society Ltd v, [1996] 1 71; (HL), [1997] 1 52
Armgrade Ltd, Parkside Clubs (Nottingham) Ltd v, [1995] 2 96
Armitage v South Oxfordshire District Council, (QB), [1998] 2 121
Armitage & Guest, Brocklesby v, [2001] 1 67
Artesian Residential Investments Ltd v Beck, (CA), [1999] 2 30
Ashurst (trustee in bankruptcy), Pollard v, [2000] 2 29
Ashworth Frazer Ltd v Gloucester City Council, (CA), [1997] 1 104
Ashworth Frazer Ltd v Gloucester City Council, (CA), [2000] 1 44
Ashworth Frazer Ltd v Gloucester City Council (No 2), (HL) [2002] 1 15
Associated Newspapers Property Ltd v Master and Wardens and Brethren and Sisters of the Guild or Fraternity of the Blessed Mary the Virgin of the Mystery of Drapers in the City of London, (CA), [1997] 2 88
Aston Cantlow Parochial Church Council v Wallbank, (Ch), [2000] 2 149
Attenborough (Barbara) Associates, Hindcastle Ltd v, [1996] 1 94
Attwell, Wandsworth London Borough Council v, [1996] 1 57
Attwood v Bovis Homes Ltd, (Ch), [2000] 3 139
Aubergine Enterprises Ltd v Lakewood International Ltd, (Ch), [2001] 3 71
Austin, Windsor Life Assurance Co Ltd v, [1996] 2 169
Austin Gray (a firm), Raja v, [2002] 3 61
Austin Gray (a firm), Raja (administratrix of the estate of Raja (deceased)) v, [2003] 1 91
Australian Mutual Provident Society, Melanesian Mission Trust Board v, [1997] 2 128

B&Q plc, GS Fashions Ltd v, [1995] 1 62
B&Q plc v Liverpool & Lancashire Properties Ltd, (Ch), [2001] 1 92
BHP Petroleum Ltd, Beegas Nominees Ltd v, [1998] 2 57
BHP Petroleum Great Britain Ltd v Chesterfield Properties Ltd, (Ch), [2001] 2 11
BHP Petroleum Great Britain Ltd v Chesterfield Properties Ltd, (CA), [2002] 1 121
BNP Mortgages Ltd v Barton Cook & Sams, (Off Ref), [1996] 1 239
BNP Mortgages Ltd v Key Surveyors Nationwide Ltd, (CA), [1995] 1 129
BP Exploration Operating Co Ltd, Ottewell (VO) v, [1996] 1 203
BP Oil UK Ltd v Kent County Council, (CA), [2003] 3 1
BRS Northern Ltd v Templeheights Ltd, (Ch), [1998] 2 182
BSE Trading Ltd v Hands, (CA), [1996] 2 214
BSE Trading Ltd, McDougalls Catering Foods Ltd v, [1997] 2 65
Babb, Merrett v, [2001] 1 145
Bacchiocchi v Academic Agency Ltd, (CA), [1998] 3 157
Badcock, Walker v, [1997] 2 163
Baines, Church Commissioners for England v, [1998] 1 73
Bairstow, Christopher Moran Holdings Ltd v, [1999] 1 1
Bairstow, Moran (Christopher) Holdings Ltd v, [1996] 2 49; (CA), [1997] 2 36
Bairstow Eves London Central Ltd v Smith, (QB), [2004] 2 25
Baker, Sanctuary Housing Association v, [1998] 1 42
Bakewell Management Ltd, Brandwood v, [2003] 1 17
Bakewell Management Ltd v Brandwood, (HL), [2004] 2 15
Ballard (Kent) Ltd v Oliver Ashworth (Holdings) Ltd, (Ch), [1998] 3 60; (CA), [1999] 2 23
Ballard, Linkproud Ltd v, [1997] 1 181

Bank of Ireland Home Mortgages v **South Lodge Developments, (Ch), [1996]** 1 **91**
Banks v **Kokkinos, (Ch), [1999]** 3 **133**
Banks, Paragon Finance plc v, **[2003]** 1 **85**
Bankway Properties Ltd v **Penfold-Dunsford, (CA), [2001]** 2 **36**
Bannocks v **Secretary of State for Transport, (LT), [1995]** 2 **157**
Banque Bruxelles Lambert SA v **Eagle Star Insurance Co Ltd, (CA), [1995]** 1 **129**
Bar Joseph, Maryland Estates Ltd v, **[1998]** 2 **47**
Barclays Bank plc v **Bean, (Ch), [2004]** 3 **71**
Barclays Bank plc v **Bee, (CA), [2001]** 3 **41**
Barclays Bank plc v **Boulter, (HL), [1999]** 3 **88**
Barclays Bank plc v **Daejan Investments (Grove Hall) Ltd, (Ch), [1995]** 1 **68**
Barclays Bank plc, Harries v, **[1997]** 2 **15**
Barclays Bank plc v **Kent County Council, (CA), [1998]** 2 **14**
Barclays Bank plc v **Layton Lougher & Co, (CA), [1998]** 3 **103**
Barclays Bank plc v **NE Hopkin John & Co, (CA), [1998]** 3 **103**
Barclays Bank plc v **Prudential Assurance Co Ltd, (Ch), [1998]** 1 **44**
Barclays Bank plc v **Savile Estates Ltd, (CA), [2002]** 1 **16**
Barclays Bank plc v **Weeks Legg & Dean, (CA), [1998]** 3 **103**
Barking and Dagenham London Borough Council, Chapman v, **[1997]** 2 **141**
Barnes, Severn Trent Water Ltd v, **[2004]** 2 **95**
Barnett v **O'Sullivan, (CA), [1995]** 1 **93**
Barratt (C Brian) & Son Ltd, Wallace v, **[1997]** 2 **1**
Barratt Homes Ltd, Lewin v, **[2000]** 1 **77**
Barrett v **Mole Valley District Council, (LT), [1997]** 1 **167**
Barrett v **Morgan, (HL), [2000]** 1 **8**
Barrett v **Morgan (Ch), [1997]** 1 **1; (CA), [1998]** 3 **3**
Barrett v **Powell (Ch), [1998]** 2 **117**
Barrington Court Developments Ltd v **Barrington Court Residents' Association, (LT), [2001]** 2 **99**
Barrington Court Residents' Association, Barrington Court Developments Ltd v, **[2001]** 2 **99**
Barry v **Heathcote Ball & Co (Commercial Auctions) Ltd, (CA), [2000]** 3 **7**
Bartholomew, Metropolitan Properties Co (Regis) Ltd v, **[1995]** 1 **65; (CA), [1996]** 1 **79**
Bartlett, Bristol & West plc v, **[2003]** 1 **85**
Barton Cook & Sams, BNP Mortgages Ltd v, **[1996]** 1 **239**
Basildon District Council, Tomkins v, **[2002]** 3 **33**
Basingstoke and Deane Borough Council v **Paice, (CA), [1995]** 2 **9**
Bass Holdings Ltd, Bisichi Mining Ltd v, **[2002]** 2 **4**
Bass Holdings Ltd v **Brodie, (QB), [1998]** 1 **51**
Batchelor v **Marlow, (Ch), [2001]** 1 **119**
Bateman v **Lancashire County Council, (LT), [1999]** 2 **203**
Batt v **Adams, (Ch), [2001]** 2 **92**
Battersea Freehold & Leasehold Co Ltd v **Wandsworth London Borough Council, (Ch), [2001]** 2 **75**
Batty v **Burfoot (QB), [1995]** 2 **142**
Batty v **Merriman (QB), [1995]** 2 **142**
Baxall Securities Ltd v **Sheard Walshaw Partnership, (CA), [2002]** 1 **73**
Baxter v **Camden London Borough Council, (CA), [1998]** 2 **29; (HL), [1999]** 3 **35**
Baygreen Properties Ltd v **Gil, (CA), [2002]** 3 **42**
Baylis (Gloucester) Ltd and another, Secretary of State for the Environment, Transport and the Regions v, **[2000]** 2 **13**
Bean, Barclays Bank plc v, **[2004]** 3 **71**
Beanby Estates Ltd v **Egg Stores (Stamford Hill) Ltd, (Ch), [2003] 3 85**
Beck, Artesian Residential Investments Ltd v, **[1999]** 2 **30**
Becker Properties Ltd v **Garden Court NW8 Property Co Ltd, (LT), [1998]** 1 **121**
Bee, Barclays Bank plc v, **[2001]** 3 **41**
Beegas Nominees Ltd v **BHP Petroleum Ltd, (CA), [1998]** 2 **57**
Beegas Nominees Ltd v **Decco Ltd, (Ch), [2003]** 3 **25**
Begg, WX Investments Ltd v, **[2002]** 3 **47**
Bell v **General Accident Fire & Life Assurance Corporation Ltd, (CA), [1998]** 1 **69**
Bell, de Rothschild v, **[1999]** 1 **35**
Belle Vue Gardens Ltd, Raymere Ltd v, **[2003]** 3 **40**
Bellhurst Ltd, Wellcome Trust Ltd v, **[2002]** 2 **57**
Bellshore Property Investments Ltd, High Street Investments Ltd v, **[1996]** 2 **40**
Bellshore Property Investments Ltd, Stapel v, **[2001]** 2 **7**
Belvedere Court Management Ltd v **Frogmore Developments Ltd, (CA), [1996]** 1 **59**
Benhams Ltd, *ex parte*, R v Director General of Fair Trading, *ex parte* Steed **[2001]** 1 **21**
Benjamin (VO) v **Anston Properties Ltd, (LT), [1998]** 2 **147**
Beresford (on the application of) R, City of Sunderland v, **[2004]** 1 **94**
Berkowitz v **MW (St John's Wood) Ltd, (CA), [1995]** 1 **29**
Berkshire Capital Funding Ltd v **Street, (CA), [1999]** 2 **92**
Berry v **Newport Borough Council, (CA), [2000]** 2 **26**
Berrycroft Management Co Ltd v **Sinclair Gardens Investments (Kensington) Ltd, (CA), [1997]** 1 **47**
Berthon Boat Co Ltd v **Hood Sailmakers Ltd, (QB), [2000]** 1 **39**
Bestley v **North West Water Ltd, (LT), [1998]** 1 **187**
Bevan Ashford (a firm), Portman Building Society v, **[2000]** 1 **81**
Bhogal v **Cheema (Ch), [1998]** 2 **50**
Bickenhall Engineering Co Ltd v **Grandmet Restaurants Ltd, (CA), [1995]** 1 **110**
Bickers, Tandridge District Council v, **[1998]** 3 **31**
Bircham & Co Nominees (No 2) Ltd, Tiffany Investments Ltd v, **[2004]** 2 **31**
Bircham & Co Nominees (No 2) Ltd v **Worrell Holdings Ltd, (CA), [2001]** 3 **83**
Bird, Routestone Ltd v, **[1997]** 1 **123**
Birmingham Midshires Building Society v **Wretham, (QB:TCC), [1999]** 1 **69**
Birrane, Westminster (Duke of) v, **[1995]** 1 **89**
Bishopsgate Foundation, The v **Curtis, (CC), [2004]** 3 **57**
Bishopsgate Space Management Ltd v **London Underground Ltd, (LT), [2004]** 2 **175**
Bisichi Mining Ltd v **Bass Holdings Ltd, (Ch), [2002]** 2 **4**
Bistern Estate Trust's Appeal, (LT), [2000] 2 **91**
Black v **Eton College, (LVT), [1995]** 1 **223**
Blacker v **Wimbledon and Putney Commons Conservators, (CC), [2004]** 3 **67**
Blackstone Investments Ltd v **Middleton-Dell Management Co Ltd, (LT), [1997]** 1 **185**
Blaenau Gwent County Borough Council (No 2), Williams v, **[1999]** 2 **195**
Blake, Medforth v, **[1999]** 2 **75**
Bland v **Ingrams Estates Ltd, (CA), [2001]** 2 **23**
Bland v **Ingrams Estates Ltd, (Ch), [1999]** 2 **49**
Bland v **Ingrams Estates Ltd (No 2), (CA), [2001]** 3 **34**
Blaquiere, Mutual Place Property Management Ltd v, **[1996]** 2 **78**
Blaquiere, Maunder Taylor v, **[2003]** 1 **52**
Blaquire, Maunder Taylor v, **[2002]** 2 **35**
Blendcrown Ltd v **Church Commissioners for England, (LT), [2004]** 1 **28**
Blockbuster Entertainment Ltd v **Leakcliff Properties Ltd, (Ch), [1997]** 1 **28**
Bluestorm Ltd v **Portvale Holdings Ltd, (CA), [2004]** 2 **38**
Blunden v **Frogmore Investments Ltd, (CA), [2002]** 2 **29**
Bocholt Developments Ltd, Salford Van Hire (Contracts) Ltd v, **[1995]** 2 **50**
Bodentien-Meyrick, Yeoman's Row Management Ltd v, **[2002]** 2 **39**
Bolton v **North Dorset District Council, (LT), [1997]** 2 **180**
Bolton v **Southern Electric plc, (LT), [1999]** 1 **177**
Booker, Cheltenham & Gloucester plc v, **[1997]** 1 **142**
Books etc Ltd, Sight & Sound Education Ltd v, **[1999]** 3 **45**
Boots the Chemist, Postel Properties Ltd v, **[1996]** 2 **60**
Boulden, Massey v, **[2003]** 1 **24**
Boulter, Barclays Bank plc v, **[1999]** 3 **88**
Bourke, Wase v, **[1996]** 1 **164**
Bovis Homes Ltd, Attwood v, **[2000]** 3 **139**
Bowerman & Partners, Mortgage Express Ltd v, **[1996]** 1 **126**
Bowring Services Ltd v **Scottish Widow's Fund & Life Assurance Society, (Ch), [1995]** 1 **158**
Boyle v **Verrall, (CA), [1997]** 1 **25**
Brackley, Notting Hill Housing Trust v, **[2001]** 3 **11**
Bradford Metropolitan District Council, Dunn v, **[2002]** 3 **104**
Brandwood v **Bakewell Management Ltd, (CA), [2003]** 1 **17**
Brandwood, Bakewell Management Ltd v, **[2004]** 2 **15**
Braintree District Council, R v, *ex parte* Halls, **[2000]** 3 **19**
Brazier, North Devon Homes Ltd v, **[2003]** 2 **14**
Brent London Borough Council, Burrows v, **[1997]** 1 **32**
Bridgegrove Ltd v **Smith, (CA), [1997]** 2 **40**
Briggs v **McCusker, (Ch), [1996]** 2 **197**
Brighton & Hove City Council v **Collinson, (CA), [2004]** 2 **65**
Bristol City Council, Burroughs Day v, **[1996]** 1 **167**
Bristol City Council, *ex parte* Everett, R v, **[1998]** 3 **25**

Bristol & West plc v **Bartlett, (CA), [2003]** 1 **85**
Britannia LAS Direct Ltd, Lemmerbell Ltd v, **[1998]** 3 **67**
British Glass Manufacturers' Confederation v **University of Sheffield, (Ch), [2004]** 1 **40**
British Home Stores (Jersey) Ltd, Epoch Properties Ltd v, **[2004]** 3 **34**
British Railways Board v **Ringbest Ltd, (Ch), [1996]** 2 **82**
British Railways Board, Friends Provident Life Office v, **[1995]** 2 **55**
British Telecom plc v **Sun Life Assurance Society plc, (CA), [1995]** 2 **44**
British Waterways Board v **London Power Networks plc, (Ch), [2003]** 1 **111**
Brixton plc, Procter & Gamble Technical Centres Ltd v, **[2003]** 2 **24**
Broadgate Square plc v **Lehman Brothers Ltd, (CA), [1995]** 1 **97**
Broadgate Square plc v **Lehman Brothers Ltd (No 2), (Ch), [1995]** 2 **5**
Broadwalk House Residents Ltd, Iperion Investments Corporation v, **[1995]** 2 **47**
Brocklesby v **Armitage & Guest, (CA), [2001]** 1 **67**
Brodie, Bass Holdings Ltd v, **[1998]** 1 **51**
Bromley London Borough Council, Dayani v, **[1999]** 3 **144**
Broomhead's Application, (LT), [2003] 2 **157**
Brown & Root Technology Ltd v **Sun Alliance & London Assurance Co Ltd, (CA), [1997]** 1 **39**
Brown v **City of London Corporation, (Ch), [1996]** 1 **139**
Brown v **Gloucester City Council, (CA), [1998]** 1 **95**
Bruton v **London & Quadrant Housing Trust, (CA), [1997]** 2 **91**
Bruton v **London & Quadrant Housing Trust, (HL), [1999]** 2 **59**
Bryan v **United Kingdom, (ECHR), [1996]** 2 **123**
Buckinghamshire County Council, R (on the application of Laing Homes Ltd) v, **[2003]** 3 **69**
Buckle v **Holderness Borough Council, (LT), [1996]** 2 **133**
Bullard, Kinch v, **[1998]** 3 **112**
Bullock & Lees, DNB Mortgages v, **[2000]** 1 **92**
Burfoot, Batty v, **[1995]** 2 **142**
Burford Estates & Property Co Ltd, Jarrett v, **[1999]** 1 **181**
Burford Midland Properties Ltd v **Marley Extrusions Ltd, (Ch), [1995]** 2 **15**
Burman v **Mount Cook Land Ltd, (CA), [2002]** 1 **61**
Burman v **Mount Cook Land Ltd, (CC), [2001]** 1 **62**
Burr v **Eyre (CC), [1998]** 2 **92**
Burrells Wharf Freeholds Ltd v **Galliard Homes Ltd,** (TCC), **[1999]** 2 **81**
Burroughs Day v **Bristol City Council, (QB), [1996]** 1 **167**
Burrows v **Brent London Borough Council, (HL), [1997]** 1 **32**
Burton v **Camden London Borough Council, (HL), [2000]** 1 **49**
Butcher Robinson & Staples Ltd v **London Regional Transport, (QB: TCC), [1999]** 3 **63**
Butterwick, McLeod v, **[1998]** 3 **147**
Bybrook Barn Garden Centre Ltd v **Kent County Council, (QB), [2000]** 2 **63**
Byland Close (Winchmore Hill) Ltd, Verkan & Co Ltd v, **[1998]** 2 **139**
Byrne v **Hall Pain & Foster, (CA), [1999]** 1 **73**
Byrne v **Inntrepreneur Beer Supply Co Ltd, (CA), [1999]** 2 **145**

CA Webber (Transport) Ltd v **Railtrack plc, (CA), [2004]** 1 **49**
CGU Life Assurance Ltd, Wallis Fashion Group Ltd v, **[2000]** 2 **49**
CIN Properties Ltd (Scotland), Dollar Land (Cumbernauld) Ltd v, **[1998]** 3 **79**
CIN Properties Ltd v **Rawlins, (CA), [1995]** 2 **130**
CPL Distribution Ltd, Starmark Enterprises Ltd v, **[2000]** 3 **37**
Cadogan, Durley House v, **[2000]** 1 **60**
Cadogan, Loder Dyer v, **[2001]** 3 **149**
Cadogan Estates Ltd v **McMahon, (HL), [2001]** 1 **47**
Cadogan Estates Ltd v **McMahon, (CA), [1999]** 2 **67**
Cadogan Estates Ltd v **Shahgholi, (LT), [1999]** 1 **189**
Cadogan Estates Ltd, *ex parte*. R v London Rent Assessment Panel, *ex parte* Cadogan Estates Ltd (QB), **[1997]** 2 **134**
Cadogan Estates, Historic Houses Hotels Ltd v, **[1995]** 1 **117**
Cadogan v **McGirk, (CA), [1996]** 2 **75**
Cadogan v **Morris, (CA), [1999]** 1 **59**
Cadogan v **Royal Brompton Hospital National Health Trust, (Ch), [1996]** 2 **115**
Cadogan (Earl) v **Strauss, (CA), [2004]** 2 **69**
Cadogan Estates Ltd, Norfolk Capital Group Ltd v, **[2004]** 2 **50**
Calcott v **JS Bloor (Measham) Ltd, (CA), [1998]** 3 **1**
Calcott, JS Bloor (Measham) Ltd v, **[2002]** 1 **1**
Callaghan, Stanton v, **[1998]** 3 **165**
Camden London Borough Council, Burton v, **[2000]** 1 **49**
Camden London Borough Council v **Gunby, (QB), [1999]** 3 **13**
Camden London Borough Council, Baxter v, **[1998]** 2 **29; [1999]** 3 **35**
Campbell, Sanctuary Housing Association v, **[1999]** 2 **20**
Campbell, Solent Pines Residents Association Ltd v, **[1996]** 1 **213**
Camtel Investments SA, Observatory Hill Ltd v, **[1997]** 1 **140**
Canary Wharf Investments (Three) v **Telegraph Group Ltd, (Ch), [2003]** 3 **31**
Caradon District Council v **Paton, (CA), [2000]** 3 **57**
Cargill plc, Shortlands Investments Ltd v, **[1995]** 1 **51**
Carl v **Grosvenor Estate Belgravia, (LT), [2000]** 3 **79**
Carmarthenshire County Council, Faraday v, **[2004]** 2 **5**
Carmel Ltd, Jolley v, **[2000]** 3 **68**
Carmel Ltd, Jolley v, **[2000]** 2 **153**
Carrel v **London Underground Ltd, (LT), [1996]** 1 **179**
Carrick, Lloyds Bank plc v, **[1997]** 1 **160**
Casey, York v, **[1998]** 2 **25**
Casswell, Welby v, **[1995]** 2 **1**
Castlegroom Ltd v **Enoch, (CC), [2003]** 2 54
Castlegroom Ltd v **Enoch (No 2), (LVT), [2003]** 3 **46**
Castles Estate Agents Ltd, Enfield London Borough Council v, **[1996]** 2 **21**
Caton's Application, In re, (LT), [1999] 3 **121**
Cavendish Funding Ltd v **Spencer (Henry) & Sons Ltd, (CA), [1998]** 1 **104**
Central Estates Ltd, Heron Maple House Ltd v, **[2002]** 1 **35**
Central Estates Ltd v **Secretary of State for the Environment, (CA), [1997]** 1 **239**
Central London Commercial Estates Ltd v **Kato Kagaku Ltd and another, (Ch), [1998]** 3 **55**
Central Manchester Development Corporation, Sceneout Ltd v, **[1995]** 2 **179**
Chaffe v **Kingsley, (CA), [2000]** 1 **104**
Chairman of the Greater Manchester & Lancashire Rent Assessment Committee, Spath Holme Ltd v, **[1995]** 2 **80**
Chairman of the London Rent Assessment Committee, Curtis v, **[1998]** 1 **79**
Chairman of the London Rent Assessment Committee, Northumberland & Durham Property Trust Ltd v, **[1998]** 3 **85**
Challock Parish Council v **Shirley, (Ch), [1995]** 2 **137**
Chamberlayne, South v, **[2001]** 3 **54**
Chancebutton Ltd v **Compass Services UK & Ireland Ltd, (Ch), [2004]** 2 **47**
Channel Hotels & Properties (UK) Ltd v **Al Tamimi, (Ch), [2004]** 1 **16**
Channel Hotels & Properties (UK) Ltd, First Penthouse Ltd v, **[2004]** 1 **16**
Chapman v **Barking and Dagenham London Borough Council, (QB), [1997]** 2 **141**
Charlegrove Securities Ltd, Gilje v, **[2002]** 1 **41**
Charlegrove Securities Ltd, Gilje v, **[2000]** 3 **89**
Charlegrove Securities Ltd (No 2), Gilje v, **[2003]** 3 **9**
Charlton, Howard v, **[2002]** 3 **65**
Chartered Society of Physiotherapy v **Simmonds Church Smiles, (Off Ref), [1995]** 1 **155**
Chartered Trust plc v **Davies, (CA), [1997]** 2 **83**
Chattey v **Farndale Holdings Inc, (CA), [1997]** 1 **153**
Checkpoint Ltd v **Strathclyde Pension Fund, (CA), [2003]** 1 **1**
Checkpoint Ltd v **Strathclyde Pension Fund, (Ch), [2002]** 2 **97**
Cheema, Bhogal v, **[1998]** 2 **50**
Chelsea Yacht & Boat Co Ltd v **Pope, (CA), [2000]** 2 **23**
Cheltenham & Gloucester plc v **Booker, (CA), [1997]** 1 **142**
Cheltenham Builders Ltd (on the application of) R, South Gloucester District Council v, **[2004]** 1 **85**
Cheong, Veena SA v, **[2003]** 1 **175**
Chester City Council, Crown House Developments Ltd v, **[1997]** 1 **169**
Chesterfield Properties Ltd, BHP Petroleum Great Britain Ltd v, **[2002]** 2 **121**
Chesterfield Properties Ltd, BHP Petroleum Great Britain Ltd v, **[2001]** 2 **155**
Chester-le-Street District Council, Co-operative Wholesale Society Ltd v, **[1996]** 2 **143; [1998]** 3 **11**
Chesterton plc, Gibson Investments Ltd v, **[2003]** 1 **142**
Cheverell Estates Ltd v **Harris, (QB), [1998]** 1 **27**
Cheverell Estates Ltd, Willison v, **[1996]** 1 **116**
Childers (JW) Trustees v **Anker, (CA), [1996]** 1 **1**
Childs, Gilbert v, **[1995]** 2 **142**
Chiswick Quay Freeholds Ltd, Langevad v, **[1999]** 1 **61**
Christopher Moran Holdings Ltd v **Bairstow, (HL), [1999]** 1 **1**
Christos v **Secretary of State for Transport, (CA), [2004]** 1 **5**
Chubb, White v, **[2002]** 1 **9**

Church Commissioners for England, Blendcrown Ltd v, **[2004] 1 143**
Church Commissioners for England v **Baines, (CA), [1998]** 1 **73**
Church Commissioners for England v **Ibrahim, (CA), [1997]** 1 **13**
Church Commissioners for England v **Mummery (VO), (LT), [1998]** 3 **209**
City of Bradford Metropolitan District Council v **Yorkshire Water Services Ltd, (QB: Admin), [2002]** 1 **85**
City of London Corporation, Brown v, **[1996]** 1 **139**
City of London Corporation, Davy's of London (Wine Merchants) Ltd v, **[2004]** 3 **39**
City of London Council, *ex parte* Mystery of the Barbers of London, R v, **[1996]** 2 **128**
City of London Real Property Co Ltd, Paragon Finance plc v, **[2002]** 1 **97**
City of Sunderland, R (on the application of Beresford) v, **[2004]** 1 **94**
City & St James, Guardcliffe Properties Ltd v, **[2003]** 2 **16**
Clark (administrator), Environment Agency v, **[2000]** 1 **113**
Clarke v **Lloyds TSB Bank plc, (QB: TCC) [2002]** 3 **93**
Clements v **Simmonds, (QB), [2002]** 3 **22**
Clickex Ltd v **McCann, (CA), [1999]** 2 **63**
Clinton Cards (Essex) Ltd v **Sun Alliance & London Assurance Co Ltd, (Ch), [2002]** 3 **19**
Clwyd Alyn Housing Association Ltd v **Welsh Office, (LT), [1996]** 1 **184**
Clydebank District Council, Royal Bank of Scotland plc v, **[1995]** 1 **229**
Coal Authority, Hepworth Building Projects v, **[1999]** 3 **99**
Cohen Arnold & Co, HIT Finance Ltd v, **[1998]** 1 **140**
Colherne Court Freehold Ltd, Latifi v, **[2003]** 1 **78**
Collins v **Howard de Walden Estates Ltd, (CA), [2003]** 3 **35**
Collins v **Howard de Walden Estates Ltd, (CC), [2002]** 2 **61**
Collins, Uratemp Ventures Ltd v, **[2001]** 3 **93**
Collins, Uratemp Ventures Ltd v, **[2000]** 1 **156**
Collinson, Brighton & Hove City Council v, **[2004]** 2 **65**
Coltrane, Day v, **[2003]** 2 **21**
Comet Properties Ltd, Romulus Trading Co Ltd v, **[1996]** 2 **70**
Commercial Union Life Assurance Co Ltd v **Moustafa, (QB), [1999]** 2 **44**
Commercial Union Life Assurance Co Ltd v **Woolworths plc, (Ch), [1996]** 1 **237**
Commission for the New Towns v **Cooper (Great Britain) Ltd, (CA), [1995]** 2 **113**
Commission for New Towns, Evis v, **[2002]** 2 **167**
Commissioners of Customs & Excise v **Moorish, (QB), [1998]** 3 **117**
Commissioners of Inland Revenue, Starke v, **[1996]** 1 **157**
Commissioners of Inland Revenue, Walton's Exors v, **[1996]** 1 **159**
Commons Commissioners, R (on the application of Whitmey) v, **[2004]** 3 **1**
Compass Services UK & Ireland Ltd, Chancebutton Ltd v, **[2004]** 2 **47**
Confederation Life Insurance Co (UK) Ltd, H Turner & Son Ltd v, **[2003]** 2 **11**
Conley, Goldstein v, **[1999]** 1 **95**
Connie Investments Ltd, Kened Ltd v, **[1997]** 1 **21**
Cook v **Winchester City Council, (LT), [1995]** 1 **179**
Cooper (Great Britain) Ltd, Commission for the New Towns v, **[1995]** 2 **113**
Co-operative Insurance Society Ltd v **Argyll Stores (Holdings) Ltd, (CA), [1996]** 1 **71; (HL), [1997]** 1 **52**
Co-operative Retail Services, Dandara Holdings Ltd v, **[2004]** 2 **163**
Co-operative Wholesale Society Ltd v **Chester-le-Street District Council, (LT), [1996]** 2 **143; (CA), [1998]** 3 **11**
Co-operative Wholesale Society Ltd v **National Westminster Bank plc, (CA), [1995]** 1 **97**
Cooper-Smith, Escalus Properties Ltd v, **[1995]** 2 **23**
Cork v **Cork, (Ch), [1997]** 1 **5**
Corke, Rother District Investments v, **[2004]** 1 **47**
Country Landowners' Association, *ex parte*. R v Minister of Agriculture, Fisheries and Food, *ex parte* Country Landowners' Association (Eur Ct), **[1996]** 2 **1**
Countryside Residential (North Thames) Ltd v **Tugwell, (CA), [2000]** 2 **59**
Countrywide Surveyors Ltd, Oswald v, **[1996]** 2 **104**
Courage Ltd v **Crehan, (CA), [1999]** 2 **145**
Covent Garden Group Ltd v **Naiva, (CA), [1995]** 1 **243**
Coventry Building Society v **William Martin & Partners, (QB), [1997]** 2 **146**
Craneheath Securities Ltd v **York Montague Ltd, (CA), [1996]** 1 **130**
Craven (Builders) Ltd v **Secretary of State for Health, (Ch), [2000]** 1 **128**
Crean Davidson Investments Ltd v **Earl Cadogan, (CC), [1998]** 2 **96**
Credit Lyonnais SA v **Russell Jones & Walker, (Ch), [2002]** 2 **65**
Crehan, Courage Ltd v, **[1999]** 2 **145**
Crehan v **Inntrepreneur Pub Co (CPC), (CA), [2004]** 3 **128**
Creska Ltd v **Hammersmith and Fulham London Borough Council, (CA), [1998]** 3 **35**
Cressingham Properties Ltd, Re, (LT), [1999] 2 **117**
Crest Nicholson Residential (South) Ltd v **McAllister, (Ch), [2003]** 1 **165**
Crest Nicholson Residential (South) Ltd v **McAllister, (CA), [2004]** 2 **79**
Crewe Services & Investment Corporation v **Silk, (CA), [1998]** 2 **1**
Cricket Ltd v **Shaftesbury plc, (Ch), [1999]** 2 **57**
Cromwell Developments Ltd v **Godfrey, (CA), [1998]** 2 **62**
Crossman Investments, Ganton House Investments v, **[1995]** 1 **239**
Crown Estate Commissioners v **Signet Group plc, (Ch), [1996]** 2 **200**
Crown Estate Commissioners, Aberdeen Steak Houses Group Ltd v, **[1997]** 2 **107**
Crown Estate Commissioners, Dream Factory Ltd v, **[2000]** 3 **107**
Crown Estate Commissioners, Hackney London Borough Council v, **[1996]** 1 **151**
Crown House Developments Ltd v **Chester City Council, (LT), [1997]** 1 **169**
Croydon Unique Ltd v **Wright, (CA), [1999]** 3 **28**
Currys Group plc v **Martin, (QB), [1999]** 3 **165**
Curtis v **Chairman of the London Rent Assessment Committee, (CA), [1998]** 1 **79**
Curtis, The Bishopsgate Foundation v, **[2004]** 3 **57**
Cussins Property Group plc, Minja Properties Ltd v, **[1998]** 2 **52**
Custins, Peacock v, **[2001]** 1 **87**

Daejan Investments (Grove Hall) Ltd, Barclays Bank plc v, **[1995]** 1 **68**
Daejan Properties Ltd, R (on the application of), London Leasehold Valuation Tribunal v, **[2001]** 3 **28**
Daejan Properties Ltd, *ex parte*, R v London Leasehold Valuation Tribunal, **[2000]** 3 **44**
Daejan Properties Ltd v **Mahoney, (CA), [1995]** 2 **75**
Daejan Properties Ltd v **Weeks, (LT), [1998]** 3 **125**
Dairywise Farms Ltd, Swift v, **[2001]** 3 **101**
Dalziel, Speedwell Estates Ltd v, **[2002]** 1 **55**
Dandara Holdings Ltd v **Co-operative Retail Services, (Ch), [2004]** 2 **163**
Dangoor, Iceland Foods plc v, **[2002]** 2 **5**
Das v **Linden Mews Ltd, (CA), [2002]** 2 **76**
David J Pinder plc, UCB Bank plc v, **[1998]** 2 **203**
David Wilson Homes Ltd, Martin v, **[2004]** 3 **77**
Davies' application, In, (LT), [2001] 1 **111**
Davies, Chartered Trust plc v, **[1997]** 2 **83**
Davies v **Eckroyd (H&R) Ltd, (Ch), [1996]** 2 **5**
Davies, Johnson v, **[1997]** 1 **42; (CA), [1998]** 3 **72**
Davina Ltd, Dong Bang Minerva (UK) Ltd v, **[1995]** 1 **41; (CA), [1996]** 2 **31**
Davy's of London (Wine Merchants) Ltd v **City of London Corporation, (Ch), [2004]** 3 **39**
Day v **Coltrane, (CA), [2003]** 2 **21**
Dayani v **Bromley London Borough Council, (QB:TCC), [1999]** 3 **144**
DB Ramsden & Co Ltd, Nurdin & Peacock plc v, **[1999]** 1 **119**
DB Ramsden & Co Ltd (No 2), Nurdin & Peacock plc v, **[1999]** 1 **15**
de Rothschild v **Bell, (CA), [1999]** 1 **35**
Dean and Chapter of the Cathedral and Metropolitan Church of Christ Canterbury v **Whitbread plc, (Ch), [1995]** 1 **82**
Dean, Roadrunner Properties Ltd v, **[2004]** 1 **73**
Debtors no 13A10 and 14A10 of 1994, In re v **(Ch), [1995]** 2 **33**
Decco Ltd, Beegas Nominees Ltd v, **[2003]** 3 **25**
Delaware Mansions Ltd v **Westminster City Council, (CA), [1999]** 3 **68**
Den Danske Bank A/S v **Skipton Building Society, (QB), [1998]** 1 **155**
Dennis, Escalus Properties Ltd v, **[1995]** 2 **23**
Dennis v **Ministry of Defence, (QB), [2003]** 2 **121**
Department of the Environment, Transport and the Regions, Wagstaff v, **[1999]** 2 **108**
Department of Transport, Norman v, **[1996]** 1 **190**
Derby City Council, Jelson Ltd v, **[1999]** 3 **91**
Design Progression Ltd v **Thurloe Properties Ltd, (Ch), [2004]** 1 **121**
Deverill, Aldavon Co Ltd v, **[1999]** 2 **69**
Devonshire Reid Properties Ltd v **Trenaman, (LT), [1997]** 1 **45**
Di Luca v **Juraise (Springs) Ltd (CA), [1998]** 2 **125**
Dickinson v **Enfield London Borough Council, (CA), [1996]** 2 **88**
Diggens' Application, In, (LT), [2000] 3 **87**
Diggens' Application (No 2) Re, (LT), [2001] 2 **163**
Dingleside Development Co Ltd v **Powys County Council, (LT), [1995]** 1 **183**

Director of Buildings and Lands v **Shun Fung Ironworks Ltd, (PC), [1995] 1 19**
Director General of Fair Trading, *ex parte* Benhams Ltd, R v, **[2001]** 1 **21**
DNB Mortgages v **Bullock & Lees, (CA), [2000]** 1 **92**
Docklands Estates Ltd, R v, **[2000]** 3 **17**
Dolgellau Golf Club v **Hett, (CA), [1998]** 2 **75**
Dollar Land (Cumbernauld) Ltd v **CIN Properties Ltd (Scotland), (HL), [1998]** 3 **79**
Donath v **Second Duke of Westminister's Trustees, (LVT), [1997]** 1 **203**
Dong Bang Minerva (UK) Ltd v **Davina Ltd, (Ch), [1995]** 1 **41; (CA), [1996]** 2 **31**
Donohue, Harrow London Borough Council v, **[1995]** 1 **257**
Doorbar v **Alltime Securities Ltd, (CA), [1996]** 2 **33**
Dorset County Council, King v, **[1997]** 1 **245**
Dovecorn Reversions Ltd, Ibrahim v, **[2001]** 2 **46**
Downie, Woodridge Ltd v, **[1997]** 2 **193**
Dowsett, Macklin v, **[2004]** 2 **75**
Dream Factory Ltd v **Crown Estate Commissioners, (Ch), [2000]** 3 **107**
Drew-Morgan v **Hamid-Zadeh, (CA), [1999]** 2 **13**
Drury v **Secretary of State for the Environment, Food and Rural Affairs, (CA), [2004]** 3 **85**
Dudley Magistrates' Court, *ex parte* Hollis, R v, **[1998]** 2 **19**
Dugan-Chapman v **Grosvenor Estates, (CC), [1997]** 1 **96**
Dukeminster (Ebbgate House One) Ltd v **Somerfield Properties Co Ltd, (CA), [1997]** 2 **125**
Dun & Bradstreet Software Services (England) Ltd v **Provident Mutual Life Assurance Association, (CA), [1998]** 2 **175**
Dunn v **Bradford Metropolitan District Council, (CA), [2002]** 3 **104**
Dunraven Ltd, Hemingway Securities Ltd v, **[1995]** 1 **61**
Durley House v **Cadogan, (Ch), [2000]** 1 **60**
Dutton, Manchester Airport plc v, **[1999]** 1 **147**
Dutton Mirrors Ltd, Snell & Prideaux Ltd v, **[1995]** 1 **259**

e.surv Ltd (formerly known as GA Valuation & Survey Ltd), McKinnon v, **[2003]** 2 **57**
Eagle Star Insurance Co Ltd, Banque Bruxelles Lambert SA v, **[1995]** 1 **129**
Eagle Star Life Assurance Co Ltd, Mannai Investment Co Ltd v, **[1996]** 1 **69; (HL), [1997]** 1 **57**
Earl Cadogan, Crean Davidson Investments Ltd v, **[1998]** 2 **96**
Earl Cadogan v **Search Guarantees plc, (CA), [2004]** 3 **55**
Earnshaw v **Hartley, (CA), [1999]** 2 **100**
East Crown Ltd, Inntrepreneur Pub Co v, **[2000]** 3 **31**
Eaton Square Properties Ltd v **Ogilvie, (RAC), [2000]** 1 **73**
Ebied v **Hopkins, (CA), [1998]** 1 **73**
Eckroyd (H&R) Ltd, Davies v, **[1996]** 2 **5**
Edison First Power Ltd (on the application of) R, Secretary of State for the Environment, **[2003]** 2 **133**
Edward Erdman Group Ltd, Nykredit Mortgage Bank plc v, **[1996]** 1 **119**
Edward Erdman Group Ltd, Nykredit Mortgage Bank plc v, **[1995]** 1 **129**
Edward Erdman Group Ltd (No 2), Nykredit Mortgage Bank plc v, **[1998]** 1 **99**
Edward Symmons & Partners, Lloyds TSB Bank plc v, **[2003]** 2 **95**
Edwin Evans, Hamlin v, **[1996]** 2 **106**
Edwin Shirley Productions Ltd v **Workspace Management Ltd, (Ch), [2001]** 2 **16**
Egan Lawson Ltd v **Standard Life Assurance Co, (CA), [2001]** 1 **27**
Egg Stores (Stamford Hill) Ltd, Beanby Estates Ltd v, **[2003]** 3 **85**
86 Lordship Road Ltd, Phyllis Trading Ltd v, **[2001]** 2 **85**
Elcock v **Newham London Borough Council, (LT), [1996]** 2 **137**
Electricity Supply Nominees Ltd v **National Magazine Co Ltd, (CC), [1999]** 1 **130**
Elghanian, Shina v, **[1999]** 3 **103**
Elitestone Ltd v **Morris, (HL), [1997]** 2 **115**
Emden, Kennedy v, **[1997]** 2 **137**
Enfield London Borough Council v **Castles Estate Agents Ltd, (QB:DC), [1996]** 2 **21**
Enfield London Borough Council, Dickinson v, **[1996]** 2 **88**
Enoch, Castlegroom Ltd v, **[2003]** 2 **54**
Enoch (No 2) Ltd, Castlegroom Ltd v, **[2003]** 3 **46**
Environment Agency v **Clark (administrator), (CA), [2000]** 1 **113**
Environment Agency, Matthews v, **[2002]** 3 **168**
Environment Agency, Official Receiver (as liquidator of Celtic Extraction Ltd & Bluestone Chemicals Ltd) v, **[1999]** 3 **21**
Epoch Properties Ltd v **British Home Stores (Jersey) Ltd, (CA: Ch Islands), [2004]** 3 **34**
Eren v **Tranmac Ltd (CA), [1997]** 2 **211**
Erewash Borough Council, Sportoffer Ltd v, **[1999]** 3 **136**
Eronpark Ltd v **Secretary of State for Transport, (LT), [2000]** 2 **165**
Eronpark Ltd v **Secretary of State for the Environment, Transport and the Regions (No 2), (LT), [2001]** 3 **133**
Ervin, Mattey Securities Ltd v, **[1998]** 2 **66**
Escalus Properties Ltd v **Cooper-Smith, (CA), [1995]** 2 **23**
Escalus Properties Ltd v **Dennis, (CA), [1995]** 2 **23**
Escalus Properties Ltd, Moore v, **[1997]** 1 **200**
Escalus Properties Ltd v **Robinson, (CA), [1995]** 2 **23**
Espley v **Williams (CA), [1997]** 1 **9**
Esselte AB v **Pearl Assurance plc, (Ch), [1995]** 2 **61; (CA), [1997]** 1 **73**
Eton College, Black v, **[1995]** 1 **223**
Europe Mortgage Co Ltd v **GA Property Services Ltd, (QB), [1999]** 3 **53**
Evans, James v, **[2000]** 3 **1**
Everett, *ex parte.* R v Bristol City Council, *ex parte* Everett (QB), **[1998]** 3 **25**
Evis v **Commission for New Towns, (LT), [2002]** 2 **167**
Exmos Cover Ltd, Wroe v, **[2000]** 1 **66**
Eyre Estate Trustees v **Shack, (LVT), [1995]** 1 **213**
Eyre, Burr v, **[1998]** 2 **92**
Ezekiel v McDade (CA), [1995] 2 **107**

Factorset Ltd v **Selby District Council, (LT), [1995]** 2 **190**
Faircharm Investments Ltd, Khazanchi v, **[1998]** 3 **147**
Faraday v **Carmarthenshire County Council, (CA), [2004]** 2 **5**
Farley v **Skinner, (HL), [2001]** 3 **57**
Farley v **Skinner, (CA), [2000]** 2 **125**
Farndale Holdings Inc, Chattey v, **[1997]** 1 **153**
Farris (VO), Monsanto plc v, **[1999]** 1 **199**
Fat Cat Café Bars Ltd, Lady Manor Ltd v, **[2001]** 2 **88**
Fennessy v **London City Airport Ltd, (LT), [1995]** 2 **167**
Fernandez, McDonald v, **[2003]** 3 **22**
Ferrishurst Ltd v **Wallcite Ltd, (CA), [1999]** 1 **85**
Fetherstonhaugh, Living Waters Christian Centres Ltd v, **[1999]** 2 **1**
Field v **Leeds City Council, (CA), [2000]** 1 **54**
Field Palmer (a firm), Izzard v, **[2000]** 1 **177**
Filross Securities Ltd v **Midgeley, (CA), [1998]** 3 **43**
First National Commercial Bank plc v **Humberts, (CA), [1995]** 1 **142**
First Penthouse Ltd v **Channel Hotels & Properties (UK) Ltd, (Ch), [2004]** 1 **16**
First Post Homes Ltd v **Johnson, (CA), [1996]** 1 **175**
First Property Growth Partnership LP v **Royal & Sun Alliance Property Services Ltd, (CA), [2003]** 1 **39**
First Property Growth Partnership LP v **Royal & Sun Alliance Property Services Ltd, (Ch), [2002]** 2 **11**
First Secretary of State , Ainsdale Investments Ltd v, **[2004]** 2 **9**
Flairline Properties Ltd v **Hassan, (QB), [1999]** 1 **138**
Fletcher Estates (Harlescott) Ltd, Secretary of State for the Environment v, **[1998]** 3 **13; [2000]** 1 **13**
Fletcher King Services Ltd, Theodore Goddard v, **[1997]** 2 **131**
Floyer-Acland v **Osmond, (CA), [2000]** 2 **1**
Fluor Daniel Properties Ltd v **Shortlands Investments Ltd, (Ch), [2001]** 2 **103**
Footwear Corporation Ltd v **Amplight Properties Ltd, (Ch), [1998]** 2 **38**
Forcelux Ltd v **Sweetman, (LT), [2001]** 2 **173**
Fox & Widley v **Gurham, (QB), [1998]** 1 **91**
Frances Holland School v **Wassef, (CC), [2001]** 2 **88**
Fraser Pipestock Ltd v **Gloucester City Council, (Ch), [1995]** 2 **90**
Freehold Properties Ltd, Swainland Builders Ltd v, **[2002]** 2 **71**
Freeman,, Kasseer v, **[2002]** 1 **9**
Friends Provident Life Office v **British Railways Board, (CA), [1995]** 2 **55**
Frogmore Developments Ltd, Belvedere Court Management Ltd v, **[1996]** 1 **59**
Frogmore Developments Ltd v **Shirayama Shokusan Co Ltd, (Ch), [2000]** 1 **121**
Frogmore Investments Ltd, Blunden v, **[2002]** 2 **29**
Fuller v **Happy Shopper Markets Ltd, (Ch), [2001]** 2 **32**

GA Property Services Ltd, Europe Mortgage Co Ltd v, **[1999]** 3 **53**
GS Fashions Ltd v **B&Q plc, (Ch), [1995]** 1 **62**
Gafford v **Graham, (CA), [1999]** 3 **75**
Galliard Homes Ltd, Burrells Wharf Freeholds Ltd v, **[1999]** 2 **81**
Ganton House Investments v **Crossman Investments, (CC), [1995]** 1 **239**

Garden Court NW8 Property Co Ltd, Becker Properties Ltd v, [1998] 1 121
Gardner v **Marsh & Parsons, (CA), [1997]** 1 **111**
Garner (Inspector of Taxes) v **Pounds Shipowners & Shipbreakers Ltd, (HL), [2000]** 2 **69**
Garston v **Scottish Widows Fund & Life Assurance Society, (CA), [1998]** 2 **73**
Garston v **Scottish Widows Fund & Life Assurance Society, (Ch), [1996]** 1 **113**
Gatwick Parking Services Ltd v **Sargent, (CA), [2000]** 2 **45**
Gemmell, Gibbs Mew plc v, **[1999]** 1 **43**
General Accident Fire & Life Assurance Corporation Ltd, Bell v, **[1998]** 1 **69**
General Accident Fire & Life Assurance Corporation Ltd, Ultraworth Ltd v, **[2000]** 2 **115**
George, John v, **[1995]** 1 **9; (CA), [1996]** 1 **7**
George, Parsons v, **[2004]** 3 **49**
Geotechnical Engineering (Northern) Ltd, Scarborough Borough Council v, **[1997]** 2 **213**
Ghaidan v **Godin-Mendoza, (HL), [2004]** 2 **132**
Gibbs Mew plc v **Gemmell, (CA), [1999]** 1 **43**
Gibson Investments Ltd v **Chesterton plc, (Ch), [2003]** 1 **142**
Gil, Baygreen Properties Ltd v, **[2002]** 3 **42**
Gilbert v **Childs, (QB), [1995]** 2 **142**
Gilje v **Charlegrove Securities Ltd, (CA), [2002]** 1 **41**
Gilje v **Charlegrove Securities Ltd, (LT), [2000]** 3 **89**
Gilje v **Charlegrove Securities Ltd (No 2), (Ch), [2003]** 3 **9**
Gillman, Hair v, **[2000]** 3 **74**
Girls' Day School Trust (1872) Application, In the, (LT), [2002] 2 **89**
Global Grange Ltd, Ivorygrove Ltd v, **[2003]** 2 **87**
Global Grange Ltd, Marazzi v, **[2003]** 2 **42**
Globalgrange Ltd, Willingale v, **[2000]** 2 **55**
Gloucester City Council, Ashworth Frazer Ltd v, **[2000]** 1 **44**
Gloucester City Council, Ashworth Frazer Ltd v, **[1997]** 1 **104**
Gloucester City Council, Brown v, **[1998]** 1 **95**
Gloucester City Council, Fraser Pipestock Ltd v, **[1995]** 2 **90**
Gloucester City Council (No 2), Ashworth Frazer Ltd v, **[2002]** 1 **15**
Glyn Owen & Co, Williams v, **[2004]** 1 **67**
Go West Ltd v **Spigarolo, (CA), [2003]** 1 **133**
Godfrey, Cromwell Developments Ltd v, **[1998]** 2 **62**
Godfrey v **Simm (VO), (LT), [2000]** 3 **85**
Godin-Mendoza, Ghaidan v, **[2004]** 2 **132**
Gojra, Railtrack plc v, **[1998]** 1 **63**
Goldings, Sneesby v, **[1995]** 2 **102**
Goldmile Properties Ltd v **Lechouritis, (CA), [2003]** 1 **60**
Goldstein v **Conley (LT), [1999]** 1 **95**
Gooden v **Northamptonshire County Council, (CA), [2002]** 1 **137**
Gosforth Allotments & Gardens Association, *ex parte.* R v Secretary of State for the Environment, *ex parte* Gosforth Allotments & Gardens Association (QB), **[1996]** 1 **147; (CA), [1996]** 2 **117**
Gotts, Yaxley v, **[1999]** 2 **181**
Gough, Mulvaney v, **[2002]** 3 **72**
Gould & Swayne, Halifax plc v, **[1998]** 3 **177**
Graham, Gafford v, **[1999]** 3 **75**
Graham, JA Pye (Oxford) Ltd v, **[2001]** 2 **69**
Graham, JA Pye (Oxford) Ltd v, **[2000]** 2 **137**
Grammer v **Lane, (CA), [2000]** 1 **1**
Grammer v **Stone, (CA), [2000]** 1 **1**
Grandmet Restaurants Ltd, Bickenhall Engineering Co Ltd v, **[1995]** 1 **110**
Grant, Halifax plc v, **[2003]** 1 **85**
Graysim Holdings Ltd v **P&O Property Holdings Ltd, (HL), [1996]** 1 **109**
Greater London Employer's Association Ltd, Provident Mutual Life Assurance Association v, **[1996]** 1 **106**
Greater Manchester Passenger Transport Executive, Jones v, **[1998]** 3 **121**
Greenalls Management Ltd v **Smith, (CA), [1999]** 2 **145**
Greenbank v **Pickles, (CA), [2001]** 1 **1**
Greenham Trading Ltd, Prudential Nominees Ltd v, **[1995]** 1 **97**
Greenwich London Borough Council, Welsh v, **[2000]** 3 **41**
Greenwood, Skinns v, **[2002]** 2 **53**
Gregory v **Shepherds (a firm), (CA), [2000]** 3 **49**
Gregory v **Turner, (CA), [2003]** 3 **129**
Griffin, Wandsworth London Borough Council v, **[2000]** 2 **105**
Griffiths (VO), Sharp v, **[1999]** 3 **113**
Grossman v **Hooper, (CA), [2001]** 2 **82**
Grosvenor Estate Belgravia, Carl v, **[2000]** 3 **79**
Grosvenor Estate (Mayfair) London's Application, Re v, **(LVT), [1995]** 2 **202**
Grosvenor Estates, Dugan-Chapman v, **[1997]** 1 **96**
Grosvenor West End Properties v **Whiston Properties Ltd, (LT), [2002]** 2 **87**
Groundpremium Property Management Ltd v **Longmint Ltd, (LVT), [1998]** 1 **131**
Guardcliffe Properties Ltd v **City & St James, (Ch), [2003]** 2 **16**
Guinness Ltd, Railtrack plc v, **[2003]** 1 **124**
Gulliksen v **Pembrokeshire County Council, (CA),** [2002] 3 **9**
Gullis, Stevens v, **[1999]** 3 **71**
Gunby, Camden London Borough Council v, **[1999]** 3 **13**
Gunmark Ltd, March Estates plc v, **[1996]** 2 **38**
Gurham, Fox & Widley v, **[1998]** 1 **91**

H Turner & Son Ltd v **Confederation Life Insurance Co (UK) Ltd, (Ch), [2003]** 2 **11**
HSBC Bank plc, Westminster City Council v, **[2003]** 1 **62**
Hackett, Inverugie Investments Ltd v, **[1996]** 1 **149**
Hackney London Borough Council v **Crown Estate Commissioners, (Ch), [1996]** 1 **151**
Hackney London Borough Council, R (on the application of Lemon Land Ltd) v, **[2002]** 1 **81**
Hackney London Borough Council, Shewu v, **[1998]** 2 **232; [1999]** 3 **1**
Hackney London Borough Council, Pankhania v, **[2004]** 1 **135**
Hackney London Borough Council, *ex parte* Structadene Ltd, *R v* **[2001]** 1 **15**
Hair v **Gillman, (CA), [2000]** 3 **74**
Halcrow Ltd, Scarlett v, **[1995]** 1 **219**
Halifax plc v **Gould & Swayne, (CA), [1998]** 3 **177**
Halifax plc v **Grant, (CA), [2003]** 1 **85**
Halifax (SW) Ltd, UCB Bank plc v, **[1996]** 2 **102**
Halifax (SW) Ltd, UCB Corporate Services Ltd v, **[2000]** 1 **87**
Hall Pain & Foster, Byrne v, **[1999]** 1 **73**
Hall, Ravenseft Properties Ltd v, **[2002]** 1 **9**
Hallam Land Management Ltd v **UK Coal Mining, (CA), [2002]** 2 **80**
Hallows v **Welsh Office, (LT), [1995]** 1 **191**
Halls, *ex parte*, R v Braintree District Council, **[2000]** 3 **19**
Halstead v **Manchester City Council, (CA), [1998]** 1 **1**
Hamer, Taylor v, **[2003]** 1 **103**
Hamid-Zadeh, Drew-Morgan v, **[1999]** 2 **13**
Hamlin v **Edwin Evans (CA), [1996]** 2 **106**
Hamlyn Taylor Neck, Portman Building Society v, **[1998]** 2 **113**
Hammad, Wellcome Trust Ltd v, **[1998]** 1 **73**
Hammersmith and Fulham London Borough Council, Creska Ltd v, **[1998]** 3 **35**
Hampshire County Council, Hurst v, **[1997]** 2 **165**
Handoll v **Warner Goodman & Streat, (CA), [1995]** 1 **173**
Hands, BSE Trading Ltd v, **[1996]** 2 **214**
Hannon v **169 Queen's Gate Ltd, (Ch), [2000]** 1 **40**
Happy Shopper Markets Ltd, Fuller v, **[2001]** 2 **32**
Harben Style Ltd v **Rhodes Trust, (Ch), [1995]** 1 **118**
Hargreaves, Zubaida v, **[1995]** 1 **127**
Harley House (Marylebone) Ltd, Michaels v, **[1997]** 2 **44**
Harries v **Barclays Bank plc, (CA), [1997]** 2 **15**
Harris Distribution Ltd, Amsprop Trading Ltd v, **[1997]** 2 **78**
Harris v **Welsh Development Agency, (LT), [1999]** 3 **207**
Harris, Cheverell Estates Ltd v, **[1998]** 1 **27**
Harris, Jervis v, **[1996]** 1 **78**
Harrow London Borough Council v **Donohue, (CA), [1995]** 1 **257**
Harrow London Borough Council v **Qazi, (HL), [2003]** 3 **109**
Harrow London Borough Council, Wildtree Hotels Ltd v, **[2000]** 2 **5**
Harrow London Borough Council, Wildtree Hotels Ltd v, **[1998]** 3 **133**
Harry, Sykes v, **[2001]** 1 **53**
Harry Lester Ltd v **Southwark London Borough Council, (LT), [1999]** 3 **179**
Hartley, Earnshaw v, **[1999]** 2 **100**
Harper, Hayling v, **[2003]** 3 **5**
Hawkesbrook Leisure Ltd v **The Reece-Jones Partnership, (Ch), [2004]** 2 **61**
Hayling v Harper, (CA), [2003] 3 5
Harwood (Michael) T/A RSBS Group v **Smith, (CA), [1998]** 1 **5**
Hassan, Flairline Properties Ltd v, **[1999]** 1 **138**
Hatch & Fielding, Higgins v, **[1996]** 1 **133**
Hatt, Newman (HMIT) v, **[2002]** 1 **89**

Hawkins, Lloyds Bank plc v, **[1998]** 3 **109**
Hazel v **Akhtar, (CA), [2002]** 1 **45**
Heathcote v **Marks (David) & Co, (QB), [1996]** 1 **123**
Heathcote Ball & Co (Commercial Auctions) Ltd, Barry v, **[2000]** 3 **7**
Hedworth, UCB Group Ltd v, **[2002]** 3 **76**
Hellings v **Parker Breslin Estates, (Ch), [1995]** 2 **99**
Hemingway Securities Ltd v **Dunraven Ltd, (Ch), [1995]** 1 **61**
Henry, Lambeth London Borough Council v, **[2000]** 1 **33**
Henry Smith's Charity Trustees, Mainwaring v, **[1996]** 2 **25**
Henry Smith's Charity Trustees (No 2), Mainwaring v, **[1997]** 1 **93**
Hepworth Building Projects v **Coal Authority, (CA), [1999]** 3 **99**
Heron Maple House Ltd v **Central Estates Ltd, (CC), [2002]** 1 **35**
Hett, Dolgellau Golf Club v, **[1998]** 2 **75**
Higgins v **Hatch & Fielding, (CA), [1996]** 1 **133**
High Street Investments Ltd v **Bellshore Property Investments Ltd, (CA), [1996]** 2 **40**
Highland & Universal Properties Ltd v **Safeway Properties Ltd, (Ct Sess: Inner House), [2000]** 3 **110**
Highways Agency, Nicholls v, **[2000]** 2 **81**
Highways Agency, Shraff Tip Ltd v, **[1999]** 2 **205**
Highways Agency, Shraff Tip Ltd v, (No 2), **[1999]** 3 **224**
Hillingdon London Borough Council v **ARC Ltd, (CA), [1998]** 3 **18**
Hillingdon London Borough Council v **ARC Ltd (No 2), (CA), [2000]** 3 **97**
Hillingdon London Borough Council v **ARC Ltd (No 2), (Ch), [1999]** 3 **125**
Hindcastle Ltd v **Attenborough (Barbara) Associates, (HL), [1996]** 1 **94**
Hirshorn, Viscount Chelsea v, **[1998]** 2 **90**
Historic Houses Hotels Ltd v **Cadogan Estates, (CA), [1995]** 1 **117**
HIT Finance Ltd v **Cohen Arnold & Co, (QB), [1998]** 1 **140**
Hoare v **National Trust, (LT), [1997]** 2 **229; (CA), [1999]** 1 **155**
Hodgetts (VO), R v, *ex parte* Nationwide Building Society, **[2000]** 1 **97**
Holbeck Hall Hotel Ltd v **Scarborough Borough Council, (Off Ref), [1997]** 2 **213**
Holderness Borough Council, Buckle v, **[1996]** 2 **133**
Holding & Barnes plc, *ex parte*. R v Secretary of State for the Environment, Transport and the Regions, *ex parte* Holding & Barnes plc, **[2001]** 1 **33**
Hollis, *ex parte* R v Dudley Magistrates' Court, *ex parte* Hollis (QB), **[1998]** 2 **19**
Homebase Ltd, Allied Dunbar Assurance plc v, **[2002]** 2 **23**
Hood Sailmakers Ltd, Berthon Boat Co Ltd v, **[2000]** 1 **39**
Hooper, Grossman v, **[2001]** 2 **82**
Hopkins, Ebied v, **[1998]** 1 **73**
Horden v **Viscount Chelsea, (LVT), [1997]** 1 **195**
Hospital Plan Insurance Services Ltd v **Persaud (VO), (LT), [1998]** 3 **189**
House, Tye v, **[1997]** 2 **171**
Houssein, Savva v, **[1996]** 2 **65**
Howard v **Charlton, (CA), [2002]** 3 **65**
Howard de Walden Estates Ltd, Collins v, **[2003]** 3 **35**
Howard de Walden Estates Ltd, Collins v, **[2002]** 2 **61**
Howard de Walden Estates Ltd, Malekshad v, **[2003]** 1 **151**
Howard de Walden Estates Ltd, Malekshad v, **[2001]** 3 **47**
Howard de Walden Estates Ltd v **Pasta Place Ltd, (Ch), [1995]** 1 **79**
HSBC Bank plc, McDonald's Property Co Ltd v, **[2001]** 3 **19**
Huelin (VO), R v, *ex parte* Murphy Ltd, **[2000]** 1 **97**
Humberts, First National Commercial Bank plc v, **[1995]** 1 **142**
Humberts, Spencer-Ward v, **[1995]** 1 **123**
Hunts Refuse Disposals Ltd v **Norfolk Environmental Waste Services Ltd, (CA), [1997]** 1 **16**
Hurst v **Hampshire County Council, (CA), [1997]** 2 **165**
Hutchinson, Parc Battersea Ltd v, **[1999]** 2 **33**
Hynes v Twinsectra Ltd (CA), **[1995]** 2 **69**

Ibrahim, Church Commissioners for England v, **[1997]** 1 **13**
Ibrahim v **Dovecorn Reversions Ltd, (Ch), [2001]** 2 **46**
Iceland Foods Ltd v **Dangoor, (Ch), [2002]** 2 **5**
Ikarian Reefer
Indiana Investments Ltd v **Taylor, (CC), [2004]** 3 **63**
Ingle v **Scarborough Borough Council, (CA), [2002]** 2 **161**
Ingrams Estates Ltd (No 2), Bland v, **[2001]** 3 **34**
Ingrams Estates Ltd, Bland v, **[2001]** 2 **23**
Ingrams Estates Ltd, Bland v, **[1999]** 2 **49**
Inland Revenue Commissioners, LM Tenancies 1 plc v, **[1996]** 2 **119**
Inntrepreneur Beer Supply Co Ltd, Byrne v, **[1999]** 2 **145**
Inntrepreneur Beer Supply Co Ltd v **Langton, (CA), [1999]** 2 **145**
Inntrepreneuer Estates (CPC) Ltd v **Worth, (Ch), [1996]** 1 **84**
Inntrepreneur Pub Co (CPC), Crehan v, **[2004]** 3 **128**
Inntrepreneur Pub Co v **East Crown Ltd, (Ch), [2000]** 3 **31**
Inntrepreneur Pub Co v **Langton, (Ch), [2000]** 1 **34**
Inntrepreneur Pub Co, Star Rider Ltd v, **[1998]** 1 **53**
Inntrepreneur Pub Co (CPC) v **Sweeney, (CA), [2002]** 2 **132**
Insley, Alan Wibberley Building Ltd v, **[1998]** 2 **107; [1999]** 2 **89**
International Computers Ltd, P&O Property Holdings Ltd v, **[1999]** 2 **13**
International Traders Ferry Ltd v **Adur District Council, (CA), [2004]** 2 **89**
Inverugie Investments Ltd v **Hackett, (PC), [1996]** 1 **149**
Iperion Investments Corporation v **Broadwalk House Residents Ltd, (CA), [1995]** 2 **47**
Iqbal v **Thakrar, (CA), [2004]** 3 **21**
Irving Brown & Daughter v **Smith (VO), (LT), [1996]** 2 **183**
Islington London Borough Council, Rogers v, **[1999]** 3 **17**
Islington London Borough Council, Mean Fiddler Holdings Ltd v, **[2003]** 2 **7**
Islington London Borough Council (No 2), Mean Fiddler Holdings Ltd v, **[2003]** 3 **62**
Ivory Gate Ltd v Spetale (CA), **[1998]** 2 **43**
Ivorygrove Ltd v **Global Grange Ltd, (Ch), [2003]** 2 **87**
Iza Ltd, Target Home Loans Ltd v, **[2000]** 1 **23**
Izzard v **Field Palmer (a firm), (CA), [2000]** 1 **177**

J Murphy & Sons Ltd v **Railtrack plc, (CA), [2002]** 2 **48**
JA Pye (Oxford) Ltd v **Graham, (CA), [2001]** 2 **69**
JA Pye (Oxford) Ltd v **Graham, (Ch), [2000]** 2 **137**
JA Pye (Oxford) Ltd v **Kingswood Borough Council, (CA), [1998]** 2 **159**
JS Bloor (Measham) Ltd v **Calcott, (Ch), [2002]** 1 **1**
JS Bloor (Measham) Ltd, Calcott v, **[1998]** 3 **1**Jackson, King v, **[1998]** 1 **30**
Jaggard v **Sawyer, (CA), [1995]** 1 **146**
James v **Evans, (CA), [2000]** 3 **1**
James Beauchamp (a firm), Nationwide Building Society v, **[2001]** 3 **6**
Jarrett v **Burford Estates & Property Co Ltd, (LT), [1999]** 1 **181**
Jarvis Homes Ltd v **Marshall, (CA), [2004]** 3 **81**
Jeffrey Rogers plc, Petra Investments Ltd v, **[2000]** 3 **120**
Jelson Ltd v **Derby City Council, (QB), [1999]** 3 **91**
Jennings, Royal Bank of Scotland plc v, **[1995]** 2 **87; (CA), [1997]** 1 **101**
Jephson Homes Housing Association v **Moisejevs, (CA), [2001]** 3 **14**
Jerome v **Kelly (HMIT), (CA), [2003]** 2 **73**
Jervis v **Harris (CA), [1996]** 1 **78**
Jewell v **McGowan, (CA), [2002]** 3 **87**
Jillas' application, In, v **(LT), [2000]** 2 **99**
Jobson v **Record, (CA), [1998]** 1 **113**
John v **George, (Ch), [1995]** 1 **9; (CA), [1996]** 1 **7**
John D Wood & Co (Residential & Agricultural) Ltd v **Knatchbull, (QB), [2003]** 1 **33**
John Laing Construction Ltd v **Amber Pass Ltd, (Ch), [2004]** 2 **128**
John Lyon's Charity v **Shalson, (HL), [2003]** 2 **49**
John Lyon's Charity v **Shalson, (CA), [2002]** 2 **55**
Johnsey Estates (1990) Ltd v **Secretary of State for the Environment, (CA), [2001]** 2 **128**
Johnson v **Davies, (CA), [1998]** 3 **72**
Johnson, First Post Homes Ltd v, **[1996]** 1 **175**
Johnson Security Ltd, Orchard Trading Estate Management Ltd v, **[2002]** 2 **1**
Jolley v **Carmel Ltd, (CA), [2000]** 3 **68**
Jolley v **Carmel Ltd, (Ch), [2000]** 2 **153**
Jones v **Greater Manchester Passenger Transport Executive (LT), [1998]** 3 **121**
Jones v **Morgan, (CA), [2002]** 1 **125**
Jones v **Owen (Ch), [1997]** 2 **5**
Jones v **Stones, (CA), [1999]** 3 **81**
Jones, Lower Street Properties Ltd v, **[1996]** 2 **67**
Jones, Twinsectra Ltd v, **[1998]** 2 **129**
Jones, Williams v, **[2002]** 3 **69**
Joseph, Maryland Estates Ltd v, **[1997]** 2 **96**
Juraise (Springs) Ltd, Di Luca v, **[1998]** 2 **125**
Jury's Hotel Management (UK) Ltd, Amec Developments Ltd v, **[2001]** 1 **81**

Kasseer v **Freeman, (CA), [2002]** 1 **9**
Karet, Long Acre Securities Ltd v, **[2004]** 2 **121**
Kataria v **Safeland plc, (CA), [1998]** 1 **39**
Kato Kagaku Ltd and another, Central London Commercial Estates Ltd v, **[1998]** 3 **55**
Kay-Green v **Twinsectra Ltd, (LVT), [1997]** 1 **219**

Kay-Green v **Twinsectra Ltd, (CA), [1996]** 2 **43**
Kaye, Ambrose v, **[2002]** 1 **49**
Keelwalk Properties Ltd v **Waller, (CA), [2002]** 3 **79**
Keepers and Governors of John Lyon Grammar School v **Secchi, (CA), [1999]** 3 **49**
Keepers and Governors of the Possessions Revenues and Goods of the Free v **Mayhew, (CA), [1997]** 1 **88**
Keepers and Governors of the Possessions Revenues and Goods of the Free Grammar School of John Lyon, Vignaud v, **[1996]** 2 **179**
Kelly (HMIT), Jerome v, **[2003]** 2 **73**
Kened Ltd v **Connie Investments Ltd, (CA), [1997]** 1 **21**
Kennedy v **Emden, (CA), [1997]** 2 **137**
Kennet Properties Ltd Application, Re, (LT), [1996] 2 **163**
Kent County Council, BP Oil UK Ltd v, **[2003]** 3 **1**
Kent County Council, Barclays Bank plc v, **[1998]** 2 **14**
Kent County Council, Bybrook Barn Garden Centre Ltd v, **[2000]** 2 **63**
Kent County Council, Smith v, **[1995]** 2 **196**
Kent County Council, Wards Construction (Medway) Ltd v, **[1999]** 1 **111**
Kettering Borough Council v **Anglian Water Services Ltd, (LT), [2001]** 2 **157**
Key Surveyors Nationwide Ltd, Abbey National Mortgages plc v, **[1995]** 2 **134**
Key Surveyors Nationwide Ltd, BNP Mortgages Ltd v, **[1995]** 1 **129**
Keymood Ltd, Lambert v, **[1997]** 2 **70**
Khatun (on the application of) R, Newham London Borough Council v, **[2004]** 1 **34**
Khazanchi v **Faircharm Investments Ltd, (CA), [1998]** 3 **147**
Kibata, Newham London Borough Council v, **[2004]** 1 **56**
Kiley (t/a CK Supermarkets Ltd), Williams v, **[2003]** 1 **46**
Kinch v **Bullard, (Ch), [1998]** 3 **112**
King v **Dorset County Council, (LT), [1997]** 1 **245**
King v **Jackson, (CA), [1998]** 1 **30**
Kingsley, Chaffe v, **[2000]** 1 **104**
Kingston upon Thames Royal London Borough Council v **Marlow, (QB), [1996]** 1 **101**
Kingswood Borough Council, JA Pye (Oxford) Ltd v, **[1998]** 2 **159**
Klein v **London Underground, (LT), [1996]** 1 **249**
Knatchbull, John D Wood & Co (Residential & Agricultural) Ltd v, **[2003]** 1 **33**
Knipe, National Trust for Places of Historic Interest or Natural Beauty v, **[1997]** 2 **9**
Kokkinos, Banks v, **[1999]** 3 **133**
Kreidi, Melbury Road Properties 1995 Ltd v, **[1999]** 3 **108**

LE Jones (Insurance Brokers) Ltd v **Portsmouth City Council, (CA), [2003]** 1 **99**
LM Tenancies 1 plc v **Inland Revenue Commissioners, (Ch), [1996]** 2 **119**
Ladbroke Hotels Ltd v **Sandhu, (Ch), [1995]** 2 **92**
Lady Manor Ltd v **Fat Cat Café Bars Ltd, (CC), [2001]** 2 **88**
Laing Homes (on the application of) R, Buckinghamshire County Council v, **[2003]** 3 **69**
Lakewood International Ltd, Aubergine Enterprises Ltd v, **[2001]** 3 **71**
Lamba Trading Co Ltd v **Salford City Council, (LT), [1999]** 3 **186**
Lambert Smith Hampton Group Ltd, Montlake v, **[2004]** 3 **149**
Lambeth London Borough Council v **Henry, (CA), [2000]** 1 **33**
Lambeth London Borough Council v **Rogers, (CA), [2000]** 1 **28**
Lambeth London Borough Council v **Vincent, (Ch), [2000]** 2 **73**
Lambert & Co, Mortgage Corporation v, **[1999]** 3 **59**
Lambert v **Keymood Ltd, (QB), [1997]** 2 **70**
Lamond, Surrey County Council v, **[1999]** 1 **32**
Lancashire County Council, Bateman v, **[1999]** 2 **203**
Lancaster City Council, Tague v, **[1999]** 2 **103**
Land Securities plc, Pumperninks of Piccadilly Ltd v, **[2002]** 2 **147**
Land Securities plc v **Westminster City Council (No 2), (Ch), [1995]** 1 **245**
Landmaster Properties Ltd v **Thackeray Property Services, (QB), [2003]** 2 **30**
Lands Tribunal, R (on the application of Sinclair Gardens Investments (Kensington) Ltd) v, **[2004]** 3 **15**
Landy, Rodway v, **[2001]** 3 **78**
Lane, Grammer v, **[2000]** 1 **1**
Lane Fox & Partners Ltd, McCullagh v, **[1996]** 1 **35**
Langevad v **Chiswick Quay Freeholds Ltd, (CA), [1999]** 1 **61**
Langton, Inntrepreneur Beer Supply Co Ltd v, **[1999]** 2 **145**
Langton, Inntrepreneur Pub Co v, **[2000]** 1 **34**
Latifi v **Colherne Court Freehold Ltd, (QB), [2003]** 1 **78**
Layton Lougher & Co, Barclays Bank plc v, **[1998]** 3 **103**
Layton Lougher & Co, Sullivan v, **[1995]** 2 **111**
Leach Pension (Trustees) Ltd, Oakstead Garages Ltd v, **[1996]** 1 **26**
Leakcliff Properties Ltd, Blockbuster Entertainment Ltd v, **[1997]** 1 **28**
Lechouritis, Goldmile Properties Ltd v, **[2003]** 1 **60**
Lee v **Leeds City Council, (CA), [2002]** 1 **103**
Leeds City Council, Field v, **[2000]** 1 **54**
Leeds City Council, Lee v, **[2002]** 1 **103**
Leeds City Council, Marston v, **[2002]** 3 **104**
Lehman Brothers Ltd, Broadgate Square plc v, **[1995]** 1 **97**
Lehman Brothers Ltd (No 2), Broadgate Square plc v, **[1995]** 2 **5**
Leicester City Council v **Leicestershire County Council, (LT), [1995]** 2 **169**
Leicester City Council, Ullah v, **[1996]** 1 **244**
Leicestershire County Council, Leicester City Council v, **[1995]** 2 **169**
Lemmerbell Ltd v **Britannia LAS Direct Ltd, (CA), [1998]** 3 **67**
Lemon Land Ltd (on the application of) R, Hackney London Borough Council v, **[2002]** 1 **81**
Lesquende Ltd v **Planning and Environment Committee of the State of Jersey (PC), [1998]** 1 **137**
Lester (R on the application of), London Rent Assessment Committee, v, **[2003]** 1 **58**
Lester, 134 Shirland Road Management Co Ltd v, **[1997]** 2 **207**
Lewin v **Barratt Homes Ltd, (QB:DC), [2000]** 1 **77**
Lewisham Investment Partnership Ltd v **Morgan, (Ch), [1997]** 2 **150**
Lewisham London Borough Council v **Masterson, (CA), [2000]** 1 **134**
Limegate Investment Co Ltd, St Mary's Mansions Ltd v, **[2003]** 1 **41**
Lincoln Grange Management Ltd, Adami v, **[1998]** 1 **58**
Linden Mews Ltd, Das v, **[2002]** 2 **76**
Linkproud Ltd v **Ballard, (LT), [1997]** 1 **181**
Liverpool City Council, Mallick v, **[1999]** 2 **7**
Liverpool City Council v **Walton Group plc, (Ch), [2002]** 1 **149**
Liverpool & Lancashire Properties Ltd, B&Q plc v, **[2001]** 1 **92**
Living Waters Christian Centres Ltd v **Fetherstonhaugh, (CA), [1999]** 2 **1**
Llanelec Precision Engineering Co Ltd v **Neath Port Talbot County Borough Council, (LT), [2000]** 3 **158**
Lloyd Jones v **T-Mobile (UK) Ltd, (CA), [2003]** 3 **55**
Lloyds Bank plc v **Carrick, (CA), [1997]** 1 **160**
Lloyds Bank plc v **Hawkins, (CA), [1998]** 3 **109**
Lloyds Bank plc v **Rogers, (CA), [1999]** 3 **83**
Lloyds TSB Bank plc, Clarke v, **[2002]** 3 **93**
Lloyds TSB Bank plc v **Edward Symmons & Partners, (TCC), [2003]** 2 **95**
Lloyds TSB Bank Ltd, Starling v, **[2000]** 1 **101**
Lloyds TSB Bank plc, Raja v, **[2001]** 2 **78**
Lock (on the application of) R, Secretary of State for the Environment, Transport and the Regions v, **[2002]** 3 **5**
Loder Dyer v **Cadogan, (LT), [2001]** 3 **149**
Lodge v **Wakefield Metropolitan City Council, (CA), [1995]** 2 **124**
Lomax Leisure Ltd, In re, (Ch), [1999] 2 **37**
London & Leeds Estates Ltd v **Paribas Ltd (No 2), (QB), [1995]** 1 **102**
London & Quadrant Housing Trust, Bruton v, **[1999]** 2 **59**
London & Quadrant Housing Trust, Bruton v, **[1997]** 2 **91**
London Baggage Co (Charing Cross) Ltd v **Railtrack plc (No 2), (Ch), [2003]** 1 **141**
London City Airport Ltd, Fennessy v, **[1995]** 2 **167**
London Fire and Civil Defence Authority, Wakeley v, **[1996]** 2 **148**
London Leasehold Valuation Tribunal, R v, *ex parte* Daejan Properties Ltd, **[2000]** 3 **44**
London Leasehold Valuation Tribunal, R (on the application of Daejan Properties Ltd) v, **[2001]** 3 **28**
London Power Networks plc, British Waterways Board v, **[2003]** 1 **111**
London Regional Transport, Butcher Robinson & Staples Ltd v, **[1999]** 3 **63**
London Regional Transport, Ryde International plc v, **[2001]** 1 **101**
London Regional Transport, Ryde International plc v, **[2004]** 2 **1**
London Rent Assessment Committee, R (on the application of Lester) v, **[2003]** 1 **58**
London Rent Assessment Committee, R (on the application of Morris) v, **[2002]** 2 **11**
London Rent Assessment Committee, R (on the application of Morris) v, **[2001]** 2 **51**

London Rent Assessment Committee, Northumberland & Durham Property Trust Ltd v, **[1997]** 1 **236**
London Rent Assessment Committee (No 2), Northumberland & Durham Property Trust Ltd v, **[1998]** 2 **99**
London Rent Assessment Committee, R (on the application of Wolters) v, **[2003]** 3 **17**
London Rent Assessment Panel, *ex parte* Cadogan Estates Ltd, R v, **[1997]** 2 **134**
London Underground Ltd, Bishopsgate Space Management Ltd v, **[2004]** 2 **175**
London Underground, Klein v, **[1996]** 1 **249**
London Underground Ltd, Carrel v, **[1996]** 1 **179**
London Underground Ltd, Platt v, **[2001]** 2 **121**
Long, McMeekin v, **[2003]** 2 **81**
Long v **Southwark London Borough Council, (CA), [2002]** 3 **37**
Long v **Tower Hamlets London Borough Council (Ch), [1997]** 1 **78**
Long Acre Securities Ltd v **Karet, (Ch), [2004]** 2 **121**
Longmint Ltd, Groundpremium Property Management Ltd v, **[1998]** 1 **131**
Longmint Ltd v **Marcus, (LT), [2004]** 3 **171**
Lotteryking Ltd v **AMEC Properties Ltd, (Ch), [1995]** 2 **13**
Louis v **Sadiq (CA), [1997]** 1 **136**
Lovatt, Northampton Borough Council v, **[1998]** 1 **15**
Lower Street Properties Ltd v Jones (CA), **[1996]** 2 **67**
Luminar Leisure Ltd v **Apostole, (Ch), [2001]** 3 **23**
Lynari Properties Ltd, Shortdean Place (Eastbourne) Residents' Association Ltd v, **[2003] 3 147**

M25 Group Ltd, National Grid Co plc v, **[1998]** 2 **85; [1999]** 1 **65**
M25 Group Ltd, Tudor v, **[2004]** 1 **23**
Macepark (Whittlebury) Ltd, Sargeant v, **[2004]** 3 **26**
Macklin v **Dowsett, (CA), [2004]** 2 **75**
Macleod v **National Grid Co plc, (LT), [1998]** 2 **217**
Macob Civil Engineering Ltd v **Morrison Construction Ltd, (QB:TCC), [1999]** 3 **7**
Mahoney, Daejan Properties Ltd v, **[1995]** 2 **75**
Maidstone Borough Council v **Secretary of State for the Environment, (CA), [1996]** 1 **29**
Mainwaring v **Henry Smith's Charity Trustees (No 2), (CA), [1997]** 1 **93**
Mainwaring v **Henry Smith's Charity Trustees, (CA), [1996]** 2 **25**
Malekshad v **Howard de Walden Estates Ltd, (HL), [2003]** 1 **151**
Malekshad v **Howard de Walden Estates Ltd, (CA), [2001]** 3 **47**
Mallick v **Liverpool City Council, (CA), [1999]** 2 **7**
Manchester Airport plc v **Dutton, (CA), [1999]** 1 **147**
Manchester City Council, Halstead v, **[1998]** 1 **1**
Manchester City Council, Wallace v, **[1998]** 3 **38**
Manel v **Memon, (CA), [2000]** 2 **40**
Mannai Investment Co Ltd v **Eagle Star Life Assurance Co Ltd, (CA), [1996]** 1 **69; (HL), [1997]** 1 **57**
Manuel, Wandsworth London Borough Council v, **[2002]** 2 **128**
Manyfield Ltd, 63 Perham Road Ltd v, **[1995]** 2 **206**
March Estates plc v **Gunmark Ltd, (Ch), [1996]** 2 **38**
Marcic v **Thames Water Utilities, (QB:TCC), [2001]** 3 **111**
Marcus, Longmint Ltd v, **[2004]** 3 **171**
Maribella Ltd, Staszewski v, **[1998]** 1 **34**
Marklands v **Virgin Retail Ltd, (Ch), [2004]** 2 **43**
Marks & Spencer plc, Universities Superannuation Scheme Ltd v, **[1999]** 1 **13**
Marks (David) & Co, Heathcote v, **[1996]** 1 **123**
Marley Extrusions Ltd, Burford Midland Properties Ltd v, **[1995]** 2 **15**
Marlow, Batchelor v, **[2001]** 1 **119**
Marlow, Kingston upon Thames Royal London Borough Council v, **[1996]** 1 **101**
Marlton v **Turner, (QB), [1998]** 3 **185**
Marsh & Parsons, Gardner v, **[1997]** 1 **111**
Marsh v **Powys County Council, (LT), [1997]** 2 **177**
Marshall, Jarvis Homes Ltd v, **[2004]** 3 **81**
Marshall v **Rubypoint Ltd, (CA), [1997]** 1 **69**
Marshall, TSB Bank plc v, **[1998]** 3 **100**
Marston v Leeds City Council, (CA), [2002] 3 **104**
Marazzi v **Global Grange Ltd, (Ch), [2003]** 2 **42**
Martin v **David Wilson Homes Ltd, (CA), [2004]** 3 **77**
Martin v **Maryland Estates Ltd, (CA), [1999]** 2 **53**
Martin v **Maryland Estates Ltd, (CA), [1998]** 2 **8**
Martin, Currys Group plc v, **[1999]** 3 **165**
Maryland Estates Ltd v **63 Perham Road Ltd, (LT), [1997]** 2 **198**
Maryland Estates Ltd v **Abbathure Flat Management Co Ltd, (LT), [1999]** 1 **100**
Maryland Estates Ltd v **Bar Joseph, (CA), [1998]** 2 **47**
Maryland Estates Ltd v **Joseph, (CC), [1997]** 2 **96**
Maryland Estates Ltd, Martin v, **[1999]** 2 **53**
Maryland Estates Ltd, Martin v, **[1998]** 2 **8**
Mason v **Totalfinaelf UK Ltd, (Ch), [2003]** 3 **91**
Massey v **Boulden, (CA), [2003]** 1 **24**
Master and Wardens and Brethren and Sisters of the Guild or Fraternity of the Blessed Mary the Virgin of the Mystery of Drapers in the City of London, Associated Newspapers Property Ltd v, **[1997]** 2 **88**
Masterson, Lewisham London Borough Council v, **[2000]** 1 **134**
Matthews v **Environment Agency, (LT), [2002] 3 168**
Mattey Securities Ltd v **Ervin, (CA), [1998]** 2 **66**
Maunder Taylor v **Blaquiere, (CA), [2003]** 1 **52**
Maunder Taylor v **Blaquire, (CC), [2002]** 2 **35**
Max Factor Ltd v **Wesleyan Assurance Society, (Ch), [1995]** 2 **38**
May v **Woollcombe Beer & Watts (QB), [1998]** 3 **94**
Mayhew, Keepers and Governors of the Possessions Revenues and Goods of the Free Grammar School of John Lyon v, **[1997]** 1 **88**
McAdams Homes Ltd v **Robinson, (Ch), [2004]** 3 **93**
McAllister, Crest Nicholson Residential (South) Ltd v, **[2003]** 1 **165**
McAllister, Crest Nicholson Residential (South) Ltd v, **[2004]** 2 **79**
McCann, Clickex Ltd v, **[1999]** 2 **63**
McCaughey, Walker Cain v, **[1999]** 2 **145**
McCullagh v **Lane Fox & Partners Ltd, (CA), [1996]** 1 **35**
McCusker, Briggs v, **[1996]** 2 **197**
McDade, Ezekiel v, **[1995]** 2 **107**
McDonald v **Fernandez, (CA), [2003] 3 22**
McDonald's Property Co Ltd v **HSBC Bank plc, (Ch), [2001]** 3 **19**
McDougalls Catering Foods Ltd v **BSE Trading Ltd, (CA), [1997]** 2 **65**
McGirk, Cadogan v, **[1996]** 2 **75**
McGowan, Jewell v, **[2002]** 3 **87**
McIntosh, Southwark London Borough Council v, **[2002]** 1 **25**
McKinnon v **e.surv Ltd (formerly known as GA Valuation & Survey Ltd), (Ch), [2003]** 2 **57**
McLeod v **Butterwick, (CA), [1998]** 3 **147**
McMahon, Cadogan Estates Ltd v, **[2001]** 1 **47**
McMahon, Cadogan Estates Ltd v, **[1999]** 2 **67**
McMeekin v **Long, (QB), [2003]** 2 **81**
Meadfield Properties Ltd v **Secretary of State for the Environment, (Ch), [1995]** 1 **39**
Mean Fiddler Holdings Ltd v **Islington London Borough Council, (CA), [2003]** 2 **7**
Mean Fiddler Holdings Ltd v **Islington London Borough Council, (No 2), (LT), [2003]** 3 **61**
Medforth v **Blake, (CA), [1999]** 2 **75**
Mehta v **Royal Bank of Scotland, (QB), [1999]** 3 **153**
Melanesian Mission Trust Board v **Australian Mutual Provident Society, (PC), [1997]** 2 **128**
Melbury Road Properties 1995 Ltd v **Kreidi, (CC), [1999]** 3 **108**
Memon, Manel v, **[2000]** 2 **40**
Merivale Moore plc v **Strutt & Parker, (QB), [1998]** 2 **195; (CA), [1999]** 2 **171**
Merrett v **Babb, (CA), [2001]** 1 **145**
Merriman, Batty v, **[1995]** 2 **142**
Metcalfe Hambleton & Co, Sansom v, **[1998]** 2 **103**
Metropolitan Properties Co (Regis) Ltd v **Bartholomew, (QB), [1995]** 1 **65; (CA), [1996]** 1 **79**
Michael v **Miller, (CA), [2004]** 2 **151**
Michaels v **Harley House (Marylebone) Ltd, (Ch), [1997]** 2 **44**
Microgen Holdings plc, Proudreed Ltd v, **[1996]** 1 **89**
Micrografix v **Woking 8 Ltd, (Ch), [1995]** 2 **32**
Microworld Technology Ltd, St Giles Hotel Ltd v, **[1997]** 2 **105**
Middleton Potts & Co, Scottish Amicable Life Assurance Society v, **[1995]** 1 **97**
Middleton-Dell Management Co Ltd, Blackstone Investments Ltd v, **[1997]** 1 **185**
Midgeley, Filross Securities Ltd v, **[1998]** 3 **43**
Midwood v **Morgan, (Ch), [2001]** 3 **127**
Milius' Application, Re (LT), [1996] 1 **209**
Miller, Michael v, **[2004]** 2 **151**
Mills, Southwark London Borough Council v, **[1998]** 3 **46; [1999]** 3 **35**
Milton Gate Investments Ltd, PW & Co v, **[2004]** 3 **103**
Milverton Group Ltd v **Warner World Ltd, (CA), [1995]** 2 **28**

Minister of Agriculture, Fisheries and Food, *ex parte* Country Landowners' Association, R v, **[1996]** 2 **1**
Ministry of Agriculture, Fisheries and Food, *ex parte* St Clere's Hall Farm, R v, **[1996]** 1 **15**
Ministry of Defence, Dennis v, **[2003]** 2 **121**
Minja Properties Ltd v **Cussins Property Group plc, (Ch), [1998]** 2 **52**
Minories Finance Ltd, Routestone Ltd v, **[1997]** 1 **123**
Moisejevs, Jephson Homes Housing Association v, **[2001]** 3 **14**
Mole Valley District Council, Barrett v, **[1997]** 1 **167**
Monella v **PizzaExpress (Restaurants) Ltd, (Ch), [2004]** 1 **43**
Monsanto plc v **Farris (VO), (LT), [1999]** 1 **199**
Montgomery, Sabella Ltd v, **[1998]** 1 **65**
Montlake v **Lambert Smith Hampton Group Ltd, (Comm), [2004]** 3 **149**
Moore v **Escalus Properties Ltd, (LVT), [1997]** 1 **200**
Moorish, Commissioners of Customs & Excise v, **[1998]** 3 **117**
Moran (Christopher) Holdings Ltd v **Bairstow, (Ch), [1996]** 2 **49; (CA), [1997]** 2 **36**
Morgan Sindall plc v **Sawston Farms (Cambs) Ltd, (CA), [1999]** 1 **90**
Morgan, Barrett v, **[2000]** 1 **8**
Morgan, Barrett v, **(Ch), [1997]** 1 **1; (CA), [1998]** 3 **3**
Morgan, Jones v, **[2002]** 1 **125**
Morgan, Lewisham Investment Partnership Ltd v, **[1997]** 2 **150**
Morgan, Midwood v, **[2001]** 3 **127**
Morland plc, Passmore v, **[1999]** 1 **51**
Morrells of Oxford Ltd v **Oxford United Football Club, (CA), [2001]** 1 **76**
Morrells of Oxford Ltd v **Oxford United Football Club, (Ch), [2000]** 3 **70**
Morris, Cadogan v, **[1999]** 1 **59**
Morris, Elitestone Ltd v, **[1997]** 2 **115**
Morris (on the application of) R, London Rent Assessment Committee v, **[2002]** 2 **11**
Morris (on the application of) R, North Somerset District Council v, **[2003]** **3** **129**
Morris, Viscount Chelsea v, **[1997]** 2 **100**
Morrison Construction Ltd, Macob Civil Engineering Ltd v, **[1999]** 3 **7**
Mortgage Corporation v **Lambert & Co, (Ch), [1999]** 3 **59**
Mortgage Corporation v **Shaire, (Ch), [2000]** 3 **131**
Mortgage Express Ltd v **Bowerman & Partners, (CA), [1996]** 1 **126**
Mount Charlotte Investments plc v **Prudential Assurance, (Ch), [1995]** 1 **15**
Mount Cook Land Ltd, Burman v, **[2002]** 1 **61**
Mount Cook Land Ltd, Burman v, **[2001]** 1 **62**
Mount Cook Land Ltd v **Rosen, (CC), [2003]** 1 **75**
Mount Eden Land Ltd, Prudential Assurance Co Ltd v, **[1997]** 1 **37**
Moustafa, Commercial Union Life Assurance Co Ltd v, **[1999]** 2 **44**
Mulvaney v **Gough, (CA), [2002]** 3 **72**
Mummery (VO), Church Commissioners for England v, **[1998]** 3 **209**
Murphy Ltd, *ex parte*, R v Huelin (VO), **[2000]** 1 **97**
Muscat v **Smith, (CA), [2003]** 3 **11**
Mutual Place Property Management Ltd v **Blaquiere, (CC), [1996]** 2 **78**
MW (St John's Wood) Ltd, Berkowitz v, **[1995]** 1 **29**
Mystery of the Barbers of London, *ex parte.* R v City of London Council, *ex parte* Mystery of the Barbers of London (QB), **[1996]** 2 **128**
Mytre Investments Ltd v Reynolds (QB), **[1995]** 2 **40**

Naidu, Yenula Properties Ltd v, **[2002]** 3 **28**
Naiva, Covent Garden Group Ltd v, **[1995]** 1 **243**
National Car Parks Ltd v **Trinity Development Co (Banbury) Ltd, (Ch), [2001]** 2 **43**
National Grid Co plc v **M25 Group Ltd, (Ch), [1998]** 2 **85; (CA), [1999]** 1 **65**
National Grid Co plc, Macleod v, **[1998]** 2 **217**
National Magazine Co Ltd, Electricity Supply Nominees Ltd v, **[1999]** 1 **130**
National Provincial Building Society v **Ahmed, (CA), [1995]** 2 **127**
National Trust for Places of Historic Interest or Natural Beauty v **Knipe, (CA), [1997]** 2 **9**
National Trust, Hoare v, **[1997]** 2 **229; [1999]** 1 **155**
National Trust v **Spratling (LT), [1997]** 2 **229**
National Westminster Bank plc, Co-operative Wholesale Society Ltd v, **[1995]** 1 **97**
Nationwide Building Society v **James Beauchamp (a firm), (CA), [2001]** 3 **6**
Naylor v **Southern Electricity Board, (LT), [1996]** 1 **195**
NE Hopkin John & Co, Barclays Bank plc v, **[1998]** 3 **103**
Neath Port Talbot County Borough Council, Llanelec Precision Engineering Co Ltd v, **[2000]** 3 **158**
Newman (HMIT) v **Hatt, (LT), [2002]** 1 **89**
Newell, Secretary of State for the Environment v, **[1998]** 3 **13; [2000]** 1 **13**
Newell v **Secretary of State for the Environment, (CA), [1998]** 3 **13; (HL), [2000]** 1 **13**
Newham London Borough Council, Elcock v, **[1996]** 2 **137**
Newham London Borough Council v **Kibata [2004]** 1 **56**
Newham London Borough Council, R (on the application of Khatun) v, **[2004]** 1 **34**
Newham London Borough Council, R (on the application of Zeb) v, **[2004]** 1 **34**
Newport Borough Council, Berry v, **[2000]** 2 **26**
Next plc, Ocean Accident & Guarantee Corporation v, **[1996]** 2 **84**
Niazi Services Ltd v **van der Loo, (CA), [2004]** 1 **62**
Nichols' Application, Re, (LT), [1997] 1 **144**
Nicholls v **Highways Agency, (LT), [2000]** 2 **81**
Norfolk Capital Group Ltd v **Cadogan Estates Ltd, (Ch), [2004]** 2 **50**
Norfolk Environmental Waste Services Ltd, Hunts Refuse Disposals Ltd v, **[1997]** 1 **16**
Norman v **Department of Transport, (LT), [1996]** 1 **190**
Northamptonshire County Council, Gooden v, **[2002]** 1 **137**
North British Housing Association Ltd v **Sheridan, (CA), [1999]** 2 **138**
North Devon Homes Ltd v **Brazier, (QB), [2003]** 2 **14**
North Dorset District Council, Bolton v, **[1997]** 2 **180**
North Somerset District Council, R (on the application of Morris) v, **[2003]** 3 **129**
North West Water Ltd, Bestley v, **[1998]** 1 **187**
North West Water Ltd, Taylor v, **[1995]** 1 **266**
North Western Rent Assessment, Spath Holme Ltd v, **[2001]** 3 **30**
Northampton Borough Council v **Lovatt, (CA), [1998]** 1 **15**
Northern Electric plc v **Addison, (CA), [1997]** 2 **111**
Northumberland & Durham Property Trust Ltd v **Chairman of the London Rent Assessment Committee, (QB), [1998]** 3 **85**
Northumberland & Durham Property Trust Ltd v **London Rent Assessment Committee (No 2), (QB), [1998]** 3 **99**
Northumberland & Durham Property Trust Ltd v **London Rent Assessment Committee, (QB), [1997]** 1 **236**
Northumbrian Water Ltd, *ex parte* Able UK Ltd, R v, **[1996]** 2 **15**
Norwich Union Life Insurance Society v **Shopmoor Ltd, (Ch), [1998]** 2 **167**
Notting Hill Housing Trust v **Brackley, (CA), [2001]** 3 **11**
Nurdin & Peacock plc v **DB Ramsden & Co Ltd, (Ch), [1999]** 1 **119**
Nurdin & Peacock plc v **DB Ramsden & Co Ltd (No 2), (Ch), [1999]** 1 **15**
Nykredit Mortgage Bank plc v **Edward Erdman Group Ltd, (CA), [1995]** 1 **129; (HL), [1996]** 2 **93**
Nykredit Mortgage Bank plc v **Edward Erdman Group Ltd (No 2), (HL), [1998]** 1 **99**
Nykredit Mortgage Bank plc v **Edward Erdman Group Ltd, (CA), [1996]** 1 **119**
Nynehead Developments Ltd v **RH Fibreboard Containers Ltd, (Ch), [1999]** 1 **7**

Oakstead Garages Ltd v **Leach Pension (Trustees) Ltd, (Ch), [1996]** 1 **26**
Observatory Hill Ltd v **Camtel Investments SA, (Ch), [1997]** 1 **140**
Ocean Accident & Guarantee Corporation v **Next plc, (Ch), [1996]** 2 **84**
Ocean Leisure Ltd v **Westminster City Council, (CA), [2004]** 3 **9**
Oceanic Village Ltd v **United Attractions Ltd, (Ch), [2000]** 1 **148**
O'Connor v **Old Etonian Housing Association Ltd, (CA), [2002]** 1 **38**
Official Receiver (as liquidator of Celtic Extraction Ltd & Bluestone Chemicals Ltd) v **Environment Agency, (CA), [1999]** 3 **21**
Ogilvie, Eaton Square Properties Ltd v, **[2000]** 1 **73**
Old England Properties Ltd v **Telford & Wrekin Council, (LT), [2000]** 3 **153**
Old Etonian Housing Association Ltd, O'Connor v, **[2002]** 1 **38**
Oliver Ashworth (Holdings) Ltd, Ballard (Kent) Ltd v, **[1998]** 3 **60; [1999]** 2 **23**
Omega Trust Co Ltd v **Wright Son & Pepper, (CA), [1997]** 1 **120**
134 Shirland Road Management Co Ltd v **Lester, (LVT), [1997]** 2 **207**
169 Queen's Gate Ltd, Hannon v, **[2000]** 1 **40**
Orchard Court Residents' Association v **St Anthony's Homes Ltd, (CA), [2003]** 2 **28**
Orchard Trading Estate Management Ltd v **Johnson Security Ltd, (CA), [2002]** 2 **1**

Osada v **Shepping, (CA), [2000]** 2 **38**
Osei-Bonsu, Wandsworth London Borough Council v, **[1999]** 1 **26**
O'Sullivan, Barnett v, **[1995]** 1 **93**
Osmond, Floyer-Acland v, **[2000]** 2 **1**
Oswald v **Countrywide Surveyors Ltd, (CA), [1996]** 2 **104**
Ottewell (VO) v **BP Exploration Operating Co Ltd, (LT), [1996]** 1 **203**
Overseas Investment Services Ltd v **Simcobuild Construction Ltd, (CA), [1996]** 1 **49**
Owen (No 2), *ex parte*. R v Secretary of State for Transport, *ex parte* Owen (No 2) (QB), **[1996]** 1 **52**
Owen, *ex parte*. R v Secretary of State for Transport, *ex parte* Owen (CA), **[1995]** 2 **213**
Owen, Jones v, **[1997]** 2 **5**
Oxford City Council, Oxfordshire County Council v, **[2004]** 1 **105**
Oxford City Council, Thames Water Utilities Ltd v, **[1999]** 1 **167**
Oxfordshire County Council v **Oxford City Council, (LT), [2004]** 1 **105**
Oxfordshire County Council, Rotherwick's Executors v, **[2000]** 2 **84**
Oxfordshire County Council, *ex parte* Sunningwell Parish Council, R v, **[1999]** 2 **94**
Oxford United Football Club, Morrells of Oxford Ltd v, **[2001]** 1 **76**
Oxford United Football Club, Morrells of Oxford Ltd v, **[2000]** 3 **70**
Oyston Shipways Ltd, Platform Home Loans Ltd v, **[1996]** 2 **110; (CA), [1998]** 1 **108; (HL), [1999]** 1 **77**

P&O Property Holdings Ltd v **International Computers Ltd, (Ch), [1999]** 2 **17**
P&O Property Holdings Ltd, Graysim Holdings Ltd v, **[1996]** 1 **109**
PW & Co v **Milton Gate Investments Ltd, (Ch), [2004]** 3 **103**
Paice, Basingstoke and Deane Borough Council v, **[1995]** 2 **9**
Pala, Razzaq v, **[1997]** 2 **53**
Palacegate Properties Ltd, Receiver for the Metropolitan Police v, **[2000]** 1 **63**
Palmeira Square Nos 2-6 Ltd, Saga Properties Ltd v, **[1995]** 1 **199**
Palmiero: debtor 3666 of 1999, Re, (Ch), [1999] 3 **27**
Pankhania v **Hackney London Borough Council, (Ch), [2004]** 1 **135**
Paragon Finance plc v **Banks, (CA), [2003]** 1 **85**
Paragon Finance plc v **City of London Real Property Co Ltd, (CA), [2002]** 1 **97**
Parc Battersea Ltd v **Hutchinson, (Ch), [1999]** 2 **33**
Paribas Ltd (No 2), London & Leeds Estates Ltd v, **[1995]** 1 **102**
Parker Breslin Estates, Hellings v, **[1995]** 2 **99**
Parkins v **Westminster City Council (CA), [1998]** 1 **22**
Parkside Clubs (Nottingham) Ltd v **Armgrade Ltd, (CA), [1995]** 2 **96**
Parness, Active Estates Ltd v, **[2002]** 3 **13**
Parsons v **George, (CA), [2004]** 3 **49**
Passmore v **Morland plc (CA), [1999]** 1 **51**
Pasta Place Ltd, Howard de Walden Estates Ltd v, **[1995]** 1 **79**
Patel, St Ermins Property Co Ltd v, **[1997]** 2 **61**
Paton, Caradon District Council v, **[2000]** 3 **57**
Payne, Pennell v, **[1995]** 1 **6**
Peacock v **Custins, (CA), [2001]** 1 **87**
Pearl Assurance plc, Esselte AB v, **[1995]** 2 **61; (CA), [1997]** 1 **73**
Pearl Property Ltd, West Hampstead Management Co Ltd v, **[2002]** 1 **115; [2002] 3 55**
Pemberton v **Southwark London Borough Council, (CA), [2000]** 2 **33**
Pembrokeshire County Council, Gulliksen v, **[2002]** 3 **9**
Penfold-Dunsford, Bankway Properties Ltd v, **[2001]** 2 **36**
Pennell v **Payne (CA), [1995]** 1 **6**
Pennycook v **Shaws (EAL) Ltd, (Ch), [2003]** 3 **28; (CA), [2004]** 2 **55**
Persaud (VO), Hospital Plan Insurance Services Ltd v, **[1998]** 3 **189**
Petra Investments Ltd v **Jeffrey Rogers plc, (Ch), [2000]** 3 **120**
Phoenix Timber Group plc, West Horndon Industrial Park Ltd v, **[1995]** 1 **77**
Phyllis Trading Ltd v **86 Lordship Road Ltd, (CA), [2001]** 2 **85**
Pickard UK Ltd, Sterling Estates v, **[1997]** 2 **33**
Pickles, Greenbank v, **[2001]** 1 **1**
PizzaExpress (Restaurants) Ltd, Monella v, **[2004]** 1 **43**
Planning and Environment Committee of the State of Jersey, Lesquende Ltd v, **[1998]** 1 **137**
Platform Home Loans Ltd v **Oyston Shipways Ltd, (Ch), [1996]** 2 **110; (CA), [1998]** 1 **108; (HL), [1999]** 1 **77**
Platt v **London Underground Ltd, (Ch), [2001]** 2 **121**
Plumbly v **Spencer (HMIT), (CA), [1999]** 2 **191**
Plummer v **Tibsco Ltd, (CA), [2002]** 1 **29**
Plustrade Ltd, Saeed v, **[2002]** 2 **19**
Pollard v **Ashurst (trustee in bankruptcy), (Ch), [2000]** 2 **29**
Poole, Rabett, v, **[2003] 3 143**
Pope, Chelsea Yacht & Boat Co Ltd v, **[2000]** 2 **23**
Porter v **Secretary of State for Transport, (LT), [1995]** 2 **175; (CA), [1996]** 2 **10**
Portman Building Society v **Bevan Ashford (a firm), (CA), [2000]** 1 **81**
Portman Building Society v **Hamlyn Taylor Neck, (CA), [1998]** 2 **113**
Portman Family Settled Estates Trustees, Williams v, **[1996]** 1 **218**
Portsmouth City Council, LE Jones (Insurance Brokers) Ltd v, **[2003]** 1 **99**
Portvale Holdings Ltd, Bluestorm Ltd v, **[2004]** 2 **38**
Possfund (North West) Ltd, Secretary of State for the Environment v, **[1997]** 2 **56**
Postel Properties Ltd v **Boots the Chemist, (QB), [1996]** 2 **60**
Pounds Shipowners & Shipbreakers Ltd, Garner (Inspector of Taxes) v, **[2000]** 2 **69**
Powell, Barrett v, **[1998]** 2 **117**
Powys County Council, Dingleside Development Co Ltd v, **[1995]** 1 **183**
Powys County Council, Marsh v, **[1997]** 2 **177**
Pradhan, Zarvos v, **[2003]** 2 **37**
Prielipp v **Secretary of State for the Environment, Transport and the Regions, (LT), [2002]** 3 **143**
Procter & Gamble Technical Centres Ltd v **Brixton plc, (Ch), [2003]** 2 **24**
Proudreed Ltd v **Microgen Holdings plc, (CA), [1996]** 1 **89**
Provident Mutual Life Assurance Association v **Greater London Employer's Association Ltd, (Ch), [1996]** 1 **106**
Provident Mutual Life Assurance Association, Dun & Bradstreet Software Services (England) Ltd v, **[1998]** 2 **175**
Provident Mutual Life Assurance, Trane (UK) Ltd v, **[1995]** 1 **33**
Prudential Assurance, Mount Charlotte Investments plc v, **[1995]** 1 **15**
Prudential Assurance Co Ltd, Barclays Bank plc v, [1998] 1 **44**
Prudential Assurance Co Ltd v **Mount Eden Land Ltd, (CA), [1997]** 1 **37**
Prudential Assurance Co Ltd v **Waterloo Real Estate Inc, (CA), [1999]** 2 **85**
Prudential Nominees Ltd v **Greenham Trading Ltd, (CA), [1995]** 1 **97**
Prudential Property Services Ltd, United Bank of Kuwait plc v, **[1995]** 1 **129**
Pumperninks of Piccadilly Ltd v **Land Securities plc, (CA), [2002]** 2 **147**
Purfleet Farms Ltd v **Secretary of State for Transport, Local Government and the Regions, (CA), [2003]** 1 **9**

Qazi, Harrow London Borough Council v, **[2003]** 3 **109**
Quilter v **Quilter, (Ch), [2004]** 1 **79**

Rabett v **Poole (CC), [2003]** 3 **143**
Raymere Ltd v **Belle Vue Gardens Ltd, (CA), [2003]** 3 **40**
R v **Braintree District Council, *ex parte* Halls, (CA), [2000]** 3 **19**
R v **Bristol City Council,** *ex parte* **Everett, (QB), [1998]** 3 **25**
R v **City of London Council,** *ex parte* **Mystery of the Barbers of London, (QB), [1996]** 2 **128**
R (on the application of Beresford) v **City of Sunderland, (Ch), [2004]** 1 **94**
R (on the application of Cheltenham Builders Ltd) v **South Gloucestershire District Council, (HL), [2004]** 1 **85**
R (on the application of Daejan Properties Ltd) v **London Leasehold Valuation Tribunal, (CA), [2001]** 3 **28**
R (on the application of Edison First Power Ltd) v **Secretary of State for the Environment, (HL), [2003]** 2 **133**
R (on the application of Haysport Properties Ltd) v **Rent Officer of the West Sussex Registration Area, (CA), [2001]** 2 **63**
R (on the application of Khatun) v **Newham London Borough Council, (Admin), [2004]** 1 **34**
R (on the application of Laing Homes Ltd) v **Buckinghamshire County Council, (Admin), [2003]** 3 **69**
R (on the application of Lemon Land Ltd) v **Hackney London Borough Council, (QB: Admin), [2002]** 1 **81**
R (on the application of Lester) v **London Rent Assessment Committee, (QB:Admin), [2003]** 1 **58**
R (on the application of Lock) v **Secretary of State for the Environment Transport and the Regions, (QB: Admin), [2002]** 3 **5**
R (on the application of Morris) v **London Rent Assessment Committee, (QB), [2001]** 2 **51**
R (on the application of Morris) v **North Somerset District Council, (CA), [2003]** 3 **129**
R (on the application of Sinclair Gardens Investments (Kensington) Ltd) v **Lands Tribunal, (Admin), [2004]** 3 **15**

R (on the application of Whitmey) v **Commons Commissioners, (CA), [2004]** 3 **1**
R (on the application of Wolters (London) Ltd) v **London Rent Assessment Committee, (Admin), [2003]** 3 **17**
R (on the application of Zeb) v **Newham London Borough Council, (Admin), [2004]** 1 **34**
R v **Director General of Fair Trading,** *ex parte* **Benhams Ltd, (QB), [2001]** 1 **21**
R v **Docklands Estates Ltd, (CA:CrDiv), [2000]** 3 **17**
R v **Dudley Magistrates' Court,** *ex parte* **Hollis, (QB), [1998]** 2 **19**
R v **Hackney London Borough Council,** *ex parte* **Structadene Ltd, (QB), [2001]** 1 **15**
R v **Hodgetts (VO),** *ex parte* **Nationwide Building Society, (CA), [2000]** 1 **97**
R v **Huelin (VO),** *ex parte* **Murphy Ltd, (CA), [2000]** 1 **97**

R v **London Leasehold Valuation Tribunal,** ***ex parte*** **Daejan Properties Ltd, (QB), [2000]** 3 **44**
R v **London Rent Assessment Panel,** *ex parte* **Cadogan Estates Ltd, (QB), [1997]** 2 **134**
R v **Minister of Agriculture, Fisheries and Food,** *ex parte* **Country Landowners' Association, (Eur Ct), [1996]** 2 **1**
R v **Ministry of Agriculture, Fisheries and Food,** *ex parte* **St Clere's Hall Farm, (QB), [1996]** 1 **15**
R (on the application of Morris) v **London Rent Assessment Committee, (CA), [2002]** 2 **13**
R v **Northumbrian Water Ltd,** *ex parte* **Able UK Ltd, (QB), [1996]** 2 **15**
R v **Oxfordshire County Council,** *ex parte* **Sunningwell Parish Council, (HL), [1999]** 2 **94**
R v **Secretary of State for Defence,** *ex parte* **Wilkins, (QB), [2000]** 3 **11**
R v **Secretary of State for the Environment,** *ex parte* **Gosforth Allotments & Gardens Association Ltd, (QB), [1996]** 1 **147; (CA), [1996]** 2 **117**
R v **Secretary of State for the Environment, Transport and the Regions,** *ex parte* **Holding & Barnes plc, (QB (DC)), [2001]** 1 **33**
R v **Secretary of State for the Environment, Transport and the Regions,** *ex parte* **Spath Holme Ltd, (HL), [2001]** 1 **129**
R v **Secretary of State for the Environment, Transport and the Regions,** *ex parte* **Spath Holme Ltd, (CA), [2000]** 1 **167**
R v **Secretary of State for the Environment, Transport and the Regions,** *ex parte* **Wheeler, (QB), [2000]** 3 **63**
R v **Secretary of State for Transport,** *ex parte* **Owen (No 2), (QB), [1996]** 1 **52**
R v **Secretary of State for Transport,** *ex parte* **Owen, (CA), [1995]** 2 **213**
R v **Suffolk County Council,** *ex parte* **Steed, (QB), [1995]** 2 **232**
Racal Tracs, Sun Life Assurance plc v, **[2000]** 1 **138**
Railtrack plc, CA Webber (Transport) Ltd v, **[2004]** 1 **49**
Railtrack plc v **Gojra (CA), [1998]** 1 **63**
Railtrack plc v **Guinness Ltd, (CA), [2003]** 1 **124**
Railtrack plc, J Murphy & Sons Ltd v, **[2002] 2** 48
Railtrack plc (No 2), London Baggage Co (Charing Cross) Ltd v, **[2003]** 1 **141**
Rainbow Estates Ltd v **Tokenhold Ltd, (Ch), [1998]** 2 **34**
Raja v **Austin Gray (a firm), (QB), [2002]** 3 **61**
Raja (administratrix of the estate of Raja (deceased)) v **Austin Gray (a firm), (CA), [2003]** 1 **91**
Raja v **Lloyds TSB Bank plc, (Ch), [2001]** 2 **78**
Ratcliffe v **Sandwell Metropolitan Borough Council, (CA), [2002]** 1 **103**
Ravenseft Properties Ltd v **Hall, (CA), [2002]** 1 **9**
Rawlins, CIN Properties Ltd v, **[1995]** 2 **130**
Razzaq v **Pala, (Ch), [1997]** 2 **53**
Receiver for the Metropolitan Police v **Palacegate Properties Ltd, (CA), [2000]** 1 **63**
Record, Jobson v, **[1998]** 1 **113**
Reed Personnel Services plc v **American Express Ltd, (Ch), [1997]** 1 **229**
Rees v **Skerrett, (CA), [2001]** 3 **1**
Regional Railways North East Ltd, VCS Car Park Management Ltd v, **[2000]** 1 **57**
Reigate and Banstead Borough Council, Yeandle v, **[1996]** 1 **20**
Rent Officer of the West Sussex Registration Area, R (on the application of Haysport Properties Ltd) v, **[2001]** 2 **63**
Reynolds, Mytre Investments Ltd v, **[1995]** 2 **40**
RH Fibreboard Containers Ltd, Nynehead Developments Ltd v, **[1999]** 1 **7**
Rhodes Trust, Harben Style Ltd v, **[1995]** 1 **118**
Rhondda Cynon Taff County Borough Council v **Watkins, (CA), [2003]** 1 **117**
Rhymney Valley District Council, Adams v, **[2000]** 3 **25**
Richards v **Somerset County Council, (LT), [2001]** 2 **135**
Ringbest Ltd, British Railways Board v, **[1996]** 2 **82**
Ritchings, Rodd v, **[1995]** 2 **142**
Roadrunner Properties Ltd v **Dean, (CA), [2004]** 1 **73**
Roberts v **South Gloucestershire Council, (CA), [2003]** 2 **1**
Robinson, Escalus Properties Ltd v, **[1995]** 2 **23**
Robinson, McAdams Homes Ltd v, **[2004]** 3 **93**
Rodd v **Ritchings, (QB), [1995]** 2 **142**
Rodway v **Landy, (CA), [2001]** 3 **78**
Rogan v **Woodfield Building Services Ltd, (CA), [1995]** 1 **72**
Rogers v **Islington London Borough Council, (CA), [1999]** 3 **17**
Rogers, Lambeth London Borough Council v, **[2000]** 1 **28**
Rogers, Lloyds Bank plc v, **[1999]** 3 **83**
Romain v **Scuba TV Ltd (CA), [1996]** 1 **103**
Romines, Wellcome Trust Ltd v, **[1999]** 3 **229**
Romulus Trading Co Ltd v **Comet Properties Ltd, (QB), [1996]** 2 **70**
Ropaigealach v **Allied Irish Bank plc, (CA), [2002]** 1 **83**
Rosen, Mount Cook Land Ltd v, **[2003]** 1 **75**
Rosen v **Trustees of the Camden Charities, (LT), [1999]** 2 **213**
Rosen v **Trustees of the Camden Charities, (CA), [2001]** 1 **59**
Rother District Investments v **Corke, (Ch), [2004]** 1 **47**
Rotherwick's Executors v **Oxfordshire County Council, (LT), [2000]** 2 **84**
Routestone Ltd v **Bird (Ch), [1997]** 1 **123**
Routestone Ltd v **Minories Finance Ltd, (Ch), [1997]** 1 **123**
Royal & Sun Alliance Property Services Ltd, First Property Growth Partnership LP v, **[2003]** 1 **39**
Royal & Sun Alliance Property Services Ltd, First Property Growth Partnership LP v, **[2002]** 2 **11**
Royal Bank of Scotland plc v **Clydebank District Council, (Ct Sess), [1995]** 1 **229**
Royal Bank of Scotland plc v **Jennings (Ch), [1995]** 2 **87; (CA), [1997]** 1 **101**
Royal Bank of Scotland, Mehta v, **[1999]** 3 **153**
Royal Bank of Scotland, Silven v, **[2002]** 3 **115**
Royal Bank of Scotland plc, Silven Properties v, **[2003]** 3 **49**
Royal Brompton Hospital National Health Trust, Cadogan v, **[1996]** 2 **115**
Royal Hotel (Caithness) Ltd, Alexander v, **[2001]** 1 **6**
Rubypoint Ltd, Marshall v, **[1997]** 1 **69**
Ruralpride Ltd, System Floors Ltd v, **[1995]** 1 **48**
Russell Jones & Walker, Credit Lyonnais v, **[2002]** 2 **65**
Ryde International plc v **London Regional Transport, (LT), [2001]** 1 **101; (CA), [2004]** 2 **1**

S Robinson & Sons (Holdings) Ltd, Warborough Investments Ltd v, **[2003]** 2 **149**
Sabella Ltd v **Montgomery, (CA), [1998]** 1 **65**
Sadiq, Louis v, **[1997]** 1 **136**
Saeed v **Plustrade Ltd, (CA), [2002]** 2 **19**
Safeland plc, Kataria v, **[1998]** 1 **39**
Safeway Properties Ltd, Highland & Universal Properties Ltd v, **[2000]** 3 **110**
Saga Properties Ltd v **Palmeira Square Nos 2-6 Ltd, (LT), [1995]** 1 **199**
Salford City Council, Lamba Trading Co Ltd v, **[1999]** 3 **186**
Salford Van Hire (Contracts) Ltd v **Bocholt Developments Ltd, (CA), [1995]** 2 **50**
Samuel Smith Old Brewery (Tadcaster) Ltd, Selby District Council v, **[2001]** 1 **71**
Sanctuary Housing Association v **Baker, (CA), [1998]** 1 **42**
Sanctuary Housing Association v **Campbell, (CA), [1999]** 2 **20**
Sandhu, Ladbroke Hotels Ltd v, **[1995]** 2 **92**
Sandwell Metropolitan Borough Council, Ratcliffe v, **[2002]** 1 **103**
Sansom v **Metcalfe Hambleton & Co, (CA), [1998]** 2 **103**
Saphir, Trustees of the Eyre Estate v, **[1999]** 2 **123**
Sargeant v **Macepark (Whittlebury) Ltd, (Ch), [2004]** 3 **26**
Sargent, Gatwick Parking Services Ltd v, **[2000]** 2 **45**
Sarum Properties Ltd's Application, (LT), [1999] 2 **131**
Savile Estates Ltd, Barclays Bank plc v, **[2002]** 2 **16**
Savva v **Houssein, (CA), [1996]** 2 **65**
Sawston Farms (Cambs) Ltd, Morgan Sindall plc v, **[1999]** 1 **90**
Sawyer, Jaggard v, **[1995]** 1 **146**
Scarborough Borough Council v **Geotechnical Engineering (Northern) Ltd, (Off Ref), [1997]** 2 **213**
Scarborough Borough Council, Holbeck Hall Hotel Ltd v, **[1997]** 2 **213**
Scarborough Borough Council, Ingle v, **[2002]** 2 **161**

Scarlett v **Halcrow Ltd (LVT), [1995]** 1 **219**
Sceneout Ltd v **Central Manchester Development Corporation (LT), [1995]** 2 **179**
Scottish Amicable Life Assurance Society v **Middleton Potts & Co, (CA), [1995]** 1 **97**
Scottish & Newcastle Retail Ltd v **Williams (VO), (CA), [2001]** 1 **157**
Scottish & Newcastle Retail Ltd v **Williams (VO), (LT), [2000]** 2 **171**
Scottish Widow's Fund & Life Assurance Society, Bowring Services Ltd v, **[1995]** 1 **158**
Scottish Widows Fund & Life Assurance Society, Garston v, **[1998]** 2 **73**
Scottish Widows Fund & Life Assurance Society, Garston v, **[1996]** 1 **113**
Scuba TV Ltd, Romain v, **[1996]** 1 **103**
Search Guarantees plc, Earl Cadogan v, **[2004]** 3 **55**
Secchi, Keepers and Governors of John Lyon Grammar School v, **[1999]** 3 **49**
Second Duke of Westminister's Trustees, Donath v, **[1997]** 1 **203**
Secretary of State for Defence, R v, *ex parte* Wilkins, **[2000]** 3 **11**
Secretary of State for the Environment, Addin v, **[1997]** 1 **99**
Secretary of State for the Environment, Central Estates Ltd v, **[1997]** 1 **239**
Secretary of State for the Environment, R (on the application of Edison First Power Ltd) v, **[2003]** 2 **133**
Secretary of State for the Environment v **Fletcher Estates (Harlescott) Ltd, (CA), [1998]** 3 **13; (HL) [2000]** 1 **13**
Secretary of State for the Environment, *ex parte* Gosforth Allotments & Gardens Association, R v, **[1996]** 1 **147; [1996]** 2 **117**
Secretary of State for the Environment, Johnsey Estates Ltd v, **[2001]** 2 **128**
Secretary of State for the Environment, Maidstone Borough Council v, **[1996]** 1 **29**
Secretary of State for the Environment, Meadfield Properties Ltd v, **[1995]** 1 **39**
Secretary of State for the Environment, Newell v, **[1998]** 3 **13; [2000]** 1 **13**
Secretary of State for the Environment v **Newell, (CA), [1998]** 3 **13; [2000]** 1 **13**
Secretary of State for the Environment v **Possfund (North West) Ltd, (Ch), [1997]** 2 **56**
Secretary of State for the Environment, Stanford Marsh Ltd v, **[1997]** 1 **178**
Secretary of State for the Environment, Tesco Stores Ltd v, **[1995]** 2 **147**
Secretary of State for the Environment, Food and Rural Affairs, Drury v, **[2004]** 3 **85**
Secretary of State for the Environment, Transport and the Regions v **Baylis (Gloucester) Ltd, (Ch), [2000]** 2 **13**
Secretary of State for the Environment, Transport and the Regions (No 2), Eronpark Ltd v, **[2001]** 3 **133**
Secretary of State for the Environment, Transport and the Regions, *ex parte* Holding & Barnes plc, R v, **[2001]** 1 **33**
Secretary of State for the Environment, Transport and the Regions, R (on the application of Lock) v, **[2002]** 3 **5**
Secretary of State for the Environment, Transport and the Regions, Prielipp v, **[2002]** 3 **143**
Secretary of State for the Environment, Transport and the Regions, *ex parte* Spath Holme Ltd, R v, **[2000]** 1 **167**
Secretary of State for the Environment, Transport and the Regions, *ex parte* Spath Holme Ltd, R v, **[2001]** 1 **129**
Secretary of State for the Environment, Transport and the Regions, Stayley Developments Ltd v, **[2001]** 1 **167**
Secretary of State for the Environment, Transport and the Regions, *ex parte* Wheeler, R v, **[2000]** 3 **63**
Secretary of State for Health, Craven (Builders) Ltd v, **[2000]** 1 **128**
Secretary of State for Trade and Industry, Antonelli v, **[1996]** 1 **229; [1998]** 1 **9**
Secretary of State for Transport, Bannocks v, **[1995]** 2 **157**
Secretary of State for Transport, Christos v, **[2004]** 1 **5**
Secretary of State for Transport, Eronpark Ltd v, **[2000]** 2 **165**
Secretary of State for Transport, *ex parte* Owen (No 2), R v, **[1996]** 1 **52**
Secretary of State for Transport, *ex parte* Owen, R v, **[1995]** 2 **213**
Secretary of State for Transport, Porter v, **[1995]** 2 **175; (CA), [1996]** 2 **10**
Secretary of State for Transport, Sinclair v, **[1997]** 2 **173**
Secretary of State for Transport, Yorkshire Traction Co Ltd v, **[1997]** 2 **28**
Secretary of State for Transport, Local Government and the Regions, Purfleet Farms Ltd v, **[2003]** 1 **9**
Selby District Council, Factorset Ltd v, **[1995]** 2 **190**
Selby District Council v **Samuel Smith Old Brewery (Tadcaster) Ltd, (CA), [2001]** 1 **71**
Severn Trent Water Ltd v **Barnes, (CA), [2004]** 2 **95**
Sevington Properties Ltd v **Adams (CA), [1998]** 2 **57**
Shack, Eyre Estate Trustees v, **[1995]** 1 **213**
Shaftesbury plc, Cricket Ltd v, **[1999]** 2 **57**
Shahgholi, Cadogan Estates Ltd v, **[1999]** 1 **189**
Shaire, Mortgage Corporation v, **[2000]** 3 **131**
Shalson, John Lyon's Charity v, **[2003]** 2 **49**
Shalson, John Lyon's Charity v, **[2002]** 2 **55**
Sharp v **Griffiths, (VO), (LT), [1999]** 3 **113**
Shaws (EAL) Ltd, Pennycook v, **[2003] 3 28; [2004]** 2 **55**
Sheard Walshaw Partnership, Baxall Securities Ltd v, **[2002]** 1 **73**
Shepherds (a firm), Gregory v, **[2000]** 3 **49**
Shepping, Osada v, **[2000]** 2 **38**
Sheridan, North British Housing Association Ltd v, **[1999]** 2 **138**
Shevlin v **Trafford Park Development Corporation, (LT), [1998]** 1 **115**
Shewu v **Hackney London Borough Council, (LT), [1998]** 2 **232; (CA), [1999]** 3 **1**
Shimizu (UK) Ltd v **Westminster City Council, (CA), [1995]** 1 **167**
Shina v **Elghanian, (LT), [1999]** 3 **103**
Shirayama Shokusan Co Ltd, Frogmore Developments Ltd v, **[2000]** 1 **121**
Shirley, Challock Parish Council v, **[1995]** 2 **137**
Shopmoor Ltd, Norwich Union Life Insurance Society v, **[1998]** 2 **167**
Shortdean Place (Eastbourne) Residents' Association Ltd v **Lynari Properties Ltd, (LT), [2003]** 3 **147**
Shortlands Investments Ltd v **Cargill plc, (Off Ref), [1995]** 1 **51**
Shortlands Investments Ltd, Fluor Daniel Properties Ltd v, **[2001]** 2 **103**
Shotley Point Marina (1986) Ltd v **Spalding, (CA), [1997]** 1 **233**
Shraff Tip Ltd v **Highways Agency (No 2), (LT), [1999]** 3 **224**
Shraff Tip Ltd v **Highways Agency, (LT), [1999]** 2 **205**
Shulem B Association Ltd's Appeal, In (LT), [2001] 1 **105**
Shun Fung Ironworks Ltd, Director of Buildings and Lands v, **[1995]** 1 **19**
Sight & Sound Education Ltd v **Books etc Ltd, (Ch), [1999]** 3 **45**
Signet Group plc, Crown Estate Commissioners v, **[1996]** 2 **200**
Silk, Crewe Services & Investment Corporation v, **[1998]** 2 **1**
Silven v **Royal Bank of Scotland plc, (Ch), [2002]** 3 **115**
Silven Properties v **Royal Bank of Scotland plc (CA), [2003]** 3 **49**
Simcobuild Construction Ltd, Overseas Investment Services Ltd v, **[1996]** 1 **49**
Simm (VO), Godfrey v, **[2000]** 3 **85**
Simmonds, Clements v, **[2002]** 3 **22**
Simmonds Church Smiles, Chartered Society of Physiotherapy v, **[1995]** 1 **155**
Sinclair Gardens Investments (Kensington) Ltd v **Walsh, (CA), [1995]** 2 **23**
Sinclair Gardens Investments (Kensington) Ltd, Berrycroft Management Co Ltd v, **[1997]** 1 **47**
Sinclair Gardens Investments (Kensington) Ltd (on the application of) R, Lands Tribunal v, **[2004]** 3 **15**
Sinclair v **Secretary of State for Transport, (LT), [1997]** 2 **173**
Single Horse Properties Ltd v **Surrey County Council, (CA), [2002]** 2 **43**
63 Perham Road Ltd v **Manyfield Ltd, (LVT), [1995]** 2 **206**
63 Perham Road Ltd, Maryland Estates Ltd v, **[1997]** 2 **198**
Sked v **Towngran Developments Ltd, (LVT), [1995]** 1 **216**
Skerrett, Rees v, **[2001]** 3 **1**
Skinner, Farley v, **[2001]** 3 **57**
Skinner, Farley v, **[2000]** 2 **125**
Skinns v **Greenwood, (CA), [2002]** 2 **53**
Skipton Building Society, Den Danske Bank A/S v, **[1998]** 1 **155**
Slough Estates plc v **Welwyn Hatfield District Council, (QB), [1996]** 2 **219**
Smee, Abbahall Ltd v, **[2003]** 2 **66**
Smith, Bairstow Eves London Ltd v, **[2004]** 2 **25**
Smith, Bridgegrove Ltd v, **[1997]** 2 **40**
Smith, Greenalls Management Ltd v, **[1999]** 2 **145**
Smith, Harwood (Michael) T/A RSBS Group v, **[1998]** 1 **5**
Smith v **Kent County Council, (LT), [1995]** 2 **196**
Smith, Muscat v, **[2003]** 3 **11**
Smith v **South Gloucestershire District Council, (CA), [2002]** 3 **1**
Smith (VO), Irving Brown & Daughter v, **[1996]** 2 **183**
Smith v **Spaul, (CA), [2003]** 1 **70**
Sneesby v **Goldings, (CA), [1995]** 2 **102**
Snell & Prideaux Ltd v **Dutton Mirrors Ltd, (CA), [1995]** 1 **259**
Solent Pines Residents Association Ltd v **Campbell, (LVT), [1996]** 1 **213**
Somerfield Properties Co Ltd, Dukeminster (Ebbgate House One) Ltd v, **[1997]** 2 **125**
Somerset County Council, Richards v, **[2001]** 2 **135**
South v **Chamberlayne, (CA), [2001]** 3 **54**

South Australia Asset Management Corporation v **York Montague Ltd, (QB), [1995]** 2 **219; (HL), [1996]** 2 **93**
South Gloucestershire Council, Roberts v, **[2003]** 2 **1**
South Gloucestershire District Council, R (on the application of Cheltenham Builders Ltd) v, **[2004]** 1 **85**
South Gloucestershire District Council, Smith v, **[2002]** 3 **1**
South Lodge Developments , Bank of Ireland Home Mortgages v, **[1996]** 1 **91**
South London Co-operative Building Co Ltd, Wharfland Ltd v, **[1995]** 2 **21**
South Oxfordshire District Council, Armitage v, **[1998]** 2 **121**
Southern Electric plc, Bolton v, **[1999]** 1 **177**
Southern Electricity Board, Naylor v, **[1996]** 1 **195**
Southern Water plc, Wickham Growers Ltd v, **[1997]** 1 **175**
Southwark London Borough Council, Harry Lester Ltd v, **[1999]** 3 **179**
Southwark London Borough Council, Long v, **[2002]** 3 **37**
Southwark London Borough Council v **McIntosh, (Ch), [2002]** 1 **25**
Southwark London Borough Council v **Mills, (CA), [1998]** 3 **46; (HL), [1999]** 3 **35**
Southwark London Borough Council, Pemberton v, **[2000]** 2 **33**
Spalding, Shotley Point Marina (1986) Ltd v, **[1997]** 1 **233**
Spath Holme Ltd v **Chairman of the Greater Manchester & Lancashire Rent Assessment Committee, (CA), [1995]** 2 **80**
Spath Holme Ltd v **North Western Rent Assessment, (Admin), [2001]** 3 **30**
Spath Holme Ltd, *ex parte*. R v Secretary of State for the Environment, Transport and the Regions, *ex parte* Spath Holme Ltd **[2000]** 1 **167**
Spath Home Ltd, *ex parte*. R v Secretary of State for the Environment, Transport and the Regions, *ex parte* Spath Holme Ltd **[2001]** 1 **129**
Spaul, Smith v, **[2003]** 1 **70**
Speedwell Estates Ltd, Re, (LT), [1999] 2 **121**
Speedwell Estates Ltd v **Dalziel, (CA), [2002]** 1 **55**
Spencer (Henry) & Sons Ltd, Cavendish Funding Ltd v, **[1998]** 1 **104**
Spencer (HMIT), Plumbly v, **[1999]** 2 **191**
Spencer-Ward v **Humberts, (CA), [1995]** 1 **123**
Spetale, Ivory Gate Ltd v, **[1998]** 2 **43**
Spigarolo, Go West Ltd v, **[2003]** 1 **133**
Sportoffer Ltd v **Erewash Borough Council, (Ch), [1999]** 3 **136**
Spratling, National Trust v, **[1997]** 2 **229**
Squarepoint (London) Ltd, Visible Information Packaged Systems Ltd v, **[2000]** 2 **93**
St Anthony's Homes Ltd, Orchard Court Residents' Association v, **[2003]** 2 **28**
St Clere's Hall Farm, *ex parte*. R v Ministry of Agriculture, Fisheries and Food, *ex parte* St Clere's Hall Farm (QB), **[1996]** 1 **15**
St Ermins Property Co Ltd v **Patel, (CA), [1997]** 2 **61**
St Ermins Property Co Ltd v **Tingay, (Ch), [2002]** 3 **53**
St Giles Hotel Ltd v **Microworld Technology Ltd, (CA), [1997]** 2 **105**
St Mary's Mansions Ltd v **Limegate Investment Co Ltd, (CA), [2003]** 1 **41**
Standard Life Assurance Co, Egan Lawson Ltd v, **[2001]** 1 **27**
Standard Life Assurance v **Unipath Ltd, (CA), [1997]** 2 **121**
Stanford Marsh Ltd v **Secretary of State for the Environment, (LT), [1997]** 1 **178**
Stanton v **Callaghan (CA), [1998]** 3 **165**
Stapel v **Bellshore Property Investments Ltd, (Ch), [2001]** 2 **7**
Star Rider Ltd v **Inntrepreneur Pub Co, (Ch), [1998]** 1 **53**
Starling v **Lloyds TSB Bank Ltd, (CA), [2000]** 1 **101**
Starke v **Commissioners of Inland Revenue, (CA), [1996]** 1 **157**
Starmark Enterprises Ltd v **CPL Distribution Ltd, (Ch), [2000]** 3 **37**
Staszewski v **Maribella Ltd (CA), [1998]** 1 **34**
Stayley Developments Ltd v **Secretary of State for the Environment, Transport and the Regions, (LT), [2001]** 1 **167**
Steed, *ex parte*. R v Suffolk County Council, *ex parte* Steed (QB), **[1995]** 2 **232**
Stepien Lake Gilbert & Paling, Aran Catering Ltd v, **[2002]** 1 **69**
Sterling Estates v **Pickard UK Ltd, (Ch), [1997]** 2 **33**
Stevenage Borough Council, Turner v, **[1996]** 1 **23**
Stevens v **Gullis, (CA), [1999]** 3 **71**
Stone, Grammer v, **[2000]** 1 **1**
Stone, Wilson v, **[1998]** 2 **155**
Stones, Jones v, **[1999]** 3 **81**
Strathclyde Pension Fund, Checkpoint Ltd v, **[2003]** 1 **1**
Strathclyde Pension Fund, Checkpoint Ltd v, **[2002]** 2 **97**
Strauss, Cadogan (Earl) v, **[2004]** 2 **69**
Street, Berkshire Capital Funding Ltd v, **[1999]** 2 **92**
Stretch v **United Kingdom, (ECHR), [2004]** 1 **11**
Stretch v **West Dorset District Council, (CA), [1998]** 3 **62**
Structadene Ltd, *ex parte*. R v Hackney London Borough Council, *ex parte* Structadene Ltd **[2001]** 1 **15**
Strutt & Parker, Merivale Moore plc v, **[1998]** 2 **195; [1999]** 2 **171**
Strutt & Parker, Western Trust & Savings Ltd v, **[1998]** 3 **89**
Suffolk County Council, *ex parte* Steed, R v, **[1995]** 2 **232**
Sullivan v **Layton Lougher & Co, (CA), [1995]** 2 **111**
Sun Alliance & London Assurance Co Ltd, Brown & Root Technology Ltd v, **[1997]** 1 **39**
Sun Alliance & London Assurance Co Ltd, Clinton Cards (Essex) Ltd v, **[2002]** 3 **19**
Sun Life Assurance plc v **Racal Tracs, (QB:TCC), [2000]** 1 **138**
Sun Life Assurance plc v **Thales Tracs Ltd, (CA), [2001]** 2 **57**
Sun Life Assurance Society plc, British Telecom plc v, **[1995]** 2 **44**
Sun Life Assurance Society plc v **Tantofex (Engineers) Ltd, (Ch), [1999]** 2 **135**
Surrey County Council v **Lamond (CA), [1999]** 1 **32**
Surrey County Council, Single Horse Properties Ltd v, **[2002]** 2 **43**
Swainland Builders Ltd v **Freehold Properties Ltd, (CA), [2002]** 2 **71**
Swann v **White (LT), [1996]** 1 **199**
Sweeney, Inntrepreneur Pub Co (CPC) v, **[2002]** 2 **132**
Sweetman, Forcelux Ltd v, **[2001]** 2 **173**
Swift v **Dairywise Farms Ltd, (CA), 101**
Sykes v **Harry, (CA), [2001]** 1 **53**
System Floors Ltd v Ruralpride Ltd (CA), **[1995]** 1 **48**

T-Mobile (UK) Ltd, Lloyd Jones v, **[2003]** 3 **55**
Tague v **Lancaster City Council, (LT), [1999]** 2 **103**
Taj v **Ali, (CA), [2000]** 3 **35**
Tandridge District Council v **Bickers, (CA), [1998]** 3 **31**
Tantofex (Engineers) Ltd, Sun Life Assurance Society plc v, **[1999]** 2 **135**
Target Home Loans Ltd v **Iza Ltd, (CC), [2000]** 1 **23**
Taylor v **Hamer, (CA), [2003]** 1 **103**
Taylor, Indiana Investments Ltd v, **[2004]** 3 **63**
Taylor v **North West Water Ltd (LT), [1995]** 1 **266**
Telegraph Group Ltd, Canary Wharf Investments (Three) v, **[2003] 3 31**
Telegraph Service Stations Ltd v **Trafford Borough Council, (LT), [2000]** 3 **145**
Telford & Wrekin Council, Old England Properties Ltd v, **[2000]** 3 **153**
Templeheights Ltd, BRS Northern Ltd v, **[1998]** 2 **182**
Tesco Stores Ltd v **Secretary of State for the Environment, (HL), [1995]** 2 **147**
Thackeray Property Services, Landmaster Properties Ltd v, **[2003]** 2 **30**
Thakrar, Iqbal v, **[2004]** 3 **21**
Thales Tracs Ltd, Sun Life Assurance plc v, **[2001]** 2 **57**
Thames Water Utilities, Marcic v, **[2001]** 3 **111**
Thames Water Utilities Ltd v **Oxford City Council, (Ch), [1999]** 1 **167**
Thamesmead Town Ltd v **Allotey, (CA), [1998]** 3 **97**
The Reece-Jones Partnership, Hawkesbrook Leisure Ltd v, **[2004]** 2 **61**
Theodore Goddard v **Fletcher King Services Ltd, (QB), [1997]** 2 **131**
Thurloe Properties Ltd, Design Progression Ltd v, **[2004]** 1 **121**
Tibsco Ltd, Plummer v, **[2002]** 1 **29**
Tiffany Investments Ltd v **Bircham & Co Nominees (No 2) Ltd, (CA), [2004]** 2 **31**
Tingay, St Ermins Property Co v, **[2002]** 3 **53**
Tokenhold Ltd, Rainbow Estates Ltd v, **[1998]** 2 **34**
Tomkins v **Basildon District Council, (CA), [2002]** 3 **33**
Torkington, Bolton Metropolitan Borough Council v, **[2004]** 1 **28**
Totalfinaelf UK Ltd, Mason v, **[2003]** 3 **91**
Tower Hamlets London Borough Council, Long v, **[1997]** 1 **78**
Towngran Developments Ltd, Sked v, **[1995]** 1 **216**
Trafford Borough Council, Telegraph Service Stations Ltd v, **[2000]** 3 **145**
Trafford Park Development Corporation, Shevlin v, **[1998]** 1 **115**
Trane (UK) Ltd v **Provident Mutual Life Assurance, (Ch), [1995]** 1 **33**
Tranmac Ltd, Eren v, **[1997]** 2 **211**
Trenaman, Devonshire Reid Properties Ltd v, **[1997]** 1 **45**
Trinity Development Co (Banbury) Ltd, National Car Parks Ltd v, **[2001]** 2 **43**
Truelove, Wilson v, **[2003]** 2 **63**
Trustees of the Camden Charities, Rosen v, **[1999]** 2 **213**
Trustees of the Camden Charities, Rosen v, **[2001]** 1 **59**
Trustees of the Eyre Estate v **Saphir, (LT), [1999]** 2 **123**
TSB Bank plc v **Marshall (CC), [1998]** 3 **100**
Tudor v **M25 Group Ltd, (CA), [2005]** 1 **23**

Tugwell, Countryside Residential (North Thames) Ltd v, **[2000]** 2 **59**
Turner, Gregory v, **[2003]** 3 **129**
Turner v **Stevenage Borough Council, (Ch), [1996]** 1 **23**
Turner, Marlton v, **[1998]** 3 **185**
Twinsectra Ltd, Hynes v, **[1995]** 2 **69**
Twinsectra Ltd v **Jones (LT), [1998]** 2 **129**
Twinsectra Ltd, Kay-Green v, **[1996]** 2 **43; (LVT), [1997]** 1 **219**
Tye v House (Ch), **[1997]** 2 **171**

UCB Bank plc v **David J Pinder plc, (Off Ref), [1998]** 2 **203**
UCB Bank plc v **Halifax (SW) Ltd, (QB), [1996]** 2 **102**
UCB Corporate Services Ltd v **Halifax (SW) Ltd, (CA), [2000]** 1 **87**
UCB Corporate Services Ltd, Whitbread plc v, **[2000]** 3 **60**
UCB Group Ltd v **Hedworth, (CA), [2002]** 3 **76**
UK Coal Mining Ltd, Hallam Land Management Ltd v, **[2002]** 2 **80**
Ullah v **Leicester City Council, (LT), [1996]** 1 **244**
Ultraworth Ltd v **General Accident Fire & Life Assurance Corporation Ltd, (QB:TCC), [2000]** 2 **115**
Unipath Ltd, Standard Life Assurance v, **[1997]** 2 **121**
United Attractions Ltd, Oceanic Village Ltd v, **[2000]** 1 **148**
United Bank of Kuwait plc v **Prudential Property Services Ltd, (CA), [1995]** 1 **129; (HL), [1996]** 2 **93**
United Kingdom, Bryan v, **[1996]** 2 **123**
United Kingdom, Stretch v, **[2004]** 1 **11**
University of Sheffield, British Glass Manufacturers' Confederation v, **[2004]** 1 **40**
Universities Superannuation Scheme Ltd v **Marks & Spencer plc, (CA), [1999]** 1 **13**
University of Westminster's Application, In (LT), **[1997]** 1 **191**
Uratemp Ventures Ltd v **Collins, (HL), [2001]** 3 **93**
Uratemp Ventures Ltd v **Collins, (CA), [2000]** 1 **156**

VCS Car Park Management Ltd v **Regional Railways North East Ltd, (CA), [2000]** 1 **57**
van der Loo, Niazi Services Ltd v, **[2004]** 1 **62**
Veena SA v **Cheong, (LT), [2003]** 1 **175**
Verderers of the New Forest v **Young, (CA), [2004]** 1 **1**
Verkan & Co Ltd v **Byland Close (Winchmore Hill) Ltd, (LT), [1998]** 2 **139**
Verrall, Boyle v, **[1997]** 1 **25**
Vignaud v **Keepers and Governors of the Possessions Revenues and Goods of the Free Grammar School of John Lyon, (LT), [1996]** 2 **179**
Vincent, Lambeth London Borough Council v, **[2000]** 2 **73**
Virgin Retail Ltd, Marklands v, **[2004]** 2 **43**
Viscount Chelsea v **Hirshorn, (CC), [1998]** 2 **90**
Viscount Chelsea, Horden v, **[1997]** 1 **195**
Viscount Chelsea v **Morris, (CC), [1997]** 2 **100**
Viscount Chelsea, Wax v, **[1996]** 2 **80**
Visible Information Packaged Systems Ltd v **Squarepoint (London) Ltd, (LT), [2000]** 2 **93**

WX Investments Ltd v **Begg, (Ch), [2002]** 3 **47**
Wagstaff v **Department of the Environment, Transport and the Regions, (LT), [1999]** 2 **108**
Wakefield Metropolitan City Council, Lodge v, **[1995]** 2 **124**
Wakeley v **London Fire and Civil Defence Authority, (LT), [1996]** 2 **148**
Walker Cain v **McCaughey, (CA), [1999]** 2 **145**
Walker v **Badcock (CA), [1997]** 2 **163**
Wallace v **Barratt (C Brian) & Son Ltd, (CA), [1997]** 2 **1**
Wallace v **Manchester City Council, (CA), [1998]** 3 **38**
Wallbank, Aston Cantlow Parochial Church Council v, **[2000]** 2 **149**
Wallcite Ltd, Ferrishurst Ltd v, **[1999]** 1 **85**
Waller, Keelwalk Properties Ltd v, **[2002]** 3 **79**
Wallis Fashion Group Ltd v **CGU Life Assurance Ltd, (Ch), [2000]** 2 **49**
Walsh, Sinclair Gardens Investments (Kensington) Ltd v, **[1995]** 2 **23**
Walton Group plc, Liverpool City Council v, **[2002]** 1 **149**
Walton's Exors v **Commissioners of Inland Revenue, (CA), [1996]** 1 **159**
Wandsworth London Borough Council v **Attwell, (CA), [1996]** 1 **57**
Wandsworth London Borough Council, Battersea Freehold & Leasehold Co Ltd v, **[2001]** 2 **75**
Wandsworth London Borough Council v **Griffin, (LT), [2000]** 2 **105**
Wandsworth London Borough Council v **Manuel, (Ch), [2002]** 2 **128**
Wandsworth London Borough Council v **Osei-Bonsu, (CA), [1999]** 1 **26**
Warborough Investments Ltd v **S Robinson & Sons (Holdings) Ltd, (CA), [2003]** 2 **149**
Wards Construction (Medway) Ltd v **Kent County Council, (CA), [1999]** 1 **111**
Warner Goodman & Streat, Handoll v, **[1995]** 1 **173**
Warner World Ltd, Milverton Group Ltd v, **[1995]** 2 **28**
Wase v **Bourke, (Ch), [1996]** 1 **164**
Wassef, Frances Holland School v, **[2001]** 2 **88**
Waterloo Real Estate Inc, Prudential Assurance Co Ltd v, **[1999]** 2 **85**
Waters v **Welsh Development Agency, (HL), [2004]** 2 **103; (CA), [2002]** 2 **107**
Waters v **Welsh Development Agency, (LT), [2001]** 1 **185**
Watkins, Rhondda Cynon Taff County Borough Council v, **[2003]** 1 **117**
Wax v **Viscount Chelsea, (CC), [1996]** 2 **80**
Weeks Legg & Dean, Barclays Bank plc v, **[1998]** 3 **103**
Weeks, Daejan Properties Ltd v, **[1998]** 3 **125**
Welby v **Casswell, (CA), [1995]** 2 **1**
Wellcome Trust Ltd v **Bellhurst Ltd, (CA), [2002]** 2 **57**
Wellcome Trust Ltd v **Hammad, (CA), [1998]** 1 **73**
Wellcome Trust Ltd v **Romines, (LT), [1999]** 3 **229**
Welsh v **Greenwich London Borough Council, (CA), [2000]** 3 **41**
Welsh Development Agency, Harris v, **[1999]** 3 **207**
Welsh Development Agency, Waters v, **[2004]** 2 **103; [2002]** 2 **107; [2001]** 1 **185**
Welsh Office, Clwyd Alyn Housing Association Ltd v, **[1996]** 1 **184**
Welsh Office, Hallows v, **[1995]** 1 **191**
Welwyn Hatfield District Council, Slough Estates plc v, **[1996]** 2 **219**
Wesleyan Assurance Society, Max Factor Ltd v, **[1995]** 2 **38**
West Dorset District Council, Stretch v, **[1998]** 3 **62**
West Hampstead Management Co Ltd v **Pearl Property Ltd, (LT), [2002]** 1 **115; (CA), [2002] 3 55**
West Horndon Industrial Park Ltd v **Phoenix Timber Group plc (Ch), [1995]** 1 **77**
Western Trust & Savings Ltd v **Strutt & Parker, (CA), [1998]** 3 **89**
Westminster City Council, Delaware Mansions Ltd v, **[1999]** 3 **68**
Westminster City Council v **HSBC Bank plc, (TCC), [2003]** 1 **62**
Westminster City Council, Ocean Leisure Ltd v, **[2004]** 3 **9**
Westminster City Council, Parkins v, **[1998]** 1 **22**
Westminster City Council, Shimizu (UK) Ltd v, **[1995]** 1 **167**
Westminster City Council (No 2), Land Securities plc v, **[1995]** 1 **245**
Westminster (Duke of) v **Birrane (CA), [1995]** 1 **89**
Wharfland Ltd v **South London Co-operative Building Co Ltd (QB), [1995]** 2 **21**
Wheeler, *ex parte*, R v Secretary of State for the Environment, Transport and the Regions, **[2000]** 3 **63**
Whiston Properties Ltd, Grosvenor West End Properties v, **[2002]** 2 **87**
Whitbread plc, Dean and Chapter of the Cathedral and Metropolitan Church of Christ Canterbury v, **[1995]** 1 **82**
Whitbread plc v **UCB Corporate Services Ltd, (CA), [2000]** 3 **60**
White v **Chubb, (CA), [2002]** 1 **9**
White, Swann v, **[1996]** 1 **199**
Whitmey (on the application of) R, Commons Commissioners v, **[2004]** 3 **1**
Wickham Growers Ltd v **Southern Water plc (LT), [1997]** 1 **175**
Wildtree Hotels Ltd v **Harrow London Borough Council (CA), [1998]** 3 **133**
Wildtree Hotels Ltd v **Harrow London Borough Council, (HL), [2000]** 2 **5**
Wilkins, *ex parte*, R v Secretary of State for Defence, **[2000]** 3 **11**
William Martin & Partners, Coventry Building Society v, **[1997]** 2 **146**
Williams v **Blaenau Gwent County Borough Council (No 2), (LT), [1999]** 2 **195**
Williams, Espley v, **[1997]** 1 **9**
Williams v **Glyn Owen & Co, (CA), [2004]** 1 **67**
Williams v **Jones, (CA), [2002]** 3 **69**
Williams v **Kiley (t/a CK Supermarkets Ltd), (CA), [2003]** 1 **46**
Williams v **Portman Family Settled Estates Trustees (LVT), [1996]** 1 **218**
Williams (VO), Allied Domecq Retailing Ltd v, **[2000]** 2 **171**
Williams (VO), Scottish & Newcastle Retail Ltd v, **[2000]** 2 **171**
Williams (VO), Scottish & Newcastle Retail Ltd v, **[2001]** 1 **157**
Willingale v **Globalgrange Ltd, (CA), [2000]** 2 **55**
Willis' Application, Re v **(LT), [1997]** 2 **185**
Willison v **Cheverell Estates Ltd (CA), [1996]** 1 **116**
Wilson v **Stone (LVT), [1998]** 2 **155**
Wilson v **Truelove, (Ch), [2003]** 2 **63**
Wimbledon and Putney Commons Conservators, Blacker v, **[2004]** 3 **67**
Winchester City Council, Cook v, **[1995]** 1 **179**
Windsor Life Assurance Co Ltd v **Austin (LT), [1996]** 2 **169**

Woking 8 Ltd, Micrografix v, **[1995]** 2 **32**
Wolters (on the application of), London Rent Assessment Committee v, **[2003]** 3 **17**
Woodfield Building Services Ltd, Rogan v, **[1995]** 1 **72**
Woodridge Ltd v **Downie (LT), [1997]** 2 **193**
Woodward, Agricultural Mortgage Corporation plc v, **[1995]** 1 **1**
Woollcombe Beer & Watts, May v, **[1998]** 3 **94**
Woolworths plc, Commercial Union Life Assurance Co Ltd v, **[1996]** 1 **237**
Workspace Management Ltd, Edwin Shirley Productions Ltd v, **[2001]** 2 **16**
Worrell Holdings Ltd, Bircham & Co Nominees (No 2) Ltd v, **[2001]** 3 **83**
Worth, Inntrepreneuer Estates (CPC) Ltd v, **[1996]** 1 **84**
Wretham, Birmingham Midshires Building Society v, **[1999]** 1 **69**
Wright Son & Pepper, Omega Trust Co Ltd v, **[1997]** 1 **120**
Wright, Croydon Unique Ltd v, **[1999]** 3 **28**
Wroe v Exmos Cover Ltd, (CA), **[2000]** 1 **66**

X v **A, (Ch), [2000]** 1 **19**

Yaxley *v* **Gotts, (CA), [1999]** 2 **181**
Yeandle v **Reigate and Banstead Borough Council (CA), [1996]** 1 **20**
Yenula Properties Ltd v **Naidu, (CA), [2002]** 3 **28**
Yeoman's Row Management Ltd v **Bodentien-Meyrick, (CA), [2002]** 2 **39**
York Montague Ltd, Craneheath Securities Ltd v, **[1996]** 1 **130**
York Montague Ltd, South Australia Asset Management Corporation v, **[1995]** 2 **219; [1996] 2 93**
York v **Casey (CA), [1998]** 2 **25**
Yorkshire Traction Co Ltd v Secretary of State for Transport (CA), **[1997]** 2 **28**
Yorkshire Water Services Ltd, City of Bradford Metropolitan District Council v, **[2002]** 1 **85**
Young, Verderers of the New Forsest v, **[2004]** 1 **1**

Zarvos v **Pradhan, (CA), [2003]** 2 **37**
Zeb (on the application of) R, Newham London Borough Council v, **[2004]** 1 **34**
Zubaida v **Hargreaves (CA), [1995]** 1 **127**

INDEX OF SUBJECT-MATTER

References are to year, volume number and page,
eg **[1997]** 1 **78** refers to vol 1 of 1997 at p78

ADMINISTRATION OF JUSTICE ACT 1970 — See **Mortgage** (Order for possession — Warrant for possession)

ADVERSE POSSESSION

Animus possidendi

Limitation Act 1980

Whether squatters satisfied requirements of actual possession, *animus possidendi* and adverse possession for limitation period — Whether warning off of caution and issue of the originating summons action to recover land for purposes of Limitation Act 1980 — Held, requirements for adverse possession established — Limitation period ends on commencement of action to recover possession — Only action for possession is action for purposes of s15(1) of the Limitation Act 1980

***JA Pye (Oxford) Ltd* v *Graham (Ch),* [2000]** 2 **137**

Tenancy

Tenant holding over — Ex-tenant claiming title — Para 5 of Sched 1 to 1980 Act — Whether *Powell* v *McFarlane* (1979) 38 P&CR 452 requirements applied to ex-tenant holding over without payment of rent — Whether ex-tenant acquired title — Held, *Powell* v *McFarlane* requirements did not apply to ex-tenant holding over — Ex-tenant acquired title

***Williams* v *Jones (CA),* [2002]** 3 **69**

Intention

Evidence negating intention

Keys to disputed land provided to others by alleged squatter — Whether supply of keys to third parties negativing intention to possess — Whether intention changed when keys no longer supplied —Held, supply of keys negativing intention to possess — Cessation of supply of keys did not constitute change of intention

***Battersea Freehold & Leasehold Co Ltd* v *Wandsworth London Borough Council (Ch),* [2001]** 2 **75**

Intention to exclude all others

Tenant with factual possession — Whether tenant in factual possession intended to exclude all others including landlord — Whether erection of fence equivocal — Whether upon determination of tenancy any possessory title acquired by tenant passes to landlord as against third parties — Held, tenant had insufficient intention to exclude others — Erection of fence for stock not unequivocal — Presumption that possessory title acquired by tenant passes to landlord may not apply as against third parties

***Batt* v *Adams (Ch),* [2001]** 2 **92**

Limitation period

Commencement of possession under licence — Whether intention to possess following expiration of licences established by facts — Whether squatters had intention to possess disputed land — Held, intention to possess not established — No change in intention following expiration of occupational licences — European Convention right on property does not impinge on Limitation Act 1980

***JA Pye (Oxford) Ltd* v *Graham (CA),* [2002]** 2 **69**

Lease

Registered land

Land Registration Act 1925 s75(1) — Whether on surrender of lease freeholder entitled to recover possession against squatter — Held, squatter's interest not extinguished — Squatter's interest held by leaseholder under s75(1) and binds freeholder on surrender of lease

***Central London Commercial Estates Ltd* v *Kato Kagaku Ltd and other; Axa Equity & Law Life Assurance Society plc (Third Party) (Ch),* [1998]** 3 **55**

Lease in writing

Tenancy

Landlord and tenant — Tenant not paying rent — Whether time ran from expiration of notice to quit or date when rent stopped — Whether "lease in writing" under para 5, Sched 1 to Limitation Act 1980 — Whether tenant can be in adverse possession under Landlord and Tenant Act 1954 continuation tenancy — Held, documents were not a lease in writing — No acknowledgement of title

***Long* v *Tower Hamlets London Borough Council (Ch),* [1997]** 1 **78**

Limitation period

European Convention right on protection of property

Whether acquisition of land by adverse possession breaches European Convention rights — Whether European Convention impinges upon Limitation Act 1980 — Held European Convention right on property does not impinge upon Limitation Act 1980

***JA Pye (Oxford) Ltd* v *Graham (CA),* [2002]** 2 **69**

Tenant

Animus possidendi

Non-payment of rent — Tenant wrongly believing he was paying rent — Whether tenant's possession was adverse pursuant to Limitation Act 1980, Sched 1, para 8(1) — Held, an intention to possess rather than to become owner shows *animus possidendi* and satisfies requirement of adverse possession

***Lodge* v *Wakefield Metropolitan City Council (CA),* [1995]** 2 **124**

Wall

Animus possidendi

Whether intention to claim wall sufficiently manifested to the world — Held, acts of using, cutting through and attaching things to wall sufficient conduct to the world — Wall acquired by adverse possession

***Prudential Assurance Co Ltd* v *Waterloo Real Estate Inc (CA),* [1999]** 2 **85**

AGENCY

Authority — See also **Housing** (Environment — Statutory nuisance)

Solicitors

Tenancy

Negotiations for 10–year term — Intended tenant allowed into occupation — Intended landlord dying before agreement completed — Whether tenancy granted — Whether solicitor had authority to conclude agreement to grant tenancy — Held, solicitor had no authority to conclude agreement — No apparent authority and any actual authority ended on incapacity of principal

***James* v *Evans (CA),* [2000]** 3 **1**

AGREEMENT

Breach

Lock-out agreement

Exclusivity or lock-out agreement — Damages — Lock-out agreement not to negotiate or treat with others for lock-out period — Whether breach of agreement — Whether breach cause of loss — Whether loss of chance — Whether claimant entitled to damages — Conduct of estate agents — Held, breach of agreement — Breach did not cause alleged loss — Damages not payable

***Dandara Holdings Ltd* v *Co-operative Retail Services (Ch),* [2004]** 2 **163**

Stokes principle

Valuation method

Dispute over right of way compromised by Tomlin order incorporating agreement to purchase access by reference to Stokes principle — Whether costs of development to be deducted from site value — Held, no costs to be deducted from valuation already on net basis

***Challock Parish Council* v *Shirley (Ch),* [1995]** 2 **137**

AGRICULTURAL HOLDINGS — See also **Capital gains tax** (Retirement relief — Agricultural land); **Mortgage** (Possession — Tenancy)

Agreement

Negotiations "subject to contract"

Negotiations for 10-year term — Intended tenant allowed into occupation — Intended landlord dying before agreement completed — Whether tenancy granted — Whether proprietary estoppel — Whether s2(5) of Law of Property (Miscellaneous Provisions) Act 1989 applied — Whether solicitor had authority to conclude agreement to grant tenancy — Whether summary judgment under CPR 24.2 at commencement of trial where no case management identified issues — Held, no proprietary estoppel within s2(5) of 1989 Act — Solicitor had no authority to conclude agreement — CPR 24.2 should not be used at commencement of trial — There should have been earlier case management conference to identify issues

***James* v *Evans (CA),* [2000]** 3 **1**

Agricultural tenancy

Agricultural use

Covenant restricted use of holding for "agricultural purposes only" — Tenant proposing farm visit activities on holding — Provision of open farm activities — Whether proposal contrary to user clause — Whether proposal "agricultural purposes only" — Held, proposed use contrary to user restriction — Proposed use not "agricultural purposes only"

***Jewell* v *McGowan (CA),* [2002]** 3 **87**

Proprietary estoppel

Tenant's conduct defeating claim for injunction to prevent trespass — Whether agricultural tenancy protected by statute defeasible by proprietary estoppel — Measure of damages — Held, tenancy protected by Agricultural Holdings Act 1986 capable of being defeated by proprietary estoppel — Damages for trespass based upon rental value of tenancy

***JS Bloor (Measham) Ltd* v *Calcott (Ch),* [2002]** 1 **1**

Valuation

Valuation of tenancy on dissolution of partnership — Whether any evidence of special purchasers — Whether principles in *Walton's Executors* v *Commissioners of Inland Revenue* applied — Whether valuer's evidence illogical — Held, valuation principles explained — No special purchaser evidence — Principles in *Walton's Executors* v *Commissioners of Inland Revenue* applied

***Greenbank* v *Pickles (CA),* [2001]** 1 **1**

Arbitration

Determination of rent

Rent review — Premises let originally as two holdings — Whether one holding — Whether marriage value with other land relevant — Held, tenant had two holdings — Management agreement and marriage value relevant considerations for arbitrator

***JW Childers Trustees* v *Anker (CA),* [1996]** 1 **1**

Misconduct by arbitrator

Dispute as to whether appellant held agricultural tenancy — Determination of rent — Repairs — Notices to quit — Statutory arbitrator — Whether misconduct by arbitrator in issue of vacillating directions — Whether misconduct in failing to inform parties of request to president of RICS to extend time-limit — Held, no misconduct by arbitrator

***Grammer* v *Lane; Grammer* v *Stone (CA),* [2000]** 1 **1**

Procedure

Dispute as to whether appellant held agricultural tenancy — Determination of rent — Repairs — Notices to quit — Whether landowner in dispute about existence of tenancy entitled to initiate statutory arbitrations relating to rent and repairs — Statutory arbitrator — Whether arbitrator entitled to determine appropriate course of action where existence of tenancy being disputed — Whether landowner entitled to make "without prejudice" use of arbitral provisions in Agricultural Holdings Act 1986 — Held, no reason in principle why party should not pursue "without prejudice" arbitrations

***Grammer* v *Lane; Grammer* v *Stone (CA),* [2000]** 1 **1**

Covenant to repair

Damages

Annual tenancy — Breach by tenant — Claim by landlord for damages — Assessment of damages — Whether damages should equal cost of repairs — Whether landlord had proved damage — Held, discounted cost of repair proper damages — Landlord had proved damages

***Crewe Services & Investment Corporation* v *Silk (CA),* [1998]** 2 **1**

Inheritance tax — See **Taxation** (Inheritance tax)

Milk quota — See **Milk quota**

Mortgage

Fraudulent tenancy agreement

Mortgage of agricultural holding — Grant of tenancy agreement by mortgagor to mortgagor's wife — Whether agreement to be set aside as fraudulent transaction under s423 of Insolvency Act 1986 — Whether transaction at an undervalue — Held, agreement set aside as fraud on mortgagee — Benefit in transaction to tenant showed transaction at an undervalue

***Agricultural Mortgage Corporation plc* v *Woodward (CA),* [1995]** 1 **1**

Notice to quit

Case B

Early entry clause in relation to part of holding for mineral workings — Planning permission for mineral extraction included restoration and aftercare conditions — Whether landlord entitled to rely on early entry clause — Whether "use" in Case B of Sched 3 to Agricultural Holdings Act 1986 excluded mineral extraction and restoration activities — Held, landlord entitled to rely on early entry clause — "Use" in Case B includes mineral working and restoration activities

***Floyer-Acland* v *Osmond (CA),* [2000]** 2 **1**

Case D

Notice to pay rent — S22(2)(d) of Agricultural Holdings (Scotland) Act 1991 — Whether rent due for purposes of s22(2)(d) where tenant has common law right of retention (set off) by reason of landlord's material breach of repairing obligation — Whether issue of landlord's material breach matter for arbitration under 1991 Act — Held, tenant's rights of retention against rent for material breaches of landlord's obligations unaffected by statutory scheme under 1991 Act — Notice to quit invalid if notice to pay rent invalid — Issues of whether rent due and whether landlord in material breach matters for arbiter

***Alexander* v *Royal Hotel (Caithness) Ltd (Ct Sess: Inner House),* [2001]** 1 **6**

Counternotice

Joint tenants — Notice to quit — Whether one joint tenant can compel other tenant to join in service of counternotice — Whether interlocutory mandatory injunction can be granted — Held, on arguable case that joint tenants agreed tenancy would continue injunction ordered

***Cork* v *Cork (Ch),* [1997]** 1 **5**

Enforceability

Protection from Eviction Act 1977 — Whether premises let as a dwelling — Whether notice to quit agricultural holding with dwelling must comply with 1977 Act — Held, premises let as an agricultural holding and not as a dwelling — Notice to quit without compliance with 1977 Act was valid

***National Trust for Places of Historic Interest or Natural Beauty* v *Knipe (CA),* [1997]** 2 **9**

Estoppel

Case B notice to quit — Whether common assumption that tenant would only give up possession on being provided with new buildings — Whether landlord estopped from serving notice to quit relying on planning permission — Appeal from [1995] 1 9 — Held, landlord estopped from recovering possession

***John* v *George (CA),* [1996]** 1 **7**

Head tenancy

Effect on subtenancy — Head tenants failing to serve counternotice — Whether subtenancy determined — Whether prearrangement between freeholder and head tenants a consensual transaction tantamount to a surrender — Whether *Sparkes* v *Smart* [1990] 2 EGLR 245 wrongly decided — Held, service of notice to quit by freeholder on head tenant effective to determine subtenancy — Any prearrangement with head tenant not to serve counternotice not a consensual transaction tantamount to a surrender — *Sparkes* v *Smart* [1990] 2 EGLR 245 overruled

***Barrett* v *Morgan (HL),* [2000]** 1 **8**

Subtenancy

Headlease — Notice to quit — Tenants failing to serve counternotice — Whether subtenancy determined — Whether *Sparkes* v *Smart* [1990] 2 EGLR 5 wrongly decided — Held, where collusive agreement between landlord and headtenant, notice to quit operates as a surrender of headlease — Subtenancy not destroyed where tenant fails to serve counternotice to protect headlease

***Barrett* v *Morgan (CA),* [1998]** 3 **3**

AGRICULTURAL HOLDINGS — *continued*
Notice to quit — *continued*

Validity
Case E — Subtenancy granted in breach of tenancy — Whether landlord would suffer material prejudice because on determination of tenancy landlord might be forced into direct relationship with undesirable subtenant — Held, no material prejudice because upwards notice to quit determines subtenancy
***Pennell* v *Payne (CA),* [1995]** 1 **6**

Sharing of occupation — Tenant farming through partnership — Whether tenant in breach of covenant not to part with or share possession or occupation — Held, that in case of arable land no sharing and no breach
***Wallace* v *C Brian Barratt & Son Ltd (CA),* [1997]** 2 **1**

Yearly tenancy expressed to commence on 29 September — Whether notice to quit and deliver up possession on 28 September valid — Held, notice valid whether possession demanded on 28 or 29 September
***Yeandle* v *Reigate and Banstead Borough Council (CA),* [1996]** 1 **20**

Security of tenure
Anniversary of deemed yearly tenancy
Agricultural Holdings Act 1986 — Tenancy of six months taking effect as yearly tenancy under s2(1) of Act — Whether deemed yearly tenancy from date of agreement or expiration of original contractual term — Held, by s2(1) tenancy took effect as yearly tenancy from date of agreement
***Calcott* v *JS Bloor (Measham) Ltd (CA),* [1998]** 3 **1**

Succession
Principal source of livelihood
Appellant made drawings from partnership account of which in four of seven years more than 50% was attributable to injections of capital from non-agricultural sources — Whether appellant derived his principal source of livelihood from agricultural work — Held, test in Agricultural Holdings Act 1986 s36(3) (*a*) satisfied as livelihood due to agricultural work upon holding
***Welby* v *Casswell (CA),* [1995]** 2 **1**

Tenancy outside Agricultural Holdings Act
Minister's consent
Consent granted to a tenancy of not more than five years commencing on specified date — Tenancy granted after specified date for same term — Whether tenancy outside Agricultural Holdings Act 1986 — Held, tenancy outside 1986 Act — Conditions of ministerial consent satisfied
***Jones* v *Owen (Ch),* [1997]** 2 **5**

Tenancy within Agricultural Holdings Act — See **Capital gains tax** (Retirement relief — Disposal of business)

AGRICULTURE
Common market
Livestock quotas
Organisation of livestock markets — Grant of transferable production quotas — Whether landlords entitled to compensation where holdings depreciated by transfer of quotas by tenants — Held, neither general principles of law nor the relevant regulations require member states to introduce compensation arrangements for landlords
***R* v *Minister of Agriculture, Fisheries and Food, ex parte Country Landowners' Association (Eur Ct),* [1996]** 2 **1**

ALLOTMENTS
Adequate provision
Alternative allotments
Allotments Act 1925 — Meaning of "adequate provision" in s8 — Whether alternative allotment to be equal or commensurate in fitness to allotment from which plotholders to be displaced — Appeal from [1996] 1 EGLR 147 — Held, "adequate provision" means that allotment gardening can reasonably be undertaken on alternative site
***R* v *Secretary of State for the Environment, ex parte Gosforth Allotments & Gardens Association (CA),* [1996]** 2 **117**

APPEAL — See also **Lands Tribunal**; **Procedure** (Possession proceedings — Appeal)

Civil Procedure Rules
Refusal of permission to appeal
Small claims track — Claim seeking damages for trespass — Hearing without oral evidence — Award of nominal damages only — Award of costs on refusal of permission to appeal — Whether appeal against award of costs — Whether any remedy where permission to appeal refused — Whether attorney appointed under Enduring Powers of Attorney Act 1985 entitled to conduct litigation — Whether appeal against decision refusing judicial review of decision of local planning authority not to take enforcement action — Held, no appeal against costs order — No remedy against decision to refuse permission to appeal save where complete denial of justice — Appointed attorney not entitled to conduct litigation — No appeal against decision to refuse judicial review
***Gregory* v *Turner; R (on the application of Morris)* v *North Somerset District Council (CA),* [2003]** 3 **129**

County court
Landlord and tenant
Trial judge deciding two points in possession proceedings — Proceedings dismissed by trial judge — Whether appellant tenant entitled to appeal on matter decided against him — Held, tenant entitled to appeal — Validity of notice would be relevant in subsequent proceedings — No reason why tenant should face risk of issue estoppel
***Clickex Ltd* v *McCann (CA),* [1999]** 2 **63**

ARBITRATION — See also **Agricultural holdings** (Arbitration)(Notice to quit — Case D); **Housing Grants, Construction and Regeneration Act 1996** (Construction contract — Adjudicator)
Appeal
Issue estoppel
Construction of rent review clause — Whether leave should be given to appeal arbitrator's award — Whether issue estoppel would arise at future reviews by reason of arbitrator's award on construction of clause — Held, leave refused — Unjust if landlords would be entitled to object to construction summons on point of construction decided by arbitrator throughout rest of 125-year term
***British Railways Board* v *Ringbest Ltd (Ch),* [1996]** 2 **82**

S68 of Arbitration Act 1996
Rent review — Arbitrator taking into account personal experience of specific transactions — Whether arbitrator should have given parties opportunity to comment — Whether arbitrator failing to address tenant's evidence concerning over-supply of and poor demand for comparable properties in locality — Whether arbitrator committing serious irregularity within s68(2) of Arbitration Act 1996 — Whether substantial injustice to tenant — Held, arbitrator entitled to rely upon personal knowledge and experience for which he was appointed — No serious irregularity — No substantial prejudice
***Checkpoint Ltd v Strathclyde Pension Fund (CA),* [2003]** 1 1

Appointment of arbitrator
Time limit extension
Rent review — S12 of Arbitration Act 1996 — Extension of time for appointment of arbitrator — Conditions — Time limit for application to appoint arbitrator — Time limit becoming of the essence following *Starmark Enterprises Ltd* v *CPL Distribution Ltd* [2001] EWCA Civ 1252 — Whether change of law outside contemplation of parties at date of lease — Whether just to extend time — Held, change in the law not outside parties' contemplation — Failure to make timeous application not outside such contemplation
***Monella* v *PizzaExpress (Restaurants) Ltd (Ch),* [2004]** 1 **43**

Arbitration Act 1996
S68 serious irregularity — Appeal — Landlord and tenant — Rent review — Whether serious irregularity by arbitrator — Whether leave to appeal under s69 — Held, no serious or any irregularity — Leave to appeal refused
***Checkpoint Ltd* v *Strathclyde Pension Fund (Ch),* [2002]** 2 **97**

Arbitrator
Misconduct
Application for extension of time to apply for remission of award — Whether substantial prejudice — Whether arbitrator wrong in rejecting tenant's discount for onerous nature of rent review clause — Failure to adduce evidence or make submissions on point — Held, application dismissed — No prejudice — Arbitrator entitled to refer to evidence without hearing submissions on the point
***Broadgate Square plc* v *Lehman Brothers Ltd (No 2) (Ch),* [1995]** 2 **5**

Application to remove arbitrator — Whether demand and receipt of interim payment of arbitrator's fees from one party misconduct — Held, demand of interim fees on previously agreed contractual basis not misconduct —Receipt of fees by one party not objectionable
***Turner* v *Stevenage Borough Council (Ch),* [1996]** 1 **23**

ARBITRATION — *continued*
Arbitrator — *continued*
Misconduct — continued

Arbitrator failed to inspect comparables contrary to his directions — Whether misconduct — Whether leave to apply out of time to remit or set aside — Held, misconduct — Extension of time granted and award remitted
***Oakstead Garages Ltd* v *Leach Pension (Trustees) Ltd (Ch),* [1996]** 1 **26**

Directions of arbitrator related to comparables and arbitrator's inspections — Whether misconduct of arbitrator in failing to abide by own directions — Held, arbitrator misconducted himself
***Mount Charlotte Investments plc* v *Prudential Assurance (Ch),* [1995]** 1 **15**

Time-limits
Trigger notice — Counternotice to be served within three months — Time of the essence — Whether extension of time for trigger notice under Arbitration Act 1996 s12 — Held, no extension of time — Tenant failed to satisfy requirements of s12
***Fox & Widley* v *Gurham (QB),* [1998]** 1 **91**

Award
Application to set aside or remit
Application for extension of time — Failure of arbitrator to abide by own directions relating to comparables and arbitrator's inspections — Whether delay minor and misconduct made out — Held, extension of time granted as delay minor and misconduct established
***Mount Charlotte Investments plc* v *Prudential Assurance (Ch),* [1995]** 1 **15**

Remission
Arbitrator made finding that specification of comparable building different from subject building — Whether failure to allow parties to adduce evidence of respective specifications breach of rules of natural justice — Whether award made in breach of rules capable of remission under s22 of Arbitration Act 1950 — Held, award in breach of rules capable of remission — Misconduct of arbitrator in failing to give opportunity to parties to adduce evidence
***Land Securities plc* v *Westminster City Council (No 2) (Ch),* [1995]** 1 **45**

Evidence
Comparables
Hearsay evidence — Whether misconduct by arbitrator — Whether arbitrator failed to follow his own directions — Whether arbitrator took into account hearsay evidence — Whether arbitrator required to give decision on admissibility of evidence before making award — Judicial guidance on meaning and use of comparables — Held, no misconduct — Arbitrator did not use hearsay evidence — Observations on comparables approved
***Living Waters Christian Centres Ltd* v *Fetherstonhaugh (CA),* [1999]** 2 **1**

Milk quota — See **Milk quota** (Partnership asset — Dissolution of partnership)
Rent review — See **Agricultural holdings** (Arbitration)
Subpoenas "duces tecum"
Setting aside
Subpoenas required expert witness to produce proofs in two earlier arbitrations — Whether subpoenas should be set aside — Held, where witness makes prior inconsistent statement earlier statement admissible — Necessary for fair disposal of proceedings
***London & Leeds Estates Ltd* v *Paribas Ltd (No 2) (QB),* [1995]** 1 **102**

ARREARS — See **Landlord and tenant** (Equitable set-off — Assignment of right to recover rent) (Forfeiture — County Courts Act 1984 s138) (Forfeiture — Relief) (Forfeiture — Validity) (Rent arrears)

ASSIGNMENT — See **Landlord and tenant** (Assignment — Collateral deed) (Assignment — Consent) (Assignment — Landlord and Tenant Act 1988) (Assignment — Refusal of consent) (Assignment — Right of pre-emption) (Covenants — Assignment) (Joint tenancy — Assignment) (Leasehold Reform, Housing and Urban Development Act 1993 — Assignment of tenant's notice to new lease) (Option — Enforceability) (Option to determine lease — Validity) (Rent Acts — Possession); **Leasehold enfranchisement** (Leasehold Reform Act 1967 — Assignment of statutory right of enfranchisement)

AUCTIONS — See also **Local Government** (Sale of land — S123 Local Government Act 1972)
Without reserve
Collateral contract
Lots offered without reserve — Auctioneer withdrawing lots after bids he considered inadequate — Whether auctioneer entitled to withdraw lots after bids made — Whether collateral contract between auctioneer and highest bidder that lot would be sold to highest bidder — Assessment of damages — Held, collateral contract under which auctioneer promises to sell to highest bidder where lots offered without reserve — Breach of s57(4) of Sale of Goods Act 1979 — Damages difference between what would have been auction price and value of lots
***Barry* v *Heathcote Ball & Co (Commercial Auctions) Ltd (CA),* [2000]** 3 **7**

BANKRUPTCY — See also **Insolvency** (Bankruptcy — Foreign property); **Landlord and tenant** (Rent Acts — Possession); **Rent Acts** (Rent Act 1977 — Possession)
Lease
Forfeiture
Tenant's fixtures — Whether tenant entitled to remove tenant's fixtures following peaceable re-entry — Held, no right for tenant to recover tenant's fixtures
***Palmiero: debtor 3666 of 1999, Re (Ch),* [1999]** 3 **27**

BETTING OFFICE — See **Landlord and tenant** (Landlord and Tenant Act 1954 Part II — New tenancy grant)

BISON SYSTEM — See **Negligence** (Mortgage loan report — Damages)

BLOCK EXEMPTION — See **Landlord and tenant** (Tied public houses)

BOUNDARIES
Hedge and ditch presumption
Conveyance
Conveyance by reference to Ordnance Survey plan — Hedge and ditch presumption — Whether presumption rebutted — Appeal from [1998] 2 107 — Held, conveyance by reference to Ordnance Survey field enclosure numbers did not displace hedge and ditch presumption
***Alan Wibberley Building Ltd* v *Insley (HL),* [1999]** 2 **89**

BREAK CLAUSE — See **Election** (Waiver — Landlord and tenant); **Landlord and tenant** (Break clause) (Double rent — Distress for Rent Act 1737) (Part II of Landlord and Tenant Act 1954 — Notice to terminate tenancy) (User covenant — Landlord and Tenant Act 1927)

BUILDING REGULATIONS — See **Procedure** (Civil Procedure Rules — Pre-action discovery for documents)

"BWLLFA" PRINCIPLE — See **Compensation for acquisition of land** (Valuation — Injurious affection)

CAPITAL GAINS TAX — See also **Compensation for acquisition of land** (Land Compensation Act 1961)
Option to purchase land
Expenditure
Obligation on landowner to secure release from restrictive covenants — Expenditure on obligation — Whether expenditure deductible in assessing tax on option price — Held, expenditure not deductible in calculation of tax under Capital Gains Tax Act 1979
***Garner (Inspector of Taxes)* v *Pounds Shipowners & Shipbreakers Ltd (HL),* [2000]** 2 **69**

Valuation
Half-share in property
Comparables — Anonymised comparables derived from stamp duty office returns — Whether 10% proper deduction for half-share in property — Whether anonymised comparables admissible — Guidance on admissibility of transactional evidence within Inland Revenue stamp duty office — Costs — Held, 10% proper deduction for half-share — Tribunal should normally order disclosure of evidence held by stamp duty office — Guidance on admissibility of such evidence
***Newman (HMIT)* v *Hatt (LT),* [2002]** 1 **89**

Retirement relief
Agricultural land
Family farming company — Disposal of agricultural land but not of company shares — Whether "material disposal" of business assets — Whether retired farmer entitled to retirement relief from capital gains tax — Held, there was material disposal of business assets by reason of 69(2)(b) and 4(a) of Finance Act 1988 — Retirement relief allowed
***Plumbly* v *Spencer (HMIT) (CA),* [1999]** 2 **191**

Disposal of business
Surrender of agricultural tenancy — Tenant receiving capital payment — Tenant continuing to farm under licence — Finance Act 1985, s69 — Whether surrender of tenancy constituted the disposal of the whole or part of a business — Whether tenant entitled to retirement relief on chargeable gain — Held, because tenant continued to farm under licence his business continued — No disposal of whole or part of business — No retirement relief
***Barrett* v *Powell (Ch),* [1998]** 2 **117**

CAPITAL GAINS TAX — *continued*
Retirement relief — *continued*

Sale of milk quota
Disposal of milk quota one year after sale of dairy herd — Whether taxpayer entitled to retirement relief from capital gains tax — Held, sale of quota not part of disposal of business — No entitlement to retirement relief
***Wase* v *Bourke (Ch),* [1996]** 1 **164**

CASINO — See **Landlord and tenant** (Rent review clause — Construction: general)
CHANCEL REPAIRS
Chancel Repairs Act 1932
Lay rector
Whether liability of lay rector of part of rectorial lands uncertain — European Convention on Human Rights — Art 1 of First Protocol — Arts 9 and 14 of Convention — Whether law on chancel repair liability breaches Convention — Held, law not uncertain — Lay rector of part of rectorial lands liable for chancel repairs — Liability does not involve breach of Convention
***Aston Cantlow Parochial Church Council* v *Wallbank (Ch),* [2000]** 2 **149**

CHARGING ORDER
Charging order absolute
Instalment payment order
S86 of County Courts Act 1984 — Instalment payment order made prior to charging order absolute — Whether county court having jurisdiction to make charging order absolute — Whether proper exercise of discretion — Held, county court having jurisdiction to make charging order absolute following instalment payment order — Discretion properly exercised
***Ropaigealach* v *Allied Irish Bank plc (CA),* [2002]** 1 **83**

Lease
Forfeiture
County Courts Act 1984 s138(9C) — Whether holder of charging order entitled to relief from forfeiture — Whether holder of charging order owner of interest within meaning of s138(9C) — Whether possession order should be set aside — Held, owner of charging order holds interest within s138(9C) and entitled to relief from forfeiture
***Croydon Unique Ltd* v *Wright (CA),* [1999]** 3 **28**

CIVIL PROCEDURE
Discovery — See **Appeal** (Civil Procedure Rules — Refusal of permission to appeal); **Procedure** (Civil Procedure Rules — Expert evidence)

COMMISSION — See **Estate agents** (Cold-calling letter — Entitlement to fees) (Commission — Agents acting for purchaser) (Commission — Sole selling rights) (Commission — Unfair terms)

COMMON MARKET
Agriculture
Livestock quotas
Organisation of livestock markets — Grant of transferable production quotas — Whether landlords entitled to compensation where holdings depreciated by transfer of quotas by tenants — Held, neither general principles of law nor the relevant regulations require member states to introduce compensation arrangements for landlords
***R* v *Minister of Agriculture, Fisheries and Food, ex parte Country Landowners' Association (Eur Ct),* [1996]** 2 **1**

COMMONS
See **Easement** (Acquisition by prescription — Village green); **Town or village greens** (Class (c) green — Registration) (Commons Registration Act 1965 — Registration) (Registration — Sports and pastimes as of right)

COMMONS — See also **Town and village greens** (Class (c) green — Registration)(S22 of Commons Registration Act 1965 — Registration) (Registration — Sports and pastimes as of right)
Byelaws
New Forest
Enforcement — Magistrates' court — Court of Swainmote — Jurisdiction — Commoners failing to pay marking fees — Whether magistrates having jurisdiction — Whether Court of Swainmote having sole jurisdiction — Whether Court of Swainmote compliant with Art 6 of European Convention on Human Rights — Held, magistrates had jurisdiction concurrently with Court of Swainmote — Doubtful whether Court of Swainmote compliant with Art 6
***Verderers of the New Forest* v *Young (QB),* [2004]** 1 **1**

Registration
Town or village green
Green omitted from registration pre-1970 — Whether council entitled to refuse to resister based on 20-years' use post 2 January 1970 — Locality — Appeal from [1995] 2 233 — Held, application failed on inadequate evidence of use as of right — *Quaere* whether registration creates rights
***R* v *Suffolk County Council, ex parte Steed (CA),* [1997]** 1 **131**

Rights of common
Extent of grazing rights
Trespass — Right to graze two horses — Commoner erecting electric fence to contain horses — Whether erection of electric fence on common constituting trespass — S194 of Law of Property Act 1925 — Whether erection of fence on common land with ministerial consent depriving landowner of action in trespass in common law — Whether lord of manor entitled to remedy of injunction — Held, erection of electric fence constituting trespass — Grant of ministerial consent under s194 not depriving owner of action in trespass — Injunction granted
***Rabett* v *Poole (CC),* [2003]** 3 **143**

COMMONS REGISTRATION ACT 1965 — See **Town and village greens** (Class (c) green — Registration) (S22 of Commons Registration Act 1965 — Registration) (Registration — Sports and pastimes as of right); **Village green**

COMPENSATION — See also **Compulsory purchase**; **Landlord and tenant** (Landlord and Tenant Act 1954 — Request for new tenancy) (Landlord and Tenant Act 1954 Part II); **Lands Tribunal** (Procedure — Points raised by member); **Local land charge** (Local Land Charges Act 1975 — Compensation); **Restrictive covenants** (Modification — S84 Law of Property Act 1925); **Town and country planning** (Blight notice) (Enforcement — Stop notice)
Certificate of appropriate alternative development
Relevant date and policies
Relevant date for identifying policies — Whether date of entry or date of making of compulsory purchase order — Whether only scheme of acquisition or additionally underlying facts and policies to be disregarded — Held, relevant date is date of notice under Land Compensation Act 1961 s22(2)(*a*) of making of compulsory purchase order — Under s17(4) disregard scheme only not underlying facts and policies — Assume scheme only cancelled
***Secretary of State for the Environment* v *Fletcher Estates (Harlescott) Ltd; Secretary of State for the Environment* v *Newell (CA),* [1998]** 3 **13**

Compulsory purchase — See also **Compensation for electricity lines** (Statutory basis of compensation — Valuation)
McCarthy Rules
Petrol filling station — Obstruction of access roads — Whether compensation payable under s10 — Effect of stopping-up orders — Whether obstructions sufficiently proximate — Held, stopping-up orders did not precede obstruction — *Joliffe* v *Exeter Corporation* [1967] 1 WLR 993 distinguished — Compensation payable only for proximate obstructions on land in CPO
***Wagstaff* v *Department of the Environment (LT),* [1999]** 2 **108**

Compulsory Purchase Act 1965 s10
Temporary damage
Works causing temporary damage to hotel — Whether compensation limited to direct physical interference with land — Whether compensation payable for all injurious affection — Whether compensation payable for temporay interference — Whether compensation payable for reduction in letting value during temporary interference — Held, compensation not payable for damage that is not a direct physical interference with land — No compensation for noise, dust and vibration — Compensation not payable for non-actionable injurious affection — Compensation payable for temporary damage — Measure of compensation includes reduction in letting value
***Wildtree Hotels Ltd* v *Harrow London Borough Council (HL),* [2000]** 2 **5**

Disturbance — See **Compensation for acquisition of land** (Disturbance); **Landlord and tenant** (Landlord and Tenant Act 1954 Part II — Compensation for disturbance)

Highways Act 1980 s28
Footpath creation order
Claim for compensation — Ss 9 and 39 of Limitation Act 1980 — Whether claim for compensation statute-barred — Whether compensation authority estopped from relying upon expiration of limitation period — Held, claim statute-barred — Compensating authority not estopped from relying upon expiration of limitation period
***Rotherwick's Executors* v *Oxfordshire County Council, (LT),* [2000]** 2 **84**

Land Compensation Act 1961
Article 4 direction
Direction withdrawing permitted development to run motocross meetings — Direction not confirmed by Secretary of State — Claim for compensation — Whether collateral statutory nuisance procedure to prevent motocross negates loss — Whether r (4) of s5 of Land Compensation Act 1961 relevant — Held, on facts collateral procedure unlikely to prevent use — R (4) not applicable — Compensation payable
***Bolton* v *North Dorset District Council (LT),* [1997]** 2 **180**

COMPENSATION — *continued*

Land Compensation Act 1973 Part I
Altered highway
Whether resurfacing of road an alteration — Whether evidence of acoustics experts helpful — Determination of compensation — Held, part of works resurfacing not alterations — Subjective views of claimants on effect of noise more useful than evidence of acoustic experts — Compensation at 22-23% of unaffected values and 80% of total depreciation
***King* v *Dorset County Council (LT),* [1997]** 1 **245**

Immunity
Physical factors caused by use of primary school — Depreciation in value of claimant's land — Whether immunity from nuisance actions conferred on education authority — Whether claim for depreciation can be maintained — Held, no immunity conferred by Education Act 1944 — Claim for compensation precluded by s1(6) of 1973 Act
***Marsh* v *Powys County Council (LT),* [1997]** 2 **177**

Limitation
Offer of *ex gratia* payment in full and final settlement — Whether offer accepted — Whether reference statute-barred — Held, offer of *ex gratia* payment accepted by banking of enclosed cheques — Principle in *Hillingdon London Borough Council* v *ARC Ltd* [1998] 3 EGLR 18 applied to 1973 Act — Claims time-barred six years after first claim day
***Bateman* v *Lancashire County Council (LT),* [1999]** 2 **203**

Physical factor
Construction of fire station — Affected properties included dwelling subject to regulated tenancy — Whether properties depreciated in value by noise created on and in fire station — Held, properties depreciated by 6.25% — Compensation payable based on opinion evidence of values
***Wakeley* v *London Fire and Civil Defence Authority (LT),* [1996]** 2 **148**

Validity of claim
Compensation paid to company controlled by claimant for acquisition of land for construction of motorway — Compensation included injurious affection to residue of agricultural holding including principal house — Whether claimant as owner of principal house entitled to claim compensation under Part I of Land Compensation Act 1973 — Decision, claim barred by s8 as compensation already paid for same depreciation — Claimant estopped — Corporate veil can be pierced
***Bannocks* v *Secretary of State for Transport (LT),* [1995]** 2 **157**

Whether notice of claim for "an amount in excess of £50 to be agreed" valid — Whether tribunal has jurisdiction to entertain reference — Decision, claim invalid — Tribunal had no jurisdiction
***Fennessy* v *London City Airport Ltd (LT),* [1995]** 2 **167**

Valuation
Sheltered housing let on secure tenancies — Whether planning restrictions on use and existence of secure tenancies relevant in determining value and depreciation — Held, claimants' interest affected by restrictions and tenancies is relevant interest to be valued — No evidence of depreciation
***Clwyd Alyn Housing Association Ltd* v *Welsh Office (LT),* [1996]** 1 **184**

Valuation evidence
Compensation for depreciation in value attributable to bypass — Whether two valuers necessary — Burden of proof — Whether evidence of settlements and awards nationally relevant — Whether any depreciation in value — Whether subject property benefited by bypass — Held, one valuer with local knowledge best assists tribunal — National settlements and awards irrelevant — No depreciation in value
***Hallows* v *Welsh Office (LT),* [1995]** 1 **191**

Limitation period — See also **Compulsory purchase** (Compensation — Limitation period) (Lands Tribunal — Limitation period)
Waiver
Notice to treat — Land Compensation Act 1961, s31 — Withdrawal of notice to treat — Whether claim time-barred under s9 of Limitation Act 1980 — Determination of compensation — Held, by reason of waiver and estoppel claim not time-barred — Robust award of £500
***Williams* v *Blaenau Gwent County Borough Council (No 2) (LT),* [1999]** 2 **195**

Refusal of consent to fell tree
Remedial works
Compensation for refusal of consent to fell ash tree — Tree root damage to property — Remedial works carried out in consequence — Whether cost of remediation and of appeal against refusal of consent recoverable in whole or in part — Whether interest recoverable — Held, costs recoverable in part
***Buckle* v *Holderness Borough Council (LT)* , [1996]** 2 **133**

S10 of Compulsory Purchase Act 1965
Entitlement
Obstructions to highway adjoining retail premises — Construction of bridges — Provision of hoarding to works and materials store — Actionability — Whether hoarding actionable as public nuisance in absence of statutory powers — Held, erection of hoarding actionable — Entitlement to compensation
***Ocean Leisure Ltd* v *Westminster City Council (CA),* [2004]** 3 **9**

Tree preservation order — See **Town and country planning** (Tree preservation order)

COMPENSATION FOR ACQUISITION OF LAND — See also **Compulsory purchase** (Compensation — Limitation)

Betterment
Highway
Land acquired for link road — Planning permission for development of retained land dependent on completion of link road — Whether retained land benefited by highway — Whether betterment to be set off against compensation — Whether discretion should be exercised to award agricultural value of land taken — Decision, betterment to be set off against compensation — No ground for awarding agricultural value
***Leicester City Council* v *Leicestershire County Council (LT),* [1995]** 2 **169**

Certificate of appropriate alternative development
Challenge
Certificate issued by Secretary of State for car parking — Whether s5 of Land Compensation Act 1961, r (2), relevant to considerations underlying certificate — Whether use of land for public open space by authority with compulsory powers to be disregarded under s17(4) — Held, r(2) not relevant to planning assumptions under 1961 Act — S17(4) therefore inapplicable
***Maidstone Borough Council* v *Secretary of State for the Environment (CA),* [1996]** 1 **29**

Issue estoppel
Certificate issued on appeal for residential development — Whether evidence can be adduced by acquiring authority to show that conclusions by which certificate issued should not be followed in assessing compensation for severance and injurious affection — Appeal from [1995] 2 EGLR 175 — Held, *issue estoppel* did not apply — Issue determined by Secretary of State not the same as issue to be determined by Lands Tribunal
***Porter* v *Secretary of State for Transport (CA),* [1996]** 2 **10**

Matters to be disregarded
Bypass scheme — Planning history including safeguarding of route — Relevant date for indentifying policies — Whether date of entry or date of notice of making compulsory purchase order — Matters to be disregarded by local planning authority — Whether scheme of acquisition and underlying proposal to be regarded as cancelled on relevant date — Whether all effects of proposal arising before relevant date to be additionally disregarded — Held, relevant date is date of notice of making compulsory purchase order — Only scheme of acquisition and underlying proposal to be regarded as cancelled on relevant date — No assumption to be made about effects of proposal arising prior to relevant date
***Newell* v *Secretary of State for the Environment; Fletcher Estates (Harlescott) Ltd* v *Secretary of State for the Environment (HL),* [2000]** 1 **13**

Costs
Assessment of compensation
Whether claimant entitled to costs of compensation assessment tribunal under the equivalent in Jersey of r (6) of s5 of the Land Compensation Act 1961 — Held, claimant entitled to costs reasonably incurred provided claim notified to tribunal before award
***Lesquende Ltd* v *Planning and Environment Committee of the State of Jersey (PC),* [1998]** 1 **137**

Disturbance
Adaptation works
Claimant's own time — Loss of profits — Bank bridging loan — Whether cost of adapting alternative premises compensatable — Held, cost of adapting alternative premises not compensatable as claimant had value for money and compensation for land taken — Bank bridging loan costs disallowed as claimant failed to mitigate by seeking early advance payment
***Harris* v *Welsh Development Agency (LT),* [1999]** 3 **207**

Extinguishment of business
Acquisition of laundry — Total extinguishment — Whether market value or value to owner — Appropriate years' purchase — Decision, value of business to claimant exceeded market value — Robust award of £50,000 in respect of annual adjusted profits of £28,324
***Sceneout Ltd* v *Central Manchester Development Corporation (LT),* [1995]** 2 **179**

COMPENSATION FOR ACQUISITION OF LAND — *continued*
Disturbance — *continued*
Extinguishment of business — continued

Leasehold factory premises — Goodwill — Loss on forced sale — Accountancy fees — Relevance of RICS Appraisal and Valuation Manual — Multiplier of of adjusted net profit — Effect of sealed offer two days prior to hearing — Held, RICS "Red Book" may be used — Value to owner must not be arbitrary — YP of 2.5
***Shevlin* v *Trafford Park Development Corporation (LT)*, [1998]** 1 **115**

Extinguishment or relocation
Cash and carry warehouse business — Whether total extinguishment or hypothetical relocation — Determination of hypothetical relocation costs — Held, claimant should have relocated and notclosed business down — Compensation assessed on hypothetical relocation basis
***Lamba Trading Co Ltd* v *Salford City Council (LT)*, [1999]** 3 **186**

Loss of profits
Causation — Compulsory Purchase Act 1965 s20 — Whether loss of profits and goodwill caused by dispossession — Whether causation proved — Whether tribunal can make robust award — Held, claimant failed to prove causation of loss attributable to dispossession or threat of dispossession
***Harry Lester Ltd* v *Southwark London Borough Council (LT)*, [1999] 3 179**

Loss of profits due to scheme blight — Loss of anticipated profits on additional facility not constructed — Interest as disturbance element — Calculation of interest — Determination of disturbance compensation in claim based on equivalent reinstatement — R(5) s5 of Land Compensation Act 1961 — Held, disturbance compensation payable for losses due to scheme blight — Interest payable at LIBOR rates plus 2% for increased borrowings due to loss of profits
***Eronpark Ltd* v *Secretary of State for the Environment, Transport and the Regions (No 2) (LT)*, [2001]** 3 **133**

Pre-acquisition losses
Acquisition of business premises — Intention to relocate — Whether claimant entitled to costs of relocation which exceed costs of total extinguishment — Whether reasonable businessman would have relocated — Whether original business extinguished — Held, principles of compensation restated — Compensation not in principle precluded where relocation costs in excess of extinguishment — Reasonable businessman would not have relocated
***Director of Buildings and Lands* v *Shun Fung Ironworks Ltd (PC)*, [1995]** 1 **19**

Riding school business
Appropriate basis of compensation — Whether equivalent reinstatement basis — Whether total extinguishment basis — Whether relocation basis — Relocation compensation — Whether jurisdiction to determine eviction costs — Held, land capable of being valued — No compensation under r (5) equivalent reinstatement basis — Compensation assessed upon relocation basis — Lands Tribunal having no jurisdiction to address costs of eviction under s13(3) of Compulsory Purchase Act 1965
***Prielipp* v *Secretary of State for the Environment, Transport and the Regions (LT)*, [2002]** 3 **143**

Total loss of goodwill
Acquisition of hairdressing business held on tenancy — Whether landlord could have opposed grant of new tenancy — Whether decline in business profits attributable to scheme — Factors affecting appropriate YP multiplier — Held, decline in profits not due to scheme — Other issues decided in favour of claimant — Appropriate YP multiplier was 3.25
***Klein* v *London Underground (LT)*, [1996]** 1 **249**

Disturbance compensation
"Freed-up" time
Mitigation — Compensation assessed for loss of business — Lands Tribunal making deduction for claimant's freed-up time following loss of business — Whether tribunal entitled to make deduction — Whether procedural unfairness — Held, tribunal not entitled to make deduction on its own volition without putting point to the parties and hearing submissions and evidence — Freed-up time aspect of mitigation and otherwise not necessarily exceptional
***Faraday* v *Carmarthenshire County Council (CA)*, [2004]** 2 **5**

Disturbance payment
S37 of Land Compensation Act 1973
Business premises acquired — Tenants served with notices under s25 of Landlord and Tenant Act 1954 — Whether compensation payable under s37(1)(a) where entitlement to compensation under 1954 Act — Development by third party — Whether land "held" by acquiring authority within meaning of s37(1)(c) — Whether claimants displaced in consequence of the carrying out of redevelopment of the land — Held, entitlement to compensation under 1954 Act no bar to entitlement to compensation under s37(1)(a) of 1973 Act — No entitlement to compensation under s37(1)(c) where redevelopment by third party
***Evis* v *Commission for New Towns (LT)*, [2002]** 2 **167**

Equivalent reinstatement
Disturbance losses
R5, s5 of Land Compensation Act 1961 — Whether claimant entitled to compensation for disturbance and/or loss of profits in r 5 case — Held, claimant entitled to compensation for disturbance and loss of profits due to dispossession
***Eronpark Ltd* v *Secretary of State for Transport (LT)*, [2000]** 2 **165**

Interest
R5 equivalent reinstatement compensation for church — Long delay between dispossession and reinstatement — Interest — Whether interest payable on reinstatement costs — Whether interest payable from date of dispossession to date compensation paid — Limitation Act 1980 — Determination of date of cause of action — Held, interest payable on r (5) equivalent reinstatement costs — Interest payable from dispossession until compensation paid — Cause of action for interest claim accrued when amount of compensation agreed
***Halstead* v *Manchester City Council (CA)*, [1998]** 1 **1**

Riding school business
Appropriate basis of compensation — Whether equivalent reinstatement basis — Whether total extinguishment basis — Whether relocation basis — Relocation compensation — Held, land capable of being valued — No compensation under r (5) equivalent reinstatement basis — Compensation assessed upon relocation basis
***Prielipp* v *Secretary of State for the Environment, Transport and the Regions (LT)*, [2002]** 3 **143**

General vesting declaration
Time-limits
Lands Tribunal — Reference to tribunal after expiration of six-year period of limitation — Whether claim barred by s10(3) of Compulsory Purchase (Vesting Declarations) Act 1981 — Whether acquiring authority estopped from relying on statutory limitation period — Held, acquiring authority so estopped — Appeal dismissed from [1996] 2 143
***Co-operative Wholesale Society Ltd* v *Chester-le-Street District Council (CA)*, [1998]** 3 **11**

Injurious affection — See **Compensation** (Land Compensation Act 1973 Part I — Validity of claim)

Land Compensation Act 1961
Capital gains tax liability
Claimant incurred capital gains tax liability on compensation for land taken — Whether capital gains tax liability compensatable under second limb of r (6) of s5 Land Compensation Act 1961 — Held, tax liability not compensatable as a matter of law under r (6)
***Harris* v *Welsh Development Agency (LT)*, [1999]** 3 **207**

Holding costs
Sale of property frustrated by threat of compulsory acquisition — Claimant incurring interest on borrowed money — Whether holding costs compensatable under r(6) — Held, decision in *Director of Buildings and Lands* v *Shun Fung Ironworks Ltd* [1995] 1 EGLR 19 applicable — Holding costs caused by threat of acquisition — Effect on marketing of property of compulsory acquisition foreseeable and not too remote
***Ryde International plc* v *London Regional Transport (LT)*, [2001]** 1 **101**

Statutory planning assumptions
Site of part of demolished railway station — Whether s15(3)(a) of Land Compensation Act 1961 permits assumption of planning permission for single dwelling–house — Whether hope value admissible in relation to assumed permission — Held, assumed planning permission for original building and uses on original foundations — Hope value not inadmissible in law — Hope value dependent on evidence
***Old England Properties Ltd* v *Telford & Wrekin Council (LT)*, [2000]** 3 **153**

Land taken
Structures
Shed and hardstanding on land taken for road scheme — Whether Highways Agency gave undertaking to pay equivalent reinstatement compensation — Whether compensation under r5 of s5 of Land Compensation Act 1961 payable in respect of structures on land taken — Whether cost of reinstating structures payable as disturbance compensation under r 6 — Lands Tribunal Rules simplified procedure — Unconditional offer above award — Whether acquiring authority entitled to costs under simplified procedure — Held, compensation assessed under r 2 of s5 of 1961 Act — Acquiring authority entitled to costs from date of unconditional offer under s4(1)(*a*) of 1961 Act
***Nicholls* v *Highways Agency (LT)*, [2000]** 2 **81**

Limitation period

Estoppel

Reference to Lands Tribunal after expiry of limitation period — S9 of Limitation Act 1980 — Whether acquiring authority estopped from relying upon expiry of limitation period — Held, on the evidence acquiring authority not estopped from relying upon expiry of limitation period

***Llanelec Precision Engineering Co Ltd* v *Neath Port Talbot County Borough Council (LT),* [2000]** 3 **158**

Lands Tribunal

Possession taken — Agreement — Right to compensation — Contractual rights — Reference to Lands Tribunal — Whether claim to compensation arising under contractual or statutory rights — Whether limitation period running from date of entry or date of contract — Whether claim statute barred — Held, s3 of Compulsory Purchase Act 1965 permitting acquisition of title by agreement — Agreement conferring contractually enforceable terms as to compensation — Reference within six-year limitation period

***BP Oil UK Ltd v Kent County Council (CA),* [2003]** 3 **1**

Mortgage

Date of redemption

Whether Lands Tribunal has jurisdiction to determine redemption date — Held, tribunal has no jurisdiction

***Shewu* v *Hackney London Borough Council (LT),* [1998]** 2 **232**

Notice to treat

Withdrawal

Whether acquiring authority can withdraw notice to treat after taking possession — Advance payment — Determination of parties' rights after withdrawal of notice to treat — Held, where claimant submits a statement of claim, acquiring authority can withdraw a notice to treat — An excess advance payment recoverable

***R* v *Northumbrian Water Ltd, ex parte Able UK Ltd (QB),* [1996]** 2 **15**

Open market value

Development land

Development site consisting of flats and bungalows — Developer in scheme world losing profits that would have been earned on sale of completed units — Whether value of land under r (2) can include profits on sale of units — Whether value limited to single-sale transaction — Whether lost profits compensatable under r (6) — Held, value of land compensatable under r (2) on single-sale transaction that excludes profits that developer would earn on sale of completed units — Lost profits not compensatable under r (6)

***Ryde International plc* v *London Regional Transport (CA),* [2004]** 2 **1**

Pointe Gourde scheme

Ransom value

Completion of link road — Ransom value to development site — Whether Lands Tribunal made perverse finding as to scheme — Whether tribunal overlooked differing terms of acquisition agreement imposing inhibition on development — Held, tribunal entitled to find scheme as an issue of fact — Erred in distinguishing inhibitions on development without link road

***JA Pye (Oxford) Ltd* v *Kingswood Borough Council (CA),* [1998]** 2 **159**

Purchase notice

Scheme

Acquisition under purchase notice — Scheme — Whether any scheme underlying acquisition following purchase notice — Whether no-scheme world and real world the same — Whether loss of ransom value recoverable — Held, no scheme underlying acquisition — No-scheme world same as real world

***Richards* v *Somerset County Council (LT),* [2001]** 2 **135**

Scheme

Nature reserve

Creation of barrage and acquisition of land to carry out palliative measure of providing nature reserve — Whether scheme was nature reserve or barrage — Whether necessary to discount any increase in value attributable to need to acquire land for palliative measures — Held, scheme was creation of barrage — Any value due to need for palliative measure of creating nature reserve disregarded

***Waters* v *Welsh Development Agency (LT),* [2001]** 1 **185**

S9 of Land Compensation Act 1961

Indication of likely acquisition

Acquisition under purchase notice — Whether any indication under s9 by authority with compulsory powers — Scheme — Whether any scheme underlying acquisition following purchase notice — Whether no-scheme world and real world the same — Whether loss of ransom value recoverable — Held, no s9 indication — No scheme underlying acquisition — No-scheme world same as real world

***Richards v Somerset County Council (LT),* [2001]** 2 **135**

Special suitability rule

Nature reserve

Creation of barrage and acquisition of land to carry out palliative measure of providing nature reserve — Whether r3 of s5 of Land Compensation Act 1961 applied to acquisition of land for nature reserve — Held, r3 did not apply to acquisition of land for creation of nature reserve

***Waters* v *Welsh Development Agency (LT),* [2001]** 1 **185**

Statutory planning assumptions

Planning permission for retained land

Acquisition of land for motorway link — Retained land — Whether s14(2) to (4) of Land Compensation Act 1961 required permission to be taken into account — Certificate for development on land acquired — Whether certificate assumed development on retained land — Whether planning permission would have been granted for acquired and other land — Whether planning permission could reasonably have been anticipated after valuation date — Held, s14(2) to (4) did not require permission to be taken into account — S17 certificate development cannot be assumed to be part of development on adjoining land — Limited permission would have been granted in no-scheme world

***Stayley Developments Ltd* v *Secretaryof State for the Environment, Transport and the Regions (LT),* [2001]** 1 **167**

Valuation

Established use

Effect of abandonment of use — Whether abandonment caused by scheme — Whether r 4 excluded compensation for previously established light industrial use — Held, compensation awarded for established use — Abandonment of use caused by scheme

***Ullah* v *Leicester City Council (LT),* [1996]** 1 **244**

Injurious affection

Claim for ransom value of land taken — Whether additional claim for injurious affection of retained land — Whether *Bwllfa* principle applicable to deferment — Held, claimant entitled to injurious affection in addition to ransom value of land taken — *Bwllfa* principle considered

***Dingleside Development Co Ltd* v *Powys County Council (LT),* [1995]** 1 **183**

Petrol filling station

Valuation method — Appropriate years — Effect of pricing policy — Surveyors' and accountants' valuations — Whether capital value or investment method — Comparables — Petrol sales — Held, surveyors' evidence preferred to accountants' — Petrol sales throughput and capital value per gallon — Methods of adjustment

***Telegraph Service Stations Ltd* v *Trafford Borough Council (LT),* [2000]** 3 **145**

Ransom strip

Acquisition of land for access to development site — Effect of "no-scheme world" — Ransom strip — Effect of alternative means of access — Held, alternative accesses disregarded — Single access site — 30% of development value appropriate basis

***Crown House Developments Ltd* v *Chester City Council (LT),* [1997]** 1 **169**

Acquisition of land for highway providing access to a back land development — Whether cross ransom payments to be deducted from ransom values — Held, net ransom value payable — Injurious affection to retained land payable in addition

***Dingleside Development Co Ltd* v *Powys County Council (LT),* [1995]** 1 **183**

Value of land

Scheme

Identification of scheme — Cardiff Bay barrage — Land for nature reserve required to meet conservation objections — Whether Indian case wrongly decided — Whether purpose underlying acquisition identifies scheme — Whether project a scheme only if project of acquiring authority — Whether nature reserve integral part of project — Whether temporal limitation on matters regarded as part of scheme — Whether owners entitled to "barrage inhibition value" — Held, Indian case reconcilable with other decisions — Land must be acquired for some project and need not be project of acquiring authority to constitute scheme — No temporal limitation on scope of scheme — Nature reserve part of scheme — "Barrage inhibition value" not payable

***Waters* v *Welsh Development Agency (CA),* [2002]** 2 **107**

Identification of scheme — Cardiff Bay barrage — Land for nature reserve required to meet conservation objections — Claim for ransom value attributable to barrage — Whether nature reserve integral part of project — Meaning of scheme — Whether owners entitled to "barrage inhibition value" —Held, guiding principle in identifying scheme fair compensation — Proposed works and their purpose for which they are being carried out relevant to identifying the scheme

***Waters* v *Welsh Development Agency (HL),* [2004]** 2 **103**

COMPENSATION FOR ACQUISITION OF LAND — *continued*
Value of land — *continued*

Value of tenancy
Tenancies excluded from Landlord and Tenant Act 1954 and subject to break clauses — Whether as a matter of law tenancy must be valued on basis of its termination on earliest possible date — Whether possibility of renewal of tenancy relevant — Whether subtenancy extinguished by acquisition of superior tenancy — S20 of Compulsory Purchase Act 1965 — Held, tenancies must be valued on basis of earliest date of termination ignoring any possibility of renewal in the no-scheme world
***Bishopsgate Space Management Ltd v London Underground Ltd (LT),* [2004]** 2 **175**

COMPENSATION FOR ELECTRICITY LINES
Injurious affection
Claim
Compulsory purchase order — Over-sailing lines — Determination of injurious affection to dwelling-house — Held, £4,000 is proper sum of compensation
***Naylor* v *Southern Electricity Board (LT),* [1996]** 1 **195**

Waste tip
Valuation method — Whether capitalisation of net annual receipts — Whether comparables relevant — Whether capitalisation of notional royalties — Held, capitalisation of net income method accepted
***Shraff Tip Ltd* v *Highways Agency (No 2) (LT),* [1999]** 3 **224**

Jurisdiction
Lands Tribunal
Electricity Act 1989 — Unauthorised entry for temporary and permanent installations — Whether Lands Tribunal has jurisdiction to determine compensation claim — Held, no jurisdiction where no statutory wayleave
***Bolton* v *Southern Electric plc (LT),* [1999]** 1 **177**

Statutory basis of compensation
Valuation
Certificate of lawful use — Application for certificate of appropriate alternative development — Term of wayleave — Held, statutory basis for compensation excludes Land Compensation Act 1961 (except ss2 and 4), Compulsory Purchase Act 1965 and Land Compensation Act 1973 — *Horn* principle of equivalence applies — No power to seek certificate of appropriate alternative development — Certificate of lawful use adds no value in absence of demand — Para 7 of Sched 4 to Electricity Act 1989
***Macleod* v *National Grid Co plc (LT),* [1998]** 2 **217**

COMPENSATION FOR SEA DEFENCE WORKS
Loss of profits claim
Assessment
Loss of profits and capital loss on sale of property — Management time — Surveyor's fees — Whether surveyor's fees limited to Ryde's scale — Dates from which statutory interest payable — Held, loss of profits for period of disruption — Loss on capital value on profits method valuation — Management time at £28.65 per hour — Ryde's scale rejected in favour of actual fees paid — Determination of dates from which statutory interest would run.
***Matthews and another* v *Environment Agency (LT),* [2002]** 3 **168**

COMPENSATION FOR SEWERS
Ransom value
Laying of sewer
Land with potential to grant sewer easements — Sched 12 to Water Industry Act 1991 — Sewers laid under powers in s159 of 1991 Act — Whether ransom value payable — Determination of ransom value — Held, ransom value not payable — Existence of statutory powers negating any ransom value — Determination of ransom value where more than one ransom owner
***Kettering Borough Council* v *Anglian Water Services Ltd (LT),* [2001]** 2 **157**

COMPENSATION FOR WATER MAIN CONSTRUCTION

Disturbance
Post valuation events
Apple trees grubbed out for water main — Claimant grubbed out fruit trees on adjacent land post-valuation date with benefit of recently introduced grant — Whether events occurred after date of entry relevant in assessing compensation under Water Industry Act 1991 — Held, post valuation events relevant in assessing compensation for disturbance
***Wickham Growers Ltd* v *Southern Water plc (LT),* [1997]** 1 **175**

COMPULSORY PURCHASE — See also **Compensation** (Compulsory Purchase — McCarthy Rules); **Compensation for acquisition of land** (Certificate of appropriate alternative development — Matters to be disregarded) (Disturbance — Loss of profit) (General vesting declaration — Time-limits) (Land Compensation Act 1961 — Statutory planning assumptions) (Limitation period — Lands Tribunal) (Value of land)

Agreement
Estoppel
Acquiring authority offered £0.5m "subject to contract" — Offer accepted — Representations by acquiring authority that offer remained extant relied on by claimants when purchasing new property — Acquiring authority denying it was bound by offer — Whether acquiring authority estopped from advancing value of property other than £0.5m — Held, acquiring authority not estopped by "subject to contract" offer
***Christos* v *Secretary of State for Transport (CA),* [2004]** 1 **5**

Offer to acquire
Acquiring authority offered £0.5m "subject to contract" — Offer accepted – – Acquiring authority denying that it was bound by offer — Whether binding contract — Whether suspensive condition — Whether "subject to contract" meaningless in compulsory purchase situation — Held, no binding contract
***Christos* v *Secretary of State for Transport (CA),* [2004]** 1 **5**

Certificate of appropriate alternative development — See **Compensation** (Certificate of appropriate alternative development — Relevant date and policies); **Compensation for electricity lines**

Compensation
Disturbance
Premises included bed-sitting rooms let on residential furnished tenancies — Delay in payment and underestimate of advance payment of purchase price — Inability of claimant to acquire substitute premises — Whether claimant entitled to compensation for loss of income by reason of delay in receiving full value of premises — Held, claimant received compensation for value of property that reflected the total extinguishment of the rental income business — Only remedy for delay in payment is interest at the prescribed rates
***Mallick* v *Liverpool City Council (CA),* [1999]** 2 **7**

Enforceable right to claim
General vesting declaration — Whether any enforceable right to compensation after expiration of limitation period for reference to Lands Tribunal — Held, claim to compensation was statutory — No enforceable right to claim unless application made to Lands Tribunal within limitation period or compensation agreed
***Royal Bank of Scotland plc* v *Clydebank District Council (Ct Sess),* [1995]** 1 **229**

Limitation
Reference to Lands Tribunal — Expiry of statutory time limit — Estoppel by convention — Whether conditions for estoppel by convention satisfied — Promissory estoppel — Whether acquiring authority estopped from relying upon expiry of time limit — Held, estoppel by convention not established by course of negotiations — No promissory estoppel
***Hillingdon London Borough Council* v *ARC Ltd (No 2) (CA),* [2000]** 3 **97**

Limitation period
Entry under s11 of Compulsory Purchase Act 1965 — Whether claim for compensation subject to limitation period under s9 of Limitation Act 1980 — Whether cause of action accrues on entry of acquiring authority — Whether reference to Lands Tribunal an action to recover sums — Held, cause of action under s9 of 1980 Act accrues when acquiring authority enters land — Reference to Lands Tribunal is an action to recover sums under s9
***Hillingdon London Borough Council* v *ARC Ltd (CA),* [1998]** 3 **18**

Reference to Lands Tribunal — Expiration of statutory time-limit — Estoppel by convention — Whether acquiring authority estopped from relying upon expiration of time-limit — Held, on the evidence acquiring authority estopped by convention — S9 of Limitation Act 1980 did not preclude reference to Lands Tribunal
***Hillingdon London Borough Council* v *ARC Ltd (No 2) (Ch),* [1999]** 3 **125**

Mortgage
Property mortgaged — Long delay before redemption by acquiring authority — Whether acquiring authority under any duty to redeem mortgage — Whether claimants entitled to compensation representing additional interest where authority fails to redeem timeously — Whether Lands Tribunal has jurisdiction to determine claim — Held, Lands Tribunal had no jurisdiction — Acquiring authority under no statutory or contractual duty to redeem mortgage — No compensation claim
***Shewu* v *Hackney London Borough Council (CA),* [1999]** 3 **1**

Compensation
Post-valuation date damage
Acquiring authority taking possession of property in stages — Parties agreeing first date of entry valuation date — Whether any deduction for post-valuation date damage — Held, valuation on basis of condition of property on valuation date — No deduction for post-valuation date damage
***Christos* v *Secretary of State for Transport (CA),* [2004]** 1 **5**

COMPULSORY PURCHASE — *continued*
Compensation — *continued*

Willing seller

Acquisition of rights over railway line — Characteristics of hypothetical seller — Meaning of "willing seller" in s5(2) of Land Compensation Act 1961 — Whether hypothetical seller likely to have characteristics of actual seller — Whether willing seller was railway company with similar characteristics to Railtrack and London Underground — Whether tribunal entitled to consider actual offer of one of parties — Held, no principle of law at issue — Tribunal entitled to conclude that willing seller was railway company having same functions and pressures as actual sellers — Tribunal entitled to consider evidence of actual offer

***Railtrack plc* v *Guinness Ltd (CA),* [2003]** 1 **124**

Compulsory purchase order

Validity

Acquisition under s17 Housing Act 1985 for housing purposes — Property containing ground-floor commercial use — Commercial use remaining after acquisition and refurbishment — Whether lawful to acquire commercial part — Whether order made as a last resort — Whether evidence and reasons supporting need for commercial part to avoid tenure problems in relation to future repairs — Whether violation of European Convention on Human Rights — Held, lawful compulsorily to acquire and retain property containing commercial part for housing purposes — Order made as last resort — Adequate evidence and reasons to support problems of tenure if commercial part not acquired — Fair balance of interests and no violation of Convention Rights

***Ainsdale Investments Ltd* v *First Secretary of State (Admin),* [2004]** 2 **9**

CPO of bus station — Whether valid use of power to take bus undertaking — Whether effect of CPO to inhibit competition — Whether burden on applicant to show executive had inadequate resources — Held, CPO valid — Irrelevant that effect of taking land was to close undertaking

***Yorkshire Traction Co Ltd* v *Secretary of State for Transport (CA),* [1997]** 2 **28**

Human rights — Art 8 of Convention — Inspector concluding that breach of Art 8 a possibility where occupiers had no alternative relocation property — Secretary of State accepting recommendations and obtaining undertaking from local authority to address potential breach — Whether Secretary of State had made breach of Art 8 a relevant issue — Whether Secretary of State should have decided whether Art 8 would be breached — Held, Secretary of State had concluded that CPO would not breach Art 8 — Requirement of local authority undertaking did not introduce issue not already decided

***R (on the application of Lock)* v *Secretary of State for the Environment, Transport and the Regions (QB),* [2002]** 3 **5**

Crichel Down Rules

RR9 and 14(7)

Construction of rules — Application to facts — Whether material change in character of land as a whole — Whether minister entitled to conclude that no obligation to offer land back — Whether inconsistent application of rules — Whether failure to consult — Held, r 9 applies to consideration of character of land as a whole — On facts no obligation to offer back to previous owners within meaning of r 14(7) — No inconsistency or failure to consult

***R* v *Secretary of State for Defence, ex p Wilkins (QB),* [2000]** 3 **11**

Eviction of claimant

Costs of eviction

Acquiring authority incurring costs upon eviction of claimant — Whether jurisdiction to determine eviction costs — Held, Lands Tribunal having no jurisdiction to address costs of eviction under s13(3) of Compulsory Purchase Act 1965

***Prielipp* v *Secretary of State for the Environment, Transport and the Regions (LT),* [2002]** 3 **143**

Lands Tribunal — See also **Lands Tribunal** (Award — Reopening award)

Limitation period

General vesting declaration — Whether any enforceable right to compensation after expiration of limitation period for reference to the Lands Tribunal — Held, claim to compensation was statutory — No enforceable right to claim unless application made to Lands Tribunal within limitation period or compensation agreed

***Royal Bank of Scotland plc* v *Clydebank District Council (Ct Sess),* [1995]** 1 **229**

Possession

Adverse possession

Deed poll vesting title with acquiring authority — Expropriatee in adverse possession of land prior to deed poll — S77 of Lands Clauses Consolidation Act 1845 — Whether new limitation period commencing upon execution of deed poll — Whether acquiring authority entitled to possession — Deed poll — Delay — Whether acquiring authority abandoning purpose of compulsory acquisition — Whether deed poll valid and effective — Whether acquiring authority evincing intention to abandon rights of compulsory acquisition — Whether acquiring authority entitled to make deed poll — Whether public law defence available in private law proceedings — Held, new limitation period commencing upon execution of deed poll — Acquiring authority entitled to possession on basis of valid deed poll — Public law defence can be raised to challenge validity of deed poll

***Rhondda Cynon Taff County Borough Council* v *Watkins* (CA), [2003]** 1 **117**

Statutory contract

S2(1) Law of Property (Miscellaneous Provisions) Act 1989

Compensation for the acquisition of land — Offer of compensation accepted — Contract — S2(1) of Law of Property (Miscellaneous Provisions) Act 1989 — Whether statutory contract under compulsory acquisition required to observe s2(1) — Whether compensation subject of binding contract — Held, statutory contract arising on notice to treat and agreement as to compensation not required to satisfy s2(1) — Binding contract on amount of compensation

***Llanelec Precision Engineering Co Ltd* v *Neath Port Talbot County Borough Council (LT),* [2000]** 3 **158**

CONSTRUCTIVE TRUST — See **Law of Property (Miscellaneous Provisions) Act 1989** (Proprietary estoppel)

CONTRACT — See also **Landlord and tenant** (Assignment — Consent to assign) (Tenancy — Unfair terms in consumer contracts)

Breach

Implied terms

Breach — Completion conditional upon buyer obtaining planning permission — Whether implied term that buyer would use reasonable endeavours to obtain permission — Whether implied term required implication of identifiable period of time at date of contract — Whether breach of implied term — Held, terms implied requiring buyer to use reasonable endeavours to obtain planning permission — No breach of terms where matters outside control of buyer

***Jolley* v *Carmel Ltd (CA),* [2000]** 3 **68**

Exchange of contract

Law of Property (Miscellaneous Provisions) Act 1989

Whether option to surrender lease assigned — Whether s2 of 1989 Act satisfied — Held, s2 not satisfied — No exchange of contracts.

***Commission for the New Towns* v *Cooper (Great Britain) Ltd (CA),* [1995]** 2 **113**

Land

Unfair terms in consumer contracts

Local authority letting accommodation to homeless persons — Whether Unfair Terms in Consumer Contract Regulations 1999 and Council Directive 93/13/EEC apply to contracts for the transfer of an interest in land — Whether the 1999 Regulations and directive apply to tenancies offered by public authorities — Held, 1999 Regulations and directive applying to contracts for transfer of an interest in land — 1999 Regulations and directive applying to grant of tenancies by public authorities

***R (on the application of Khatun)* v *Newham London Borough Council; R (on the application of Zeb)* v *Newham London Borough Council (Admin),* [2004]** 1 **34**

Rectification

Unconscionable conduct

Unilateral mistake — One party failed to disclose true intentions behind agreement — Whether rectification to delete terms not envisaged by innocent party — Held, rectification ordered

***Commission for the New Towns* v *Cooper (Great Britain) Ltd (CA),* [1995]** 2 **110**

Registration

Separate equitable interests

Unregistered and void against chargee — Whether chargee bound by interest of purchaser under bare trust, constructive trust or proprietary estoppel separate and distinct from contract — Held, where purchaser has interest under a contract void against chargee under s4(6) of Land Charges Act 1925 no room for other equitable interests binding chargee

***Lloyds Bank plc* v *Carrick (CA),* [1997]** 1 **160**

Solicitors — See **Negligence** (Solicitors — Conveyancing)

CONTRACT FOR SALE OF LAND — See also **Vendor and Purchaser** (Contract for sale of land — Notice to complete); (Sale of land — Property)

Conditional contract

Implied terms

Breach — Completion conditional on buyer obtaining planning permission — Change of scheme — Planning appeal — Whether implied term that buyer would use reasonable endeavours to obtain permission — Whether implied term required implication of identifiable period of time at date of contract — Whether breach of implied term — Held, terms implied requiring buyer to use reasonable endeavours to obtain planning

CONTRACT FOR SALE OF LAND — *continued*
Conditional contract — *continued*
permission — No breach of terms where buyer changed schemes and appealed planning refusal
***Jolley* v *Carmel Ltd (Ch),* [2000]** 2 **153**

Rescission
Landlord's consent
Standard conditions — Sale of lease — Whether landlord gave consent to proposed assignment — Whether "in principle" consent good consent — Whether landlord can qualify consent after giving "in principle" consent — Whether buyer entitled to rescind contract — Whether buyer entitled to return of deposit — Held, landlord's in principle consent sufficient to deny buyer right to rescind — Standard condition 8.3.4 must be given purposive construction — Landlord not entitled to qualify earlier consent — Buyer not entitled to return of deposit
***Aubergine Enterprises Ltd* v *Lakewood International Ltd (Ch),* [2001]** 3 **71**

CONVEYANCING — See **Negligence** (Solicitors — Damages); **Vendor and Purchaser**
Boundaries — See **Boundaries** (Hedge and ditch presumption — Conveyance)
Contract — See **Vendor and purchaser**
S2 Law of Property (Miscellaneous Provisions) Act 1989
Parties agreeing sale of property — Whether parties agreeing terms not in writing — Whether additional term in collateral contract — Whether agreement for sale subject to an additional term — *Tootal Clothing Ltd* v *Guineau Properties Management Ltd* [1992] 2 EGLR 80 — Whether composite arrangements can be placed in separate contracts — Whether s2 applies to land contract part of composite agreement — Held, no additional term agreed — Parts of Tootal Clothing doubted
***Grossman* v *Hooper (CA),* [2001]** 2 **82**

Solicitors — See also **Negligence** (Solicitors — Conveyancing)
Undertaking
Undertaking to bank — "Good marketable title" — Meaning of undertaking — Meaning of "good marketable title" — Whether breach of undertaking — Whether qualified or absolute liability — Held, undertaking refers to titlenot property — Liability is qualified
***Barclays Bank plc* v *Weeks Legg & Dean; Barclays Bank plc* v *Layton Lougher & Co; Barclays Bank plc* v *NE Hopkin John & Co (CA),* [1998]** 3 **103**

CO-OWNERSHIP
Tenants in common
Trusts of Land and Appointment of Trustees Act 1996
Property held by beneficial tenants in common — Trust — Whether co-owner can require partition of co-owned property — Whether trustees can order that beneficiaries shall occupy and be excluded from separate parts of trust property — Held, s13 of 1996 Act allows trustees to make scheme for partition of occupation of trust property between two or more beneficial tenants in common
***Rodway* v *Landy (CA),* [2001]** 3 **78**

COSTS — See **Highways** (Agreement — S278 Highways Act 1980); **Landlord and tenant** (Repairs — Dilapidations) (Service charges — Landlord and Tenant Act 1985 s19); **Lands Tribunal** (Costs — *Calderbank* offer to settle) (Costs — Equivalence principle) (Costs — Simplified procedure) (Service charges) (Service charges — Legal costs); **Leasehold valuation tribunal** (Appeal — Application for leave to appeal)
Compensation for acquisition of land — See **Compensation for acquisition of land** (Costs)
Development — See **Valuation** (Property — Stokes principle)
Landlord's reasonable costs — *See* **Landlord and tenant** *(*Assignment — Reasonable costs)
Lands Tribunal — See **Lands Tribunal** (Costs)
Legal — See **Landlord and tenant** (Costs)
Service charges — See **Landlord and tenant**

COUNCIL TAX — See **Rates and rating** (Council tax — Self-contained unit)

COVENANTS — See also **Landlord and tenant** (Alterations and additions — Covenant to repair) (Assignment — Landlord and Tenant (Covenants) Act 1995) (Covenant against alteration)(Enforcement of covenants — S141 Law of Property Act 1925) (Repairs); **Landlord and Tenant Act 1954 Part II** (Landlord's opposition to new tenancy) (Covenants — User); **Restrictive covenant** (Annexation — S78 Law of Property Act 1925)
Breach of — See **Damages** (Lord Cairns' Act — Assessment); **Landlord and tenant** (Assignment — Breach of covenant) (Break-clause — Condition precedent: performance of covenants) (Forfeiture — Mistake) (Quiet enjoyment); **Restrictive covenants** (Application to modify — Grounds (*a*), (*aa*) and (*c*)) (Burden — Successors in title)

Freehold
Positive covenant
Whether positive covenant enforceable against successor in title to covenantor — Whether taking incidental benefit invokes all burdens under deed — Whether successor in title can choose to take benefit — Held, taking incidental benefit does not invoke all burdens — Successor in title may choose to take benefit
***Thamesmead Town Ltd* v *Allotey (CA),* [1998]** 3 **97**

Lease — See **Landlord and tenant** (Assignment — Covenants)
Obsolete covenant — See **Restrictive covenant** (Discharge or modification — Obsolete covenant)
Repairs — See **Agricultural holdings** (Covenant to repair — Damages); **Landlord and tenant** (Repairs — Covenant)
Restraint of trade agreement — See **Estate agents**
Restrictive covenants — See **Restrictive covenant** (Burden) (Discretion or modification); **Town and country planning** (Overriding private rights)
Surety — See **Landlord and tenant** (Assignment — Liability of surety) (Forfeiture — Surety covenants) (Surety covenant)
User covenant — See **Landlord and tenant** (User covenant)

CRICHEL DOWN RULES — See also **Compulsory purchase** (Crichel Down Rules — Rr 9 and 14(7)
Disposal of land
Exemption
Requirement to offer to original owner — Application of exemption — Land to be offered for park–and–ride scheme — Whether decision to apply exemption perverse or *Wednesbury* unreasonable — Held, decision within exemption — Decision not perverse or Wednesbury unreasonable
***R* v *Secretary of State for the Environment, Transport and the Regions, ex p Wheeler (QB),* [2000]** 3 **63**

CUSTOM OF LONDON — See **Easement** (Right to light — Prescription)

DAMAGES — See **Agreement** (Breach; **Landlord and Tenant** (Dilapidations — Measure of damages) (Rent Acts — Claim for damages) (Repairs — Breach); **Misrepresentation** (Negligent misrepresentation — Assessment of damages); **Negligence** (Solicitor — Assesment of damages) (Surveyors)
Assessment of — See **Negligence** (Solicitors — Assessment of damages) (Solicitors — Damages); **Landlord and tenant** (Repairs — Damages)
Breach of covenant — See **Landlord and tenant** (Repairs — Covenant); **Restrictive covenant** (Breach — Damages) (Breach — Remedies); **Water** (Trespass)
Causation
Building works
Coincidence of damage and works — Party wall subjected to forces by Kango hammer — Damage to adjoining property — Whether coincidence a forensic tool — Whether damage caused by works on balance of probability — Held, inference can be drawn that one event (works) caused another event (damage) where interval between two events short and second event precisely that expected from the first event — On balance of probability use of Kango hammer caused damage
***Roadrunner Properties Ltd* v *Dean (CA),* [2004]** 1 **73**

Contributory negligence — See **Negligence** (Valuers — Contributory negligence) (Valuers — Duty of care in tort)
Easement — See **Easement** (Right of way)
Interest — See **Negligence** (Valuers — Duty of care in tort)
Limitation Act
Date of damage — See **Negligence** *(*Limitation Act)
Lord Cairns' Act
Assessment
Breach of restrictive covenant and trespass — Use of private road and garden — Damages not injunction appropriate remedy — Whether damages for future wrongs — Proper assessment of damages — Held, damages for present and future wrongs assessed on compensatory principles applying *Wrotham Park Estate Co Ltd* v *Parkside Homes Ltd*
***Jaggard* v *Sawyer (CA),* [1995]** 1 **146**

Measure of damages — See **Agricultural holdings** (Covenant to repair); **Landlord and tenant** (Repairs — Covenant)(Repairs — Damages); **Negligence** (Damages: measure and quantification) (Mortgage loan report)
Misrepresentation — See **Landlord and tenant** (Misrepresentation — Damages)
Mitigation — See **Negligence** (Damages: measure and quantification)
Valuation — See **Negligence** (Mortgage loan report — Damages)
Wrongful eviction — See **Landlord and tenant** (Housing Act 1988 — Damages for unlawful eviction); **Licence** (Eviction)

DEED
Construction
Meaning of "transferee"
Pre-emption right granted "for so long as transferee is beneficial owner" — Land transferred to joint transferees — Later gift by one transferee of his whole legal and beneficial interest to co-transferee — Whether sole

owner entitled to exercise right of pre-emption — Whether "for so long as" referred to retained beneficial ownership of both original transferees — Held, "transferee" meant both transferees — Pre-emption right expired on grant by one of two transferees of his whole legal and beneficial interest to the other transferee

***Midwood* v *Morgan and another (Ch),* [2001]** 3 **127**

"DELAFORCE" EFFECT — See **Landlord and tenant** (Leasehold enfranchisement — Leasehold Reform Act 1967)

DEMOLITION — See **Duty of care** (Property); **Easements** (Easement of support)

DEROGATION FROM GRANT — See also **Landlord and tenant** (Derogation from grant — Repudiation)

Lease

Shopping mall

Lease of premises in shopping mall — Adjoining unit let to pawnbroker — Pawnbroker causing nuisance — Whether tenant entitled to disclaim lease — Whether landlord in breach of covenant not to derogate — Whether landlord ought to control nuisance — Held, landlord derogated from grant — Tenant entitled to disclaim lease

***Chartered Trust plc* v *Davies (CA),* [1997]** 2 **83**

DILAPIDATIONS — See **Landlord and tenant** (Dilapidations — Damages) (Repairs — Covenant)

DISCLAIMER — See **Landlord and tenant** (Disclaimer of lease — Liability of original tenant) (Disclaimer of lease — Guarantor) (Dilapidations) (Forfeiture — Possession) (Insolvency — Disclaimer of lease) (Rent arrears — Liability of original tenant) (Repairs — Breach); **Negligence** (Duty of care and liability — Disclaimer)

DISTURBANCE — See **Compensation for acquisition of land** (Disturbance) (Disturbance payment — S37 of Land Compensation Act 1973); **Compulsory purchase** (Compensation)

Compensation — See **Landlord and Tenant** (Landlord and Tenant Act 1954 Part II — Compensation for disturbance); **Water** (Pipeline)

DUTY OF CARE

Property

Demolition

Demolition of adjoining building without the carrying out of protective works against damp to common wall — Whether property owner owing duty to prevent damp penetration to adjoining property following demolition — Held, duty of care owed to adjoining property owner in respect of failure to carry out damp-proofing works

***Rees* v *Skerrett (CA),* [2001]** 3 **1**

DWELLING-HOUSE — See **Landlord and tenant** (Assured tenancy — Dwelling-house)

EASEMENT — See also **Right of light** (Express grant — Interference); **Vendor and purchaser** (Condition of sale — Implied easements)

Acquisition

Prescription

s193(4) of Law of Property Act 1925 precluding acquisition of easement under s2 of Prescription Act 1832 or by lost modern grant — Whether *Hanning* v *Top Deck Travel Group Ltd* (1993) 68 P&CR 14 correctly decided — Held, *Hanning* wrongly decided — Landowner could make lawful grant of easement and make user lawful — Where user contrary to law can be made lawful by landowner's dispensing power, easement capable of acquisition by prescription

***Bakewell Management Ltd* v *Brandwood (HL),* [2004]** 2 **15**

Parking and storage of vehicles — Whether use without planning permission precludes prescriptive claim — Whether easement to park and store vehicles — Claim under s62 Law of Property Act 1925 — Change of use of dominant land — Whether excessive use of rights of way — Held, prescriptive easement to park — Use without planning permission can support prescriptive claim to easement — Right of way under s62 limited to degree of user before change of use of dominant land

***Batchelor* v *Marlow (Ch),* [2001]** 1 **119**

Acquisition by prescription

Village green

S34(1) of Road Traffic Act 1988 — Whether right of way acquired by prescription over village green — Whether track over village green "land of any other description" under s34(1)(a) of 1988 Act — Whether use of track criminal offence contrary to s34(1) — Whether breach of s12 of Inclosure Act 1859 — Whether breach of s29 of Commons Act 1876 — Enlargement of purported dominant tenement in 1977 — Whether any easement based on prescription restricted to pre-1977 dominant tenement — Held, use of track was contrary to s34(1) — Easement not capable of being acquired by prescription — Easement capable of being acquired under s68 of Countryside and Rights of Way Act 2000 not limited to pre-1977 dominant tenement

***Massey* v *Boulden (CA),* [2003]** 1 **24**

Common use of garden

Extent of rights

Yard and access used in common for garden cultivation, drying clothes and general enjoyment — Whether use capable of being an easement — Whether any interference of easement actionable — Whether servient owner entitled to form access drive over servient land — Extent of rights over communal gardens — Held, a right to use land as a communal garden for recreational and amenity purposes capable of being an easement — Servient owner entitled to use servient tenement subject to existence of easement

***Mulvaney* v *Gough (CA),* [2002]** 3 **72**

Creation

Prescription

Right of way — Right of way for vehicles claimed by prescription — Public footpath along route of claimed right of way — Whether user by vehicles over public footpath breaching Road Traffic Act 1988 founding right of way by prescription — Whether *Hanning* v *Top Deck Travel Group Ltd* (1993) 68 P&CR 14 applying — Whether any evidence that user commenced not less than 20 years prior to Road Traffic Act 1930 — Held, *Hanning* v *Top Deck Travel Group Ltd* applying — No vehicular rights acquired by user since 1930 — Case remitted to determine whether right acquired by 20 years' prescription prior to 1930

***Hayling* v *Harper (CA),* [2003]** 3 **5**

S62 of Law of Property Act 1925

Permission to tenant to park car — Conveyance of freehold to tenant — Whether implied grant of easement to park — Whether precarious permission to park "temporary" — Held, precarious permission to tenant not "temporary" in sense contemplated in *Wright* v *Macadam* [1949] 2 KB 744 of being for a period of time — Permission became an easement on conveyance of land in 1979

***Hair* v *Gillman (CA),* [2000]** 3 **74**

Easement of support

Extent of easement

Demolition of adjoining building providing support against effect of wind suction — Whether easement of support includes support to withstand effect of wind — Held, easement of support includes support through protection against effect of wind

***Rees* v *Skerrett (CA),* [2001]** 3 **1**

Implied reservation

General rule against implied reservation

Right claimed to construct and use road — Early conveyances and plans referring to road and intended road — Whether easement implied under exception to general rule — Whether right to construct road — Held, despite intentions to construct road, no implied reservation of a right from words used by conveyances or from surrounding facts — No necessity for reservation of right — Too much uncertainty of rights intended in grants

***Chaffe* v *Kingsley (CA),* [2000]** 1 **104**

Lost modern grant

Common land

Access over common land to adjoining houses — Conditions for application of lost modern grant fiction satisfied — Whether decision in *Hanning* v *Top Deck Travel Group Ltd* (1993) 68 P&CR 14 wrongly decided — Whether Court of Appeal entitled to depart from earlier decision — Whether easements acquired under lost modern grant fiction — Held, Hanning correctly decided — Acquisition of easement under lost modern grant precluded where use illegal

Brandwood v Bakewell Management Ltd (CA), 17

Prescription

Drainage easement

Radical alteration of dominant tenement — No increase in quantity of water through servient tenement — Whether prescriptive easement lost or suspended on radical alteration of dominant tenement — Whether strict rule in easements of way applicable to easements of drainage — Held, drainage easement not lost or suspended by alteration to dominant tenement

***Attwood* v *Bovis Homes Ltd (Ch),* [2000]** 3 **139**

Right of light

Prescription

Claim based on Prescription Act 1832, lost modern grant and at common law — Whether good cause of action — Held, claim dismissed — One-year obstruction based on obstruction notice effectively registered under Rights of Light Act 1959 — Custom of London

EASEMENT — *continued*
Right of light — *continued*
Prescription — continued

defeated lost modern grant — Building not in existence in 1189 defeated common law claim
***Bowring Services Ltd* v *Scottish Widow's Fund & Life Assurance Society (Ch),* [1995]** 1 **158**

Right granted by lease
Lease reserving right to develop other land — Whether right of light subject to right to develop other land — Whether "other land" restricted to land belonging to landlord — Whether "other land" including after-acquired land — Whether provisions in lease restricting right of light otherwise arising under Prescription Act 1932 — Held, "other land" applying only to land owned by landlord at date of grant of lease — Reservation of right to build allowing development of land belonging to landlord at date of grant — Landlord entitled to restrict right of light otherwise arising under Prescription Act 1932 by terms of reservation of right to build
***Paragon Finance plc* v *City of London Real Property Co Ltd (Ch),* [2002]** 1 **97**

Rights of Light Act 1959
Obstruction notice registered under 1959 Act accompanied by temporary Lands Tribunal certificate — Whether one-year period of notional obstruction commences on date of registration of notice or date when later definite certificate granted — Whether certificates challengeable in private or public law — Held, one-year period commences on date of original registration of obstruction notice — Certificates challengeable only by judicial review
***Bowring Services Ltd* v *Scottish Widow's Fund & Life Assurance Society (Ch),* [1995]** 1 **158**

Right of way
Damages
Whether vehicular rights abandoned — Whether court should grant mandatory injunction to require removal of obstructions — Whether damages adequate remedy — Held, despite obstructions and occasional use, vehicular rights not abandoned — Damages awarded representing diminution in value attributable to loss of rights
***Snell & Prideaux Ltd* v *Dutton Mirrors Ltd (CA),* [1995]** 1 **259**

Excessive use
Right of way for all purposes — Dominant tenement used in conjunction with other land — Whether right of way can be exercised in conjunction with land additional to dominant tenement — Width and location of right of way — Whether court entitled to have regard to pre–grant agreement — Held, right of way cannot be used to access land other than dominant tenement — Pre–conveyance contract admissible to establish precise width and location of right of way
***Peacock* v *Custins (CA),* [2001]** 1 **87**

Excessive user
Grant for agricultural use — Timber stored on dominant tenement after felling on other land — Whether removal of timber over right of way excessive user — Held, such use excessive
***Jobson* v *Record (CA),* [1998]** 1 **113**

Interference
Service access to retail warehouse — Proposed obstruction — Test of actionable interference to right of way — Whether test of necessity (or reasonable necessity) or of convenience — Whether right of way could be substantially and practically exercised with obstruction as conveniently as without it — Held, test of actionable interference is one of convenience not necessity or reasonable necessity — Proposed obstruction would constitute actionable breach
***B&Q plc* v *Liverpool & Lancashire Properties Ltd (Ch),* [2001]** 1 **92**

Remedy for trespass
Carriageway used for purpose outside scope of easement benefiting dominant tenement — Use infringing rule in *Harris* v *Flower* — Whether injunction or damages under Lord Cairns' Act — Held, injunction not necessarily appropriate remedy
***Das* v *Linden Mews Ltd (CA),* [2002]** 2 **76**

Rule in Harris v Flower
Ancillary use of dominant tenement — Easement of way to dominant tenement did not include parking — Land added to dominant tenement and used for parking — Whether parking use ancillary to use of dominant tenement — Whether injunction or damages under Lord Cairns' Act — Held, parking use of additional land not use of dominant tenement — Injunction not necessarily appropriate remedy
***Das* v *Linden Mews Ltd (CA),* [2002]** 2 **76**

Standard of construction
Grant of rights of way over road to be constructed to adoption standard — Grant until adoption — Whether grantees of rights entitled at common law to enter servant land to construct road and procure adoption — Held, immediate grant of right of way — Grantees enjoyed ancillary right to enter servient land and construct road to adoption standard
***Nationwide Building Society* v *James Beauchamp (a firm) (CA),* [2001]** 3 **6**

Rights of drainage
Alteration of dominant land
Right to drain implied under the rule in *Wheeldon* v *Burrows* [1879] LR 12 ChD 31 — Alteration of dominant land from bakery to detached houses — Whether right to drain continuing for benefit of altered dominant land — Whether radical change in character of dominant land — Whether substantial increase or alteration in burden on servient land — Held, issue depending on whether substantial change in character of dominant land and substantial increase in burden on servient land — Both questions satisfied — Radical change in character of dominant land — Substantial increase in burden — Right of drainage not continuing
***McAdams Homes Ltd* v *Robinson (CA),* [2004]** 3 **93**

ELECTION
Waiver
Landlord and tenant
Break clause — Holding over — Double rent — Distress for Rent Act 1737 — Whether landlord elected between right to double rent or contractual rent — Held, tenant's break notice valid — No right to elect lease as continuing — Landlord did not treat tenant as trespasser and not entitled to double rent
***Ballard (Kent) Ltd* v *Oliver Ashworth (Holdings) Ltd (CA),* [1999]** 2 **23**

ELECTRICITY POWER LINES — See **Compensation for electricity lines**
ELECTRICITY WAYLEAVE
Grant
Land
Application for wayleave to install cables in service tunnel under dockland water — Para 6 of Sched 4 to Electricity Act 1989 — Power to grant wayleaves on, under, or over any land — Whether "land" having wide meaning under Interpretation Act 1978 — Whether service tunnel "land" — Held, wayleave can be granted in respect of service tunnel — No contrary intention and wide meaning of "land" under Interpretation Act 1978 applied
***British Waterways Board* v *London Power Networks plc (Ch),* [2003]** 1 **111**

ENFRANCHISEMENT — See **Leasehold enfranchisement**

ENVIRONMENTAL PROTECTION ACT 1990 — See also **Landlord and tenant** (Repairs — Statutory nuisance); **Public health** (Statutory nuisance — Procedure in magistrates' court); **Trusts** (Trustees — Liabilities)
Waste-management licence
Insolvency
Administration — Leave to bring proceedings — Waste-management licence — Criminal proceedings for breach of licence conditions — Whether Environment Agency requires leave to bring and continue criminal proceedings relating to waste-management licence — Whether leave should be granted — Held, leave required under ss10 and 11 Insolvency Act 1986 to bring and continue criminal proceedings — Leave given on the facts known at date of appeal
***Environment Agency* v *Clark (administrator) (CA),* [2000]** 1 **113**

ESTATE AGENTS
Cold–calling letter
Entitlement to fees
Transaction resulting from information in cold–calling letter — Whether estate agent entitled to fee — Held, no agreement was made — Offer not capable of acceptance
***Lady Manor Ltd* v *Fat Cat Café Bars Ltd (CC),* [2001]** 2 **88**

Commission
Agents acting for purchaser
Purchaser receiving information through two agents — Whether first or second agent effective cause of the purchase — Whether agents seeking instructions from purchasers entitled to any commission in absence of express terms — Held, first agent was not effective cause of purchase — Doubtful whether agents providing information toprospective purchasers entitled to commission in absence of express contractual terms
***Egan Lawson Ltd* v *Standard Life Assurance Co (CA),* [2001]** 1 **27**

Sole selling rights
Construction of agreement — Estate Agents (Provision of Information) Regulations 1991 — Exchange of contracts after expiration of sole selling right period — Whether purchaser must be introduced by estate agent — Whether estate agents entitled to commission — Held, estate agent not entitled to commission where purchaser introduced by persons other than agent himself
***Michael Harwood T/A RSBS Group* v *Smith (CA),* [1998]** 1 **5**

ESTATE AGENTS — *continued*
Commission — *continued*

Unfair terms
Unfair Terms in Consumer Contracts Regulations 1999 — Commission agreement for sale of flat — Commission payable within 10 days of completion of sale at 1.5% — Standard commission payable after 10 days at 3% — Vendors' solicitor failing to settle commission within 10-day period — Whether agreement for commission rate of 3% excluded from unfairness provisions of 1999 regulations — Whether option or obligation on vendors to pay 1.5% commission rate — Held, obligation not option to pay 1.5% commission rate — Provision for 3% commission not excluded by reg 6(2) — 3% commission subject to unfairness provisions in reg 5(1)
***Bairstow Eves London Central Ltd* v *Smith (QB),* [2004]** 2 **25**

Duties
Protection of vendor's interest
Duty to purchaser — Subsale at inflated price to defraud mortgagee — Failure of estate agents to inform vendor — Whether vendor entitled to damages for loss of right to renegotiate sale price — Held, no breach of duty — No grounds in fact or at law that price could have been renegotiated — Public policy prevented award of damages failure to participate in proceeds of a fraud
***Berkowitz* v *MW (St John's Wood) Ltd (CA),* [1995]** 1 **29**

Breach of contract — Negligence — Property in market at £1.5m asking price — Further property in same mews in market at £1.95m asking price — Whether estate agent in breach of contract — Whether estate agent negligent in determining asking price — Whether estate agent owing duty of care to pass on information relevant to principal's instructions — Whether principal would have obtained higher price — Whether loss of chance to obtain higher price — Held, estate agent not negligent in fixing asking price — Estate agent in breach of duty of care in not passing on information about higher asking price of further property — Principal had 66% chance of obtaining higher price
***John D Wood & Co (Residential & Agricultural) Ltd* v *Knatchbull* (QB), 33**

Estate Agents Act 1979
Duty of Director General of Fair Trading
Prohibition and warning powers — Alleged facts of malpractice — Subsale at higher price — Allegations against estate agent employees — Whether a "triggering" event — Whether Director General should consider exercising power to prohibit or warn — Held, breach of information regulations shown on allegations — Director General required to consider exercise of powers
***R* v *Director General of Fair Trading, ex parte Benhams Ltd (QB),* [2001]** 1 **21**

Prohibition order
Criminal offence — Estate Agents Act 1979 s3 — Order prohibiting practice as estate agent — Conviction for arson in USA court in 1973 — Whether 1979 Act retrospective — Whether 1979 Act permits a conviction before a foreign tribunal — Whether arson an offence under s3 — Held, 1979 Act retrospective — S3 offences includes foreign conviction for arson — Appeal dismissed from [1996] 1 EGLR 229
***Antonelli* v *Secretary of State for Trade and Industry (CA),* [1998]** 1 **9**

Negligence
Duty of care and liability
Negligent misstatement by estate agent of area of gardens to house — Whether duty of care owed to purchaser — Whether disclaimer effective — Held, on particular facts no duty of care owed — If duty owed effective disclaimer — Disclaimer reasonable under Unfair Contract Terms Act 1977
***McCullagh* v *Lane Fox & Partners Ltd (CA),* [1996]** 1 **35**

Property Misdescriptions Act 1991
Statutory defence
Property without planning permission described as bungalow — Whether estate agents established statutory defence of taking all reasonable steps and due diligence to avoid misstatement — Held, estate agents had no real doubt that property had planning permission — Defence established
***Enfield London Borough Council* v *Castles Estate Agents Ltd (QB:DC),* [1996]** 2 **21**

Restraint of trade agreement
Restrictive covenants
Partnership covenant restrictive of practice within two-year two-mile radius — Whether valid and enforceable — Held, covenant protected goodwill and enforceable
***Espley* v *Williams (CA),* [1997]** 1 **9**

Trade Descriptions Act 1968
Fly boarding
Penalty — Whether fine of £22,500 plus costs appropriate penalty for illegal "fly boarding" — Whether fines of £100 per offence generally imposed by magistrates too low — Held, fines of £2,000 per offence appropriate — Magistrates imposing fines that are too low
***R* v *Docklands Estates Ltd (CA: CrDiv),* [2000]** 3 **17**

ESTOPPEL — See also **Agricultural holdings** (Notice to quit — Estoppel); **Compensation** (Highways Act 1980 s28 — Footpath creation order) (Land Compensation Act 1973 Part I — Validity of claim); (Limitation period — Waiver); **Compensation for acquisition of land** (General vesting declaration — Time-limits); **Compulsory purchase** (Agreement — Estoppel) (Compensation — Limitation period) (Statutory contract — S2(1) Law of Property (Miscellaneous Provisions) Act 1989); **Landlord and tenant** (Alterations and additions — Covenant to repair) (Break clause — Estoppel) (Break clause — Notice) (Break clause — Time-limit) (Landlord and Tenant Act 1954, Part II — Licence) (Landlord and Tenant Act 1954, Part II — Tenancy by estoppel) (Landlord and Tenant Act 1987 — Tenants' rights of first refusal) (Lease or licence — Tenancy by estoppel) (Rent Acts — Statutory tenancy) (Repairs — Leasehold Property (Repairs) Act 1938); **Law of Property (Miscellaneous Provisions) Act 1989** (Proprietary estoppel); **Trespass**

Cause of action estoppel
Legal charge
Estoppel *per rem judicatam* — Legal charge — Guarantee — All moneys charge — Debtor liable under loan and guarantee — County court judgment for possession and repayment of loan — Whether cause of action in favour of bank for moneys due under guarantee — Whether bank estopped by earlier county court judgment — Held, bank estopped by earlier cause of action
***Lloyds Bank plc* v *Hawkins (CA),* [1998]** 3 **109**

Issue estoppel — See **Compensation for acquisition of land** (Certificate of appropriate alternative development — Issue estoppel); **Landlord and tenant** (Rent review clause — Issue estoppel)

EU TREATY — See **Landlord and tenant** (Rent — Set-off) (Tied public houses)

EVICTION — See **Landlord and tenant** (Housing Act 1988 — Damages for unlawful eviction)

EVIDENCE — See also **Lands Tribunal** (Procedure)
Additional evidence — See **London Building Acts 1930-39** (Party wall — Award of third surveyor)
Admissibility — See **Negligence** (Surveyors — Evidence)
Expert — See also **Expert, appointment of** (Negligent valuation — Court-appointed expert); **Negligence** (Solicitors — Conveyancing)
Immunity of witness
Joint statement of experts — Whether cause of action in negligence against expert in connection with litigation — Whether witness immunity principle applies — Held, immunity for expert even if trial does not proceed.
***Stanton* v *Callaghan (CA),* [1998]** 3 **165**

Role
Whether experts employed by defendant council independent — Held, experts not independent and should not act as advocates or interpret law
***Burroughs Day* v *Bristol City Council (QB),* [1996]** 1 **167**

EXPERT, APPOINTMENT OF
Negligent valuation
Court appointed expert
Expert evidence — Property valuations — Whether defendant valuation surveyors entitled to call numerous experts in relation to 51 properties nationwide — Whether court should appoint expert under RSC Ord 40 — Appeal from [1995] 2 EGLR 134 — Held, court entitled to appoint expert in circumstance where court would be assisted — Expert was not merely a witness, but had no jurisdiction
***Abbey National Mortgages plc* v *Key Surveyors Nationwide Ltd (CA),* [1996]** 2 **99**

Rent review — See **Landlord and tenant** (Rent review clause — Construction: general)

EXPERTS — See also **Evidence** (Expert); **Landlord and tenant** (Rent review); **Negligence** (Duty of care and liability — Valuers) (Solicitors — Conveyancing); **Procedure**
Evidence
Employee
Local authority landlord — Part 35 of CPR — Whether employee of local authority entitled to give expert opinion evidence — Held, court entitled to see employee expert's report before permitting use of employee as expert under Part 35 of CPR — Court needs to be satisfied of issues in question and ability of employee
***Field* v *Leeds City Council (CA),* [2000]** 1 **54**

EXPERTS — *continued*

Third-party determination
Non-speaking valuation
Determination of option price — Whether expert's determination void — RICS valuation practice statements — Whether expert carried out task directed by option price formulation — Held, inferences cannot be drawn as to reasons behind a non-speaking valuation — Valuer made no mistake — Mistake in formulation of price in option agreement
***Morgan Sindall plc* v *Sawston Farms (Cambs) Ltd (CA)*, [1999]** 1 **90**

FAIR RENT — See **Landlord and tenant** (Rent Acts — Fair rent) (Rent Acts — Rent assessment)

FEES — See **Arbitration** (Arbitrator — Misconduct); **Compensation for defence works); Estate agents** (Cold-calling letter — Entitlement to fees) (Commission — Sole selling rights); **Landlord and tenant** (Leasehold enfranchisement — Surveyor's fees) (Service charges)

FIRE DAMAGE — See **Landlord and tenant** (Landlord and Tenant Act 1954 Part II — Occupation)
FIXTURES — See **Landlord and tenant** (Rent review — Fixtures)

FOOTBALL CLUB — See **Restrictive covenants** (Meaning and construction — Use)

FORFEITURE — See **Landlord and tenant** (Forfeiture)(Rent Acts — Possession); **Mortgage** (Mortgage of leasehold interest — Mortgage in possession)

FRAUD
See **Vendor and purchaser** (Sale of land — Property)
Consent to assignment — See **Landlord and tenant** (Assignment — Fraud) (Forfeiture — Waiver) (Enforcement of covenants — S141 Law of Property Act 1925)
Proceeds of a fraud — See **Estate agents** (Duties — Protection of vendor's interest)
Tenancy agreement — See **Agricultural holdings** (Mortgage — Fraudulent tenancy agreement)

GUARANTEE — See **Estoppel** (Cause of action estoppel — Legal charge)

HIGHWAYS — See also **Compensation** (Highways Act 1980 s28 — Footpath creation order); **Negligence** (Duty of care — Local authority inquiries); **Sewers and drains** (Public sewers — Meaning of "sewer")';
Adoption agreement
Enforceability against successor in title
Whether agreement under s38 of Highways Act 1980 an overriding interest under s79(1) (*a*) of Land Registration Act 1925 — Whether agreement binding on successor in title to original party to agreement — Held, agreement not an overriding interest and not binding on successor
***Overseas Investment Services Ltd* v *Simcobuild Construction Ltd (CA)*, [1996]** 1 **49**

Agreement
S278 Highways Act 1980
Whether land acquisition costs contribution can be included in s278 agreement — Compulsory purchase order — Whether CPO lawfully obtained notwithstanding s278(5) — Held, land acquisition costs contribution can be included in s278 agreement — CPO authorised notwithstanding s278(5)
***Wards Construction (Medway) Ltd* v *Kent County Council (CA)*, [1999]** 1 **111**

Blight notice — See **Town and country planning** (Blight notice — Validity of counternotice)
Blighted property
Discretionary power to acquire
Discretionary power under s246(2A) of Highways Act 1980 — Whether Secretary of State entitled to rely on owner's foreknowledge of highway scheme — Held, foreknowledge can be relied on — Secretary of State had material on foreknowledge to refuse application to acquire property
***R* v *Secretary of State for Transport, ex parte Owen (No 2) (QB)*, [1996]** 1 **52**

Discretionary power under s246(2A) of Highways Act 1980 — Whether serious effect on enjoyment of property includes diminution in value as ground for consideration — Held, highway authority to consider application in two stages — (1) Whether enjoyment of land will be seriously affected — (2) Whether to exercise discretion to purchase
***R* v *Secretary of State for Transport, ex parte Owen (CA)*, [1995]** 2 **213**

Compensation for acquisition of land — See **Compensation for acquistion of land** (Betterment)

Dedication
Acceptance
Memorandum of dedication — Whether only actual use by the public constitutes acceptance — Whether local highway authority accepted dedication on behalf of the public — Consequence of public highway status — Effect of registration of land — Whether breaches of implied covenants and contract by vendor of affected land — Held, effective dedication and acceptance — Highway authority accepted on behalf of public — Breach of contract to provide vacant possession
***Secretary of State for the Environment, Transport and the Regions* v *Baylis (Gloucester) Ltd (Ch)*, [2000]** 2 **13**

Hedges
Maintenance obligation
1812 inclosure award — Perpetual obligation to keep hedge in good repair and condition — Whether terms of inclosure award bind successors in title — Whether positive obligation enforceable — Whether plaintiffs had *locus standi* as parishioners — Held, no claim — Positive obligations cannot run with the land — Inclosure award not binding on successors in title — Plaintiffs as parishioners did not have *locus standi*
***Marlton* v *Turner (QB)*, [1998]** 3 **185**

Highway maintainable at public expense
Breach of statutory duty
S36(2)(a) of Highways Act 1980 — Public footpath on housing estate — Footpath constructed by predecessor authority — Whether constructed by authority as highway authority or as housing authority — Whether single corporate entity capable of separate existence for purposes of highways — Whether footpath maintainable at public expense — Whether breach of statutory duty under s41 of 1980 Act — Held, footpath constructed by predecessor in law — Authority one corporate entity — Footpath a highway maintainable at public expense — Breach of s41 statutory duty
***Gulliksen* v *Pembrokeshire County Council (CA)*, [2002]** 3 **9**

Shopping centre — See **Shopping centre** (Pedestrian malls)
Trees — See **Trees** (Adopted highway)

HIRE AGREEMENT — See **Landlord and tenant** (Distress — Reputed ownership)

HOME-LOSS PAYMENT
Land Compensation Act 1973
Displacement of tenant
Tenant's property in area of intended regeneration — Tenant wanting to relocate and moving out to chosen location — Claim for home-loss payment — Whether tenant "displaced" — Whether test a question of fact — Held, test of displacement a question of fact — "Wanting to move" different from "willing to move" when faced with the inevitable — Tenant wanting to move rather than being displaced
***Ingle* v *Scarborough Borough Council (CA)*, [2002]** 2 **161**

HOSPITAL — See **Restrictive covenant** (Enforcement — Statutory powers)

HOUSE, MEANING OF — See **Landlord and tenant** (Leasehold enfranchisement — "House"); **Leasehold enfranchisement** (Leasehold Reform Act 1967 — Meaning of "house")

HOUSING — See also **Landlord and tenant** (Assured tenancy — Grounds for possession) (Assured tenancy — Housing Act 1988) (Assured tenancy — S20 notice) (Housing Act 1988 — Assured tenancy)
Environment
Statutory nuisance
Environmental Protection Act 1990 — Abatement notice — Meaning of owner — Person receiving rack-rent — Whether managing agent an owner — Whether s80(2) (b) of 1990 Act includes managing agent who receives rack-rent as agent of trustee — Held, managing agent receiving rack-rent can be an owner — Abatement notice may be served on managing agent
***Camden London Borough Council* v *Gunby (QB)*, [1999]** 3 **13**

Housing Act 1985 — See also **Landlord and tenant** (Secure tenancy — Joint tenancy)
Multiple occupation
Property containing 10 bedrooms of which nine are let — Whether nine occupiers of property a single household — Whether property a house in multiple occupation — Whether all factors in *Barnes* v *Sheffield City Council* relevant — Held, property was a house in multiple occupation — Not all factors in *Barnes* always relevant
***Rogers* v *Islington London Borough Council (CA)*, [1999]** 3 **17**

Housing Act 1988 — See also **Landlord and tenant** (Housing Act 1988 — Assured tenancy) (Assured shorthold tenancy)

HOUSING — *continued*

Right to buy
Defective house type
S563(1) of Housing Act 1985 — Breach of duty — Measure of damages — Whether purchaser entitled to damages reflecting statutory financial assistance scheme — Whether damages difference between price paid and price that would have been paid with knowledge of likelihood of defect — Held, damages limited to difference between price paid and value of house with knowledge that it was a house of a type designated defective
***Berry* v *Newport Borough Council (CA),* [2000] 2 26**

Secure tenancy — See also **Landlord and tenant** (Secure tenancy)
Possession claim
Housing Act 1985 — Ground 2 possession — Nuisance and annoyance to neighbours — Conduct away from dwelling-house — Whether neighbours means adjoining occupiers — Held, neighbours not confined to adjoining occupiers — Ground 2 satisfied by misconduct amounting to nuisance and annoyance to neighbours within 100m
***Northampton Borough Council* v *Lovatt (CA),* [1998] 1 15**

Separate dwelling — Exclusive possession — Housing Act 1985 — Whether licence of bedroom and shared use of whole flat created secure tenancy — Held, no secure tenancy — Conditions of exclusive possession and essential living facilities did not both apply to flat or to any part
***Parkins* v *Westminster City Council (CA),* [1998] 1 22**

Right to buy
Order of proceedings — Possession and right to buy proceedings before court — Whether court entitled to hear possession action before right to buy proceedings — Held, court entitled to hear possession proceedings before right to buy application
***Tandridge District Council* v *Bickers (CA),* [1998] 3 31**

Secure tenancy — Whether extensive repairs and alterations to be disregarded in determining purchase price — Whether improvements have narrow construction — Held, improvements given ordinary meaning in s187
***Dickinson* v *Enfield London Borough Council (CA),* [1996] 2 88**

Status of tenant
Local authority acquired headlease subject to tenancy protected by Rent Act 1977 — Whether tenant entitled to secure tenancy under Housing Act 1985 — Held, tenant acquired status of secure tenant
***Basingstoke and Deane Borough Council* v *Paice (CA),* [1995] 2 9**

HOUSING ACT 1988 — See also **Landlord and tenant** (Assured tenancy)
Assured shorthold tenancy
S20 notices
Assured Tenancies and Agricultural Occupancies (Forms) Regulations 1988 — Whether s20 notices "substantially to same effect"— Whether s20 notice misstating commencement date valid — Whether s20 notice misstating term date of tenancy valid — Whether s20 notice misstating guidance notes valid — Held, guidance in *Mannai Investment Co Ltd* v *Eagle Star Assurance Co Ltd* [1997] AC 749 applicable — Notices "substantially to same effect"
***Ravenseft Properties Ltd* v *Hall; White* v *Chubb; Kasseer* v *Freeman (CA),* [2002] 1 9**

HOUSING ASSOCIATION — See **Landlord and tenant** (Lease or licence — Flat)

HOUSING GRANTS, CONSTRUCTION AND REGENERATION ACT 1996
Construction contract
Adjudicator
Adjudicator's pre-emptory decision — Permission given under s42 of the Arbitration Act 1996 — Contract containing valid arbitration clause — Where dispute concerns validity of decision, whether decision enforceable — Held, decision enforceable — Where dispute referred to arbitration no stay by reason of the scheme
***Macob Civil Engineering Ltd* v *Morrison Construction Ltd (QB:TCC),* [1999] 3 7**

HUMAN RIGHTS — See also **Adverse possession** (Limitation period — European Convention right on protection of property); **Compulsory purchase** (Compulsory purchase order — Validity); **Landlord and tenant** (Distress — Set-off) (Landlord and Tenant Act 1927 — Notices) (Landlord and Tenant — Counternotice) (Option ro renew — Local authority) (Repairs — Condensation) (Secure tenancy — European Convention on Human Rights); **Leasehold enfranchisement** (Leasehold Reform Act 1967 — Meaning of "house"); **Nuisance** (Sewers); **Town or village greens**

Enforcement notice
Planning appeals
Appeal against enforcement notice dismissed by inspector and High Court — Whether hearings in breach of Art 6(1) of the European Convention of Human Rights — Held, hearings by inspectors do not comply with convention as to independence because Secretary of State can revoke inspector's appointment to hear appeal
***Bryan* v *United Kingdom (ECHR),* [1996] 2 123**

European Convention on Human Rights
Chancel repairs
Chancel Repairs Act 1932 — Lay rector — Whether liability of lay rector of part of rectorial lands uncertain — European Convention on Human Rights — Art 1 of First Protocol — Arts 9 and 14 of Convention — Whether law on chancel repair liability breach of Convention— Held, law not uncertain — Lay rector of part of rectorial lands liable for chancel repairs — Liability does not involve breach of Convention
***Aston Cantlow Parochial Church Council* v *Wallbank (Ch),* [2000] 2 149**

Planning procedures
Role of Secretary of State
Compulsory purchase and works orders — Role of Secretary of State — Role of High Court on review — Article 6(1) of the European Convention of Human Rights — Right to fair hearing — Whether Secretary of State independent and impartial — Whether Secretary of State's decisions in respect of planning and transport compatible with Art 6(1) — Whether Secretary of State's exercise of decision–making powers unlawful — Held, Secretary of State not an independent and impartial tribunal — Secretary of State's decisions in respect of planning and transport not compatible with Article 6(1) — Secretary of State's exercise of decision–making powers lawful
***R* v *Secretary of State for the Environment, Transport and the Regions, ex parte Holding & Barnes plc (QB (DC)),* [2001] 1 33**

IMPROVEMENTS — See **Landlord and tenant** (Landlord and Tenant Act 1927) (Rent review clause — Disregards); **Leasehold enfranchisement** (Leasehold Reform Act 1967 — Determination of enfranchisement price) (Leasehold Reform Act 1967 — S9(1A) valuation)

INHERITANCE TAX — See **Taxation** (Inheritance tax)

INJUNCTION — See also **Easement** (Right of way — Remedy for trespass)
Interim mandatory injunction
Grounds for relief
Agreement for lease — Licence to occupy — Termination of licence — Repossession of premises — Application for interim mandatory injunction to restore occupation — Whether grounds for injunction made out — Proprietary estoppel and constructive trust — "Subject to contract" negotiations — Whether serious issue to be tried — Held, interim injunction refused — Grounds for relief not established
***Edwin Shirley Productions Ltd* v *Workspace Management Ltd (Ch),* [2001] 2 16**

INJUNCTION — *Continued*

Interlocutory — See also **Landlord and tenant** (Assignment — Breach of covenant) (Repairs — Covenant)
Lord Cairns' Act
Exercise of discretion
Breach of restrictive covenant and trespass — Use of private road and garden — Whether damages appropriate remedy — Whether damages for future wrongs — Proper assessment of damages — Held, four tests in *Shelfer* v *City of London Electric Lighting Co Ltd* fulfilled — Proper to order damages in place of injunction
***Jaggard* v *Sawyer (CA),* [1995] 1 146**

INSOLVENCY — See also **Landlord and tenant** (Insolvency); **Mortgage**
Administration
Leave to bring proceedings
Waste-management licence — Criminal proceedings for breach of licence conditions — Whether Environment Agency requires leave to bring and continue criminal proceedings relating to waste-management licence — Whether leave should be granted — Held, leave required under ss10 and 11 Insolvency Act 1986 to bring and continue criminal proceedings — Leave given on the facts known at date of appeal
***Environment Agency* v *Clark (administrator) (CA),* [2000] 1 113**

Assignment
Liability of original tenants
Assignee of term insolvent — Voluntary arrangement — Whether liability of original tenants released or discharged by insolvency voluntary arrangement made by insolvent assignee of term — Held, original tenants remained liable to pay arrears of rent — Voluntary arrangement did not prevent landlords recovering rent from original tenants
***March Estates plc* v *Gunmark Ltd (Ch),* [1996] 2 38**

INSOLVENCY — *continued*

Bankruptcy
Foreign property
Trustee in bankruptcy — Whether trustee entitled to orders relating to possession and sale of foreign property — Whether Art 16.1 of Brussels Convention applies — Whether orders sought were for rights *in rem* or *in personam* — Held, rights sought were rights *in personam* — Art 16.1 did not inhibit enforcement of trust over foreign property
***Pollard and another* v *Ashurst (trustee in bankruptcy) (Ch),* [2000]** 2 **29**

Deposit — See **Vendor and purchaser** (Deposit — Purchaser's lien)
Disclaimer of freehold
Application for vesting order
Local authority registered local land charges to protect expenses of dangerous structure notices — Whether local authority with interest in local land charges entitled to vesting order in priority to freehold registered charge — Held, expenses of local authority protected by local land charges exceeded value of property — Local authority entitled to vesting order
***Hackney London Borough Council* v *Crown Estate Commissioners (Ch),* [1996]** 1 **151**

Disclaimer of lease
Determination of loss
S178(6) of Insolvency Act 1986 — Lease at rent four times' rental value disclaimed — Compensation for landlord's loss of rent for residue of term — Whether discount for early payment of rent — Appeal from [1997] 2 36 — Held, discount on sum to reflect immediate payment of rents that would have been paid in the future
***Christopher Moran Holdings Ltd* v *Bairstow (HL),* [1999]** 1 **1**

Vesting order
Landlord and tenant — Liquidator disclaiming headlease — Whether court entitled to order that headlease should vest with freeholder — Whether sublease continuing — Held, court had no power to order vesting of headlease in freeholder where subtenants made no election to take headlease — Sublease terminated
***Sterling Estates* v *Pickard UK Ltd (Ch),* [1997]** 2 **33**

Disclaimer of licence
Waste-management licence
Environmental Protection Act 1990 — Whether waste-management licence "property" within s436 of Insolvency Act 1986 — Whether liquidator entitled to disclaim onerous waste-management licence under s178 of 1986 Act — Held, waste-management licence is property — Licence can be disclaimed — No mutual inconsistency between insolvency and environmental leglislation
***Official Receiver (as liquidator of Celtic Extraction Ltd & Bluestone Chemicals Ltd)* v *Environment Agency (CA),* [1999]** 3 **21**

Forfeiture — See **Landlord and tenant** (Forfeiture — Insolvency Act 1986)
Individual voluntary arrangement
Rent arrears
Tenant insolvent — Arrangement included future liabilities — Whether individual insolvency arrangement included future rent — Whether landlord unfairly prejudiced by inclusion of future rent — Whether material irregularity in chairman of creditors' meeting giving one year's value to future rent — Held, future liabilities included future rent — No prejudice to landlord and no material irregularity
***Doorbar* v *Alltime Securities Ltd (CA),* [1996]** 2 **33**

Surety
Duty to mitigate
Surety of lease — Company associated with tenant company in possession — Rent accepted from company in occupation — Tenant in liquidation — Lease disclaimed — Whether implied acceptance of surrender of lease — Whether surety liable for arrears of rent — Whether landlord under duty to mitigate loss by recovering possession — Held, no surrender of lease — Surety liable — No obligation on landlord to mitigate
***Bhogal* v *Cheema (Ch),* [1998]** 2 **50**

Voluntary arrangement — See also **Landlord and tenant** (Assignment)
Co-debtors
Release — Joint liability — Indemnification of liability under lease — Whether co-debtors released by terms of voluntary arrangement — Whether liability of co-debtors never effected by voluntary arrangements — Held, on construction of voluntary arrangement, co-debtors not released — No principle that co-debtors can never be released
***Johnson* v *Davies (CA),* [1998]** 3 **72**

INSURANCE — See also **Landlord and tenant** (Rent — Rent suspension clause)
Commission — See **Landlord and tenant** (Leasehold enfranchisement)

Landlord's covenant to insure — See also **Landlord and tenant** (Service charges)
Tenants' negligence
Whether under covenant to insure landlord agrees to insure for benefit if tenant — Whether benefit covers tenants' negligence — Held, *Mark Rowlands Ltd* v *Berni Inns Ltd* [1985] 2 EGLR 92 no authority that landlords' covenant to insure always includes agreement to insure for benefit of tenant
***Lambert* v *Keymood Ltd (QB),* [1997]** 2 **70**

JOINT TENANTS — See **Landlord and tenant** (Joint tenancy — Assignment) (Joint tenancy — Notice to quit) (Landlord and Tenant Act 1954 Part I — Joint tenants)
Severance
Notice
Service — Law of Property Act 1925 s196 — Notice served by post and intercepted by wife before husband received it — Whether notice served — Whether joint tenancy severed — Held, notice served at place of abode but not personally received valid
***Kinch* v *Bullard (Ch),* [1998]** 3 **112**

JUDICIAL REVIEW — See **Easement** (Right to light — Rights of Light Act 1959); **Lands Tribunal**

"KEEP OPEN CLAUSE" — See **Landlord and tenant** (User covenant — Breach)

LAND COMPENSATION ACT 1961 — See **Compensation; Compensation for acquisition of land** (Certificate of appropriate alternative development — Challenge) (Costs — Assessment of compensation) (Disturbance) (Equivalent reinstatement — Disturbance losses) (Land Compensation Act 1961 — Holding costs)(Land Compensation Act 1961 — Statutory planning assumptions) (Land taken — Structure) (Special suitability rule — Nature reserve) (Statutory planning assumptions — Planning permission for retained land); **Compensation for electricity lines** (Statutory basis of compensation — Valuation); **Lands Tribunal** (Jursidiction — Laying of sewer)

LAND COMPENSATION ACT 1973 — See **Compensation** (Land Compensation Act 1973 Part I — Valuation evidence); **Compensation fo acquisition of land** (Disturbance payment — Section 37 of Land Compensation Act 1973); **Compensation for electricity lines** (Statutory basis of compensation — Valuation); **Home-loss payment** (Land Compensation Act 1973 — Displacement of tenant)

LAND REGISTRATION — See also **Adverse possession** (Lease — registered land); **Highways** (Dedication — Acceptance)
Overriding interests
Agreement under s38 of Highways Act 1980
Whether agreement under s38 of Highways Act 1980 an overriding interest under s70(1)(*a*) of Land Registration Act 1925 as relating to public rights — Whether agreement binding on successor in title to original party to agreement — Held, interest under agreement did not relate to public rights — Agreement therefore not an overriding interest
***Overseas Investment Services Ltd* v *Simcobuild Construction Ltd (CA),* [1996]** 1 **49**

Option to purchase
Unregistered option — Overriding interest — Actual occupation — Whether occupier must be in occupation of whole of landcomprised in registered disposition — Whether person in actual occupation of part of land in registered disposition can enforce against new registered proprietor's overriding interest in whole or part of the registered land — Held, person having overriding interest need not be in actual occupation of whole of land in registered disposition
***Ferrishurst Ltd* v *Wallcite Ltd (CA),* [1999]** 1 **85**

Receipt of "rents"
Beneficiary under bare trust granting tenancy to registered proprietor — S70(1)(g) of Land Registration Act 1925 — Whether beneficiary in receipt of "rents" — Whether beneficiary entitled to overriding interest — Whether chargee entitled to vacation of caution protecting beneficiary's interest — Held, beneficiary not in receipt of rents — Beneficiary not entitled to overriding interest — Caution vacated
***UCB Group Ltd* v *Hedworth (CA),* [2002]** 3 **76**

LANDLORD AND TENANT
See also **Leasehold enfranchisement** (Leasehold Reform Act 1967) (Leasehold Reform, Housing and Urban Development Act 1993 — S20 notice) (Vesting order — Change of circumstance)
Agreement
Breach
Option to renew — Tenant exercising option to renew five-year term — Occupation under equitable five-year tenancy — Landlord refusing to grant legal title to tenancy — Landlord offering 20-year full repairing lease

LANDLORD AND TENANT — *continued*
Agreement — *continued*
Breach — *continued*

— Compromise — Surrender — Whether landlord in breach of agreement — Whether tenant entitled to damages — Assessment of damages — Held, claim for breach compromised by surrender and new lease — Damages would have been nominal only
***Plummer* v *Tibsco Ltd (CA),* [2002]** 1 **29**

Entire agreement clause
Beer tie — Whether warranty that tie would be released — Whether entire agreement clause effective — Collateral warranty — Whether any warranty or promise given — Held, entire agreement clause effective to exclude collateral warranty — No collateral warranty
***Inntrepreneur Pub Co* v *East Crown Ltd (Ch),* [2000]** 3 **31**

Law of Property (Miscellaneous Provisions) Act 1989
Negotiations for 10-year term — Intended tenant allowed into occupation — Intended landlord dying before agreement completed — Whether tenancy granted — Whether proprietary estoppel — Whether s2(5) of Law of Property (Miscellaneous Provisions) Act applied — Held, no proprietary estoppel within s2(5) of 1989 Act — Solicitor had no authority to conclude agreement
***James* v *Evans (CA),* [2000]** 3 **1**

Obligation to do building works
Agreement to grant lease on completion of building works — Provisions for termination of agreement — Whether preliminary remedial notice required — Whether termination events same as forfeitable events — Held, preliminary remedial notice requirement applied to events as if relief from forfeiture principles had application
***Dream Factory Ltd* v *Crown Estate Commissioners (Ch),* [2000]** 3 **107**

Agreement for lease
Licence to occupy
Termination of licence — Repossession of premises — Application for interim mandatory injunction to restore occupation — Whether grounds for injunction made out — Proprietary estoppel and constructive trust — "Subject to contract" negotiations — Whether serious issue to be tried — Held, interim injunction refused — Grounds for relief not established
***Edwin Shirley Productions Ltd v Workspace Management Ltd (Ch),* [2001]** 2 **16**

Misrepresentation
Beer tie — Precontract negotiations — Whether tenant bound by beer tie — Whether tenant released by collateral contract — Whether fraudulent misrepresentation by landlord — Whether negligent misrepresentation or misstatement — Whether promises gave rise to any legal liability — Whether landlord estopped from enforcing beer tie — Held, tenant not released from beer tie — No fraudulent misrepresentation — Promises not giving rise to legal liability as negligent misrepresentations or misstatements — Landlord not estopped
***Inntrepreneur Pub Co (CPC)* v *Sweeney (Ch),* [2002]** 2 **132**

Set-off
Licensee in occupation under agreement for lease — Arrears of licence fees — Whether licensee entitled to set off claim for damages against licence fees — Whether agreement void under Art 85(1) of Treaty of Rome — Held, no set-off against damages claim against licence or rent arrears — Beer tie but not agreement void
***Star Rider Ltd* v *Inntrepreneur Pub Co (Ch),* [1998]** 1 **53**

Uncertainty
Whether agreement void for uncertainty — Failure to specify commencement of term — Development agreement — Agreement for lease — Implied terms — Council promoting development scheme inconsistent with scheme underlying agreement — Whether council entitled to withhold consents and promote alternative scheme — Held, commencement of term of lease certain — If uncertain, grantee entitled to elect between two dates — Agreement subject to term that council could not reasonably withhold consent
***Liverpool City Council* v *Walton Group plc (Ch),* [2002]** 1 **149**

Agreement to grant lease
Agreement to take lease of public house — Alleged misrepresentation by landlord of annual barrelage and warranty of future barrelage — Whether representations made — Whether tenants knew of decline in sales — Whether tenants would have entered into agreement — Effect of exclusion clause — Held, no misrepresentations made — Tenants did not rely on any representations — Exclusion clause not reasonable
***Inntrepreneuer Estates (CPC) Ltd* v *Worth (Ch),* [1996]** 1 **84**

Alterations and additions — See also **Landlord and tenant** (Forfeiture — S146 notice)

Covenant to repair
Covenant to repair — Lease granted for term of 1,000 years — Covenant to erect and maintain specific buildings — Tenant proposing to demolish original buildings — Whether breach of repairing covenant — Whether anticipated breach of covenant to yield up in repair — Whether tenant estopped — Held, demolition of buildings not amounting to breach of lease — No estoppel
***British Glass Manufacturers' Confederation* v *University of Sheffield (Ch),* [2004]** 1 **40**

Dilapidations
Alterations permitted under licences — Reinstatement when required by landlord — Whether schedule of dilapidations sufficient notice of requirement to reinstate — Prohibition of alterations — Whether lease prohibiting all alterations to building as existing at date of grant of lease — Held, schedule of dilapidations sufficient notice of requirement to reinstate — Lease prohibiting all alterations to building
***Westminster City Council* v *HSBC Bank plc (TCC),* [2003]** 1 **62**

Arrears — See **Landlord and tenant** (Assignment — Liability of original tenant) (Forfeiture — Relief)

Assignment — See also **Landlord and tenant** (Forfeiture — Validity) (Option — Enforceability); **Milk quota**; **Rectification** (Lease — Bona fide purchaser)
Breach of covenant
Wrongful subletting — Whether interlocutory mandatory injunction requiring surrender of sublease — Held, injunctions granted — Plain breach of contract, inducement to breach contract and breach of restrictive covenant binding on subtenant
***Hemingway Securities Ltd* v *Dunraven Ltd (Ch),* [1995]** 1 **61**

Collateral deed
Landlord and Tenant Act 1988 — Duty to give consent to underletting — Lease requiring underlettings on stipulated terms — Tenant intending to grant underlease on same terms as headlease — Collateral deed personal to parties permitting less onerous performance and indemnifying proposed underlessee — Whether underlease and collateral deed to be read together — Whether effect of personal collateral deed rendering underlease inconsistent with stipulated terms — Whether terms inconsistent — Whether any duty on landlord under s1(3)(a) of 1988 Act — Held, collateral deed must be read into underlease — Terms of underlease then inconsistent with stipulated terms in headlease — No duty under s1(3)(a) of 1988 Act
***Allied Dunbar Assurance plc* v *Homebase Ltd (CA),* [2002]** 2 **23**

Consent
Assignee proposing use contrary to implied user clause — Whether landlord entitled to withold consent to assignment — Held, landlord having reasonable ground to withhold consent on ground that assignee might breach user clause — If *Killick* v *Second Covent Garden Property Co Ltd* [1973] 1 WLR 658 establishing any rule to the contrary, it was overruled
***Ashworth Frazer Ltd* v *Gloucester City Council (No 2) (HL),* [2002]** 1 **15**

Consent to assign
Contract for sale of land — Standard conditions — Sale of lease — Whether landlord gave consent to proposed assignment — Whether "in principle" consent good consent — Whether landlord can qualify consent after giving "in principle" consent — Whether buyer entitled to rescind contract — Whether buyer entitled to return of deposit — Held, landlord's in principle consent sufficient to deny buyer right to rescind — Standard condition 8.3.4 must be given purposive construction — Landlord not entitled to qualify earlier consent — Buyer not entitled to return of deposit
***Aubergine Enterprises Ltd v Lakewood International Ltd (Ch),* [2001]** 3 **71**

Consent, refusal of
Consent refused to safeguard grant of planning permission and proposed assignment of reversion — Whether consent unreasonably withheld — Whether landlord can consider position of proposed assignee of reversion — Held, consent not unreasonably withheld
***BRS Northern Ltd* v *Templeheights Ltd (Ch),* [1998]** 2 **182**

Landlord and tenant — Landlord and Tenant Act 1988 — Landlord's opposition to subletting premises — Whether landlord entitled to rely on reasons not put in writing within reasonable time of request — Whether reasons satisfactory — Whether test of profit greater than three times' rent — Held, landlord cannot adduce reasons before court not put in writing within reasonable time of request — No test of three times' rent
***Footwear Corporation Ltd* v *Amplight Properties Ltd (Ch),* [1998]** 2 **38**

Refusal of consent to assign, change of use and make alterations — Whether refusal reasonable — Whether landlord entitled to consider estate management objectives of its estate — Whether relief from forfeiture for deliberate breach — Held, landlord entitled to consider estate management objectives — Relief refused for deliberate breach
***Crown Estate Commissioners* v *Signet Group plc (Ch),* [1996]** 2 **200**

Covenants

Assignee covenanting to perform covenants of lease — Further assignment — Landlord releasing first assignee from covenant — Whether original tenant released from covenants under lease — Held, release of assignee did not release original tenant
***Sun Life Assurance Society plc* v *Tantofex (Engineers) Ltd (Ch),* [1999]** 2 **135**

Fraud

Consent to assignment of assured tenancy obtained by fraud — Whether assignment can be set aside as fraud unravels all — Whether landlord entitled to possession — Held, two transactions of consent to assignment and assignment — Consent obtained by fraud but assignment effective and could not be set aside — Landlord not entitled to possession
***Sanctuary Housing Association* v *Baker (CA),* [1998]** 1 **42**

Insolvency voluntary arrangement

Insolvency of tenant by assignment — Whether voluntary arrangement under Insolvency Act 1986 precluded landlord from recovering rent from original tenant and/or tenant by assignment — Held, on construction of arrangement landlord entitled to recover rent accruing after arrangement date
***Burford Midland Properties Ltd* v *Marley Extrusions Ltd (Ch),* [1995]** 2 **15**

Landlord and Tenant (Covenants) Act 1995

Agreement for lease — Agreement containing personal covenants — Assignment of reversion — Landlord serving s8 notice and request for release from covenants — Whether landlord released from personal covenant imposed in collateral agreement — Whether tenant obliged to indemnify landlord in respect of costs of complying with dangerous structure notice — Held, personal covenants not released by s8 notice under 1995 Act — Tenant obliged to indemnify landlord
***BHP Petroleum Great Britain Ltd v Chesterfield Properties Ltd (Ch),* [2001]** 2 **11**

Service of notices — Landlord and Tenant Act 1927 — Whether s17 notice validly served by recorded delivery at last known abode — Whether s17 notice that includes errors valid — Whether affidavit must identify person obtaining rent records from computer — Held, notice served by recorded delivery and undelivered validly served — S17 notice valid even when it claims more than landlord entitled to — Affidavit need not identify person obtaining rent records from computer
***Commercial Union Life Assurance Co Ltd* v *Moustafa (QB),* [1999]** 2 **44**

Land Registration Act 1925

Assignment of lease executed — Failure to register assignee as proprietor — Whether assignment for purposes of early determination clause — Whether exercise of option effective — Held, legal estate not assigned — *Gentle* v *Faulkner* applied
***Brown & Root Technology Ltd* v *Sun Alliance & London Assurance Co Ltd (CA),* [1997]** 1 **39**

Landlord and Tenant Act 1988

Application for consent to underlet and change of use refused — Premises used as nightclub and underletting to fitness club proposed — Consent would cause alleged £300,000 damage to reversionary interest — Whether refusal reasonable — Whether valuer's approach reasonable and informed — Held, valuations supporting reasons on wrong basis and ill informed — Consent unreasonably refused
***Luminar Leisure Ltd v Apostole (Ch),* [2001]** 3 **23**

Consent to assign refused — Whether landlord entitled to rely on reasons not advanced in refusing consent — Whether landlord entitled to request details of assignment including premiums — Whether consent unreasonably withheld — Held, consent unreasonably refused — Landlord not entitled to details of assignment
***Kened Ltd* v *Connie Investments Ltd (CA),* [1997]** 1 **21**

Damages — Application for consent to assign lease — Statutory duty on landlord — Whether landlord unreasonably refused consent to assignment — Whether landlord in breach of statutory duty to respond to application within reasonable time — Measure of damages — Whether tenant entitled to exemplary damages — Held, landlord in breach of statutory duty under 1988 Act — Damages awarded included loss of premium — Exemplary damages awarded
***Design Progression Ltd* v *Thurloe Properties Ltd (Ch),* [2004]** 1 **121**

Lease

Whether grant of subtenancy operated as assignment of term — Whether assignment arose on oral agreement — Whether s53 of Law of Property Act 1925 satisfied — Whether s2 of Law of Property (Miscellaneous Provisions) Act 1989 applied — Held, oral grant of tenancy effective assignment by operation of law
***Parc Battersea Ltd* v *Hutchinson (Ch),* [1999]** 2 **33**

Liability of original tenant

Assignee of term insolvent — Voluntary arrangement — Whether liability of original tenants released or discharged by insolvency voluntary arrangement made by insolvent assignee of term — Held, original tenants remained liable to pay arrears of rent — Voluntary arrangement did not prevent landlords recovering rent from original tenants
***March Estates plc* v *Gunmark Ltd (Ch),* [1996]** 2 **38**

Deed of variation between assignee of term and landlord — Rent increased — Assignee in arrears — Whether original tenant liable — Held, variation did not amount to surrender and regrant — Original tenant liable for rent initially reserved by lease — No release of liability by agreement to vary rent to higher figure
***Friends Provident Life Office* v *British Railways Board (CA),* [1995]** 2 **55**

Landlord claimed rent from original tenant after receiving sums from sureties for release from surety covenants — Whether tenant entitled to set off sums received by landlord against rents in arrears and due — Held, landlord cannot recover rents from tenant which have been paid by sureties — Landlord entitled to appropriate sums paid to future instalments when he decides
***Milverton Group Ltd* v *Warner World Ltd (CA),* [1995]** 2 **28**

Variation of lease — Liquidation of second assignee of term — Disclaimer of lease — Liability of original tenant and first assignee — Principles in *Friends Provident Life Office* v *British Railways Board* [1995] 2 EGLR 55 — Whether liable under rent review memorandum varying rents payable — Vesting order — Rent payable under lease created by vesting order — Held, original tenant and first assignee not bound by rent memorandum varying rents — On vesting order rent payable as subject to reviews under lease
***Beegas Nominees Ltd* v *BHP Petroleum Ltd; Sevington Properties Ltd* v *Adams (CA),* [1998]** 2 **57**

Whether disclaimer of lease by liquidator of assignee company discharges liability of earlier assignee and surety for arrears of rent — Held, original tenant and earlier assignee and surety not discharged from liability
***Hindcastle Ltd* v *Barbara Attenborough Associates Ltd (HL),* [1996]** 1 **94**

Whether original tenants released from liability for arrears of rent by insolvency voluntary arrangement entered into by tenant — Held, original tenants were not released from obligations under lease by voluntary arrangement
***Mytre Investments Ltd* v *Reynolds (QB),* [1995]** 2 **40**

Liability of surety

Landlord granted licences permitting variation of user of premises to assignee of term — Whether sureties to original tenant discharged by variation — Whether proviso in surety covenant revented discharge — Held, sureties discharged and release not prevented by proviso
***Howard de Walden Estates Ltd* v *Pasta Place Ltd (Ch),* [1995]** 1 **79**

Sureties party to licence to assign — Further assignment and deed of variation — Whether sureties discharged from liability — Appeal from [1995] 1 65 — Held, lease but not licence which sureties guaranteed varied — Sureties not discharged
***Metropolitan Properties Co (Regis) Ltd* v *Bartholomew (CA),* [1996]** 1 **79**

Reasonable costs

Application by tenant for consent to subletting — Whether landlord entitled to defer consideration of application until receipt from tenant of undertaking to pay all landlord's costs — Appeal from [1995] 1 41 — Held, landlord may only demand reasonable costs — Landlord had demanded all and not merely reasonable costs
***Dong Bang Minerva (UK) Ltd* v *Davina Ltd (CA),* [1996]** 2 **31**

Refusal of consent

Development lease — Unregistered option to grant sublease in preliminary agreement — Agreement to pay commission — Landlord refusing consent on grounds that tenant in breach of covenants for timeous completion of development and assignee not prepared to be bound by option and commission liability — Whether continuing or once-and-for-all breach — Whether waiver of right to forfeit — Whether refusal of consent reasonable — Whether grounds of refusal concerned with relationship of landlord and tenant — Whether obligations to grant sublease and pay commission "tenant covenants" under Landlord and Tenant (Covenants) Act 1995 and valid reason for refusal of consent to assignment as being within relationship of landlord and tenant — Held, breach once and for all — Forfeiture right waived — Unreasonable withholding of consent to assignment — Grounds for withholding outside relationship of landlord and tenant — Covenants for grant of sublease and payment of commission personal and not tenant covenants — If covenants tenant covenants then benefit passed with freehold reversion
***First Penthouse Ltd* v *Channel Hotels & Properties (UK) Ltd; Channel Hotels & Properties (UK) Ltd* v *Al Tamimi and others (Ch),* [2004]** 1 **16**

Right of pre-emption

Obligation on tenant to offer lease to lessor before disposal — Whether tenant in breach — Whether on disposal of lease without offer back lessor acquired equitable interest — Whether obligation to offer lease back invalidated by s17 of Landlord and Tenant Act 1954 — Whether lessor's equitable interest arising prior to later dispositions of lease — Held, breach of obligation to offer lease back — Lessor acquired equitable interest — Offer-back obligation not invalidated by s17 of 1954 Act — Lessor's equitable interest having priority over later dispositions

***Tiffany Investments Ltd* v *Bircham & Co Nominees (No 2) Ltd (CA),* [2004]** 2 **31**

Assured tenancy — See also **Landlord and tenant** (Housing Act 1988)

Contracting–out

Rent review to £25,000-£25,000 exceeding market rent — Whether agreement a sham — Whether an unlawful contracting out — Whether rent review repugnant to principal purpose of assured tenant — Held, rent review an unlawful contracting–out — Alternatively, rent review repugnant to main purpose of agreement — Assured tenancy continued

***Bankway Properties Ltd* v *Dunsford (CA),*[2001]** 2 **36**

County Courts Act 1984

Possession proceedings — Arrears of rent — Statutory relief under s138 of County Courts Act 1984 — Whether assured tenant entitled to relief under s138 — Held, s138 has no application to assured tenancy

***Artesian Residential Investments Ltd* v *Beck (CA),* [1999]** 2 **30**

Damages for unlawful eviction

Purported determination of tenancy — Damages for unlawful eviction Whether defence to damages — Whether correct basis of valuation — Whether tenant must elect between damages under 1988 Act or reinstatement into possession — Held, no defence — Incorrect basis of valuation cannot be raised at Court of Appeal — Tenant must elect between damages and possession at trial — Wrong basis of valuation

***Wandsworth London Borough Council* v *Osei-Bonsu (CA),* [1999]** 1 **26**

Tenant under assured shorthold tenancy agreed to leave by specified date — Landlord unlawfully evicted six days prior to specified date — Whether tenant entitled to resile from agreement to vacate — Whether damages under s28 must reflect delay of only six days before vacant possession would have occurred in absence of eviction — Held, damages must reflect the difference between the value of the premises on assumption that the tenant could reside the additional six days and the value on the assumption she could not

***King* v *Jackson (CA),* [1998]** 1 **30**

Dwelling-house

Hotel rooms — Housing Act 1988 — Room without separate kitchen — Whether occupier had assured tenancy — Cooking facilities — Whether hotel room a dwelling-house — Held, hotel room without kitchen capable of being a dwelling-house — Otherwise, agreement involving sharing of kitchen would not be capable of creating a letting of a separate dwelling

***Uratemp Ventures Ltd v Collins (HL),* [2001]** 3 **93**

Grounds for possession

Ground 7 of Sched 2 to Housing Act 1988 — Death of tenant — Proceedings to recover possession within 12 months — Whether s8 notice commencement of proceedings to recover possession — Held, ground 7 refers to court proceedings — Court proceedings must be commenced within 12 months of date when landlord learns of death of tenant

***Osada* v *Shepping (CA),* [2000]** 2 **38**

Houseboat

Houseboat — Whether letting of houseboat an assured tenancy — Whether houseboat annexed to land — Whether a letting of land — Held, houseboat not annexed to land — Letting of houseboat incapable of being a lease of land — Letting of houseboat not an assured tenancy

***Chelsea Yacht & Boat Co Ltd* v *Pope (CA),* [2000]** 2 **23**

Housing Act 1988

Possession — Ground 14 of Sched 2 to the Act as amended by Housing Act 1996 — Terms of written tenancy agreement contained provisions limiting ground 14 — Provision for minimum four weeks' notice of proceedings — Whether landlord entitled to rely on statutory provisions to obtain possession — Whether tenant entitled to benefit of written provisions — Whether implied term that agreement related to 1988 Act as amended from time to time — Held, four weeks' notice related to s8 notice — Deemed four weeks' notice dispensed with under s8

***North British Housing Association Ltd* v *Sheridan (CA),* [1999]** 2 **138**

Possession order by consent — Claim for arrears of rent — Tenant's counterclaim for damages for breach of landlord's repair covenants — Limitation on court's jurisdiction to make possession orders in s7 of Housing Act 1988 — Possession sought under ground 8 — Damages set off against rent arrears — Whether any arrears of rent under ground 8 — Whether court had jurisdiction to make possession order by consent — Held, where tenant counterclaims damages rent arrears reduced or extinguished by set-off — Rent not lawfully due for purposes of ground 8 — Court has no jurisdiction to make possession order by consent where no rent lawfully due

***Baygreen Properties Ltd* v *Gil (CA),* [2002]** 3 **42**

Several tenancies granted following assured shorthold tenancy — Whether s20(4) has continuous application such that all tenancies granted after original tenancy are assured shorthold tenancies — Whether s21(4) notice seeking possession must specify date on which possession required — Whether proceedings for possession can be commenced before expiration of s21(4) notice — Held, s20(4) applies to all tenancies subsequent to original shorthold tenancy — S21(4) need not specify date where reference made to two-month period and term date — Possession proceedings must not be commenced before s21(4) expires

***Lower Street Properties Ltd* v *Jones (CA),* [1996]** 2 **67**

Lease or licence

Hotel room — Occupiers introducing cooking facilities — Whether occupiers of hotel rooms had assured tenancies — Whether occupiers licensees — Housing Act 1988 — Whether hotel room a dwelling-house — Held, introduction of cooking facilities by occupier may indicate hotel room became dwelling-house if authorised by owner

***Uratemp Ventures Ltd* v *Collins (CA),* [2000]** 1 **156**

Notice of rent increase

Application to rent assessment committee referring notice of increase — Whether application made by tenant within time limit — Whether committee entitled to conclude that tenant failing to prove posting of notice — S13(4) of Housing Act 1988 — Requirement that tenant "refers" notice of increase to rent assessment committee — Whether "refers" meaning delivery to — Whether rent assessment committee having jurisdiction if notice not received by commencement of new rent period — European Convention on Human Rights — Art 6(1) of Convention — Whether clear code on time limits — Held, no evidence before committee as to posting of application — Committee entitled to conclude that tenant failing to prove posting — "Refers" in s13(4) meaning "delivery to" — Notice not delivered within time limit — Committee had no jurisdiction — No breach of Article 6(1) — Tenant's failure to comply with time limit cause of loss of fair hearing

***R (on the application of Lester)* v *London Rent Assessment Committee (QB: Admin),* [2003]** 1 58

Statutory periodic tenancy under Housing Act 1988 — Notice of increase of rent above £25,000 — Whether rent assessment panel had jurisdiction to determine rent above £25,000 — Held, if panel found market rent was above £25,000 limit it was obliged to determine the rent accordingly

***R* v *London Rent Assessment Panel, ex parte Cadogan Estates Ltd (QB),* [1997]** 2 **134**

Possession

Failure to serve s20 notice — Landlord required possession for husband — Whether "just and equitable" to dispense with written notice in claim for possession under ground 1 of Sched 2 Housing Act 1988 — Held, having regard to all circumstances, including tenant's knowledge of mistake to serve s20 notice and persistent delay in paying rent — Just to dispense with ground 1 written notice

***Boyle* v *Verrall (CA),* [1997]** 1 **25**

Housing Act 1988 — S20 notice — Whether notice defective — Whether notice "substantially to the same effect" as Form 7 of Assured Tenancies and Agricultural Occupancies (Forms) Regulations 1988 — Omission of bullet-point advice and warnings — Held, notice incurably defective — Notice not "substantially to the same effect"

***Manel* v *Memon (CA),* [2000]** 2 **40**

Rent — See **Landlord and tenant** (Rent Acts — Rent assessment)

Assured shorthold tenancy

Housing Act 1988 s20 notice

Disparity between dates in notice and commencement of tenancy — Whether effective creation of assured shorthold tenancy — Held, notice contained no error and neither stages in *York* v *Casey* [1998] 2 EGLR 25 satisfied — Notice ineffective

***Clickex Ltd* v *McCann (CA),* [1999]** 2 **63**

LANDLORD AND TENANT — *continued*
Assured shorthold tenancy — *continued*

Validity of notice
Assured shorthold tenancy — Errors in s20 notice relating to termination date of tenancy and identity of landlord — Whether s20 notice valid — Held, principles in *Mannai Investment Co Ltd* v *Eagle Star Life Assurance Co Ltd* [1997] 1 EGLR 57 applied — S20 notice valid
***York* v *Casey (CA),* [1998]** 2 **25**

Break clause — See also **Landlord and Tenant** (Landlord and Tenant Act 1954 Part II — S26 request for new tenancy)
Condition precedent: performance of covenants
Whether subsisting breach at termination date relevant — Distinction between breach of repairing covenant and state of disrepair — Held, notice ineffective because of subsisting breaches of covenant
***Trane (UK) Ltd* v *Provident Mutual Life Assurance (Ch),* [1995]** 1 **33**

Estoppel
Penalty rent condition — Break notice served by occupier not tenant — Whether occupier agent of tenant — Whether landlord estopped from denying validity of break notice — Whether time of the essence for payment of penalty rent — Whether landlord estopped from relying on timeous penalty rent payment — Held, service of break notice by occupier ineffective — Landlord estopped from disputing validity of notice —Timely payment of penalty of the essence — No estoppel that timeous payment suspended
***Dun & Bradstreet Software Services (England) Ltd* v *Provident Mutual Life Assurance Association (CA),* [1998]** 2 **175**

Holding over
Election — Break notice — Landlord claiming contractual rent — Validity of break notice later accepted — Whether landlord entitled to double rent under Distress for Rent Act 1737 — Whether defence of election — Held, no election — Landlord entitled to double rent
***Ballard (Kent) Ltd* v *Oliver Ashworth (Holdings) Ltd (Ch),* [1998]** 3 **60**

Landlord's intention to reconstruct
Whether notice under s25 of 1954 Act and accompanying letter effective notice to exercise break clause for development purpose — Held, requirement of desire satisfied and notice valid
***Aberdeen Steak Houses Group Ltd* v *Crown Estate Commissioners (Ch),* [1997]** 2 **107**

Notice
Break notice specifying incorrect party as tenant — Whether notice valid — Whether notice given by general agent of tenant — Whether validated by *Mannai Investment Co Ltd* v *Eagle Star Assurance Co Ltd* [1997] 1 EGLR 57 — Whether lease determined — Held, notice invalid — Not validated by *Mannai* — No estoppel
***Lemmerbell Ltd* v *Britannia LAS Direct Ltd (CA),* [1998]** 3 **67**

Whether notice incorrectly specifying 10th anniversary of date rather than commencement of term of lease valid notice — Landlord and Tenant Act 1954 — Request for new tenancy to commence earlier than date lease would determine by effluxion of time valid and effective to determine lease — Appeal from [1996] 1 113 — Held, notice valid — Request for new tenancy to commence earlier than expiration of the term of lease invalid
***Garston* v *Scottish Widows Fund & Life Assurance Society (CA),* [1998]** 2 **73**

Original lessee
Break clause exercisable by original lessee — Term assigned by original lessee and re-acquired — Whether break clause exercisable by original lessee — Appeal from [1995] 2 38 — Held, break clause no longer exercisable by original lessee
***Max Factor Ltd* v *Wesleyan Assurance Society (CA),* [1996]** 2 **210**

Possession
Yield up possession — Tenant exercising break clause — Whether tenant "yielded up possession" — Whether "yield up" requiring offer of keys — Whether lease determined — Held, offer of keys not essential requirement of "yield up" — Whether premises "yielded up" involving objective assessment of intentions of tenant as known to the landlord — Lease determined
***John Laing Construction Ltd v Amber Pass Ltd (Ch),* [2004]** 2 **128**

Pre-condition
Obligation to reasonably perform and observe the covenants — Whether precondition to determination of underlease satisfied — Whether performance short of total performance reasonable performance — Held, total failure to comply with covenants — Covenants not reasonably performed — Some performance less than total could be reasonable performance
***Reed Personnel Services plc* v *American Express Ltd (Ch),* [1997]** 1 **229**

Time limit
Whether date of break determined from commencement of term or date of execution of lease — Whether landlord estopped from asserting termination date different from that asserted by tenant — Held, break date calculated from commencement of term — No estoppel
***Trane (UK) Ltd* v *Provident Mutual Life Assurance (Ch),* [1995]** 1 **33**

S20 notice
Housing Act 1988 — Whether s20 notice can be served on agent for tenant — Whether party acting for tenant agent for purpose of receiving s20 notice — Held, s20 notice capable of being served on tenant's agent — Party acting for tenant had authority to accept s20 notice
***Yenula Properties Ltd* v *Naidu (CA),* [2002]** 3 **28**

Validity of notice
Clause required 12 months' notice to determine lease at end of fifth year of term — Letter and accompanying notice referred to clause but contained a pair of wrong termination dates — Whether notice valid — Held, notice valid as landlord would have known tenant's intention — Notice did not have to specify termination date
***Micrografix* v *Woking 8 Ltd (Ch),* [1995]** 2 **32**

Notices determining fixed term lease one day earlier than date specified in break clause — Whether notices valid — Appeal from [1996] 1 69 — Held, notices valid on test of reasonable recipient — *Hankey* v *Clavering* overruled
***Mannai Investment Co Ltd* v *Eagle Star Life Assurance Co Ltd (HL),* [1997]** 1 **57**

Business tenancies — See **Landlord and tenant** (Landlord and Tenant Act 1954 Part II)
Costs
Proceedings
Tenancy provided that tenants indemnify landlords in respect of all costs in any proceedings — Whether landlords entitled to order for costs on indemnity basis — Held, save for good reason court should reflect contractual provision for costs in any costs order
***Church Commissioners for England* v *Ibrahim (CA),* [1997]** 1 **13**

Covenant for quiet enjoyment
Rubbish chute
Noise and nuisance from rubbish chute — Whether breach of covenant for quiet enjoyment — Whether nuisance — Held, no breach of covenant for quiet enjoyment — No finding that noise from chute not contemplated at date of tenancy agreement
***Long* v *Southwark London Borough Council (CA),* [2002]** 3 **37**

Covenants — See also **Landlord and tenant** (Assignment) (Forfeiture — Mistake) (Quiet enjoyment) (Repairs) (Subletting — Consent, refusal of) (User covenant — Breach)
Assignment
Assignee proposing to use premises contrary to implied user clause — Whether landlord entitled to withhold consent to assignment — Held, landlord unable to withhold consent on ground that assignee might breach user clause
***Ashworth Frazer Ltd* v *Gloucester City Council (CA),* [2000]** 1 **44**

Beer tie
Art 85 of Treaty of Rome — Whether beer tie void under Art 85 — Whether block exemptions regulations satisfied — Whether monetary claim in respect of void beer tie — Held, block exemption regulations satisfied — No breach of Art 85 — Breach of Article does not permit monetary claim
***Gibbs Mew plc* v *Gemmell (CA),* [1999]** 1 **43**

Whether warranty or promise to release — Entire agreement clause — Whether entire agreement clause effective — Collateral warranty — Whether any warranty or promise given — Held, entire agreement clause effective to exclude collateral warranty — No collateral warranty — Beer tie effective
***Inntrepreneur Pub Co* v *East Crown Ltd (Ch),* [2000]** 3 **31**

Collateral agreement
Agreements to remedy defects and give rent concession — Whether s2 of Law of Property (Miscellaneous Provisions) Act 1989 applied — Whether obligations of collateral agreements binding on successor in title to reversion — Held, collateral contracts valid and did not dispose of an interest in land — Obligations binding on successor in title
***Lotteryking Ltd* v *AMEC Properties Ltd (Ch),* [1995]** 2 **13**

Consent for alterations
Covenant against alterations without landlord's written consent — Whether consent granted by "subject to licence" letter — Held, letter satisfied requirement of lease in giving written consent — "Subject to contract" cases involving bilateral negotiations not relevant where unilateral act of giving consent relevant
***Prudential Assurance Co Ltd* v *Mount Eden Land Ltd (CA),* [1997]** 1 **37**

Covenant by landlord
Covenant in lease restricting use of land demised to another tenant — Enforcement by tenant against another tenant of landlord — Landlord and Tenant (Covenants) Act 1995 — Restrictive covenant — Whether covenant enforceable under s3(5) of 1995 Act — Whether enforceable as restrictive covenant — Held, restrictive covenant unenforceable under s3(5) of the 1995 Act — Covenant not registrable as restrictive covenant — unenforceable notwithstanding any actual or constructive notice
***Oceanic Village Ltd* v *United Attractions Ltd (Ch),* [2000]** 1 **148**

Derogation from grant
Premises let for banking purposes — Whether landlord in breach of express covenant and/or obligation not to derogate from grant in letting adjoining premises for same use — Held, express covenant did not cover letting policy — *Port* v *Griffith* remained good law — No breach of covenant or obligation
***Romulus Trading Co Ltd* v *Comet Properties Ltd (QB),* [1996]** 2 **70**

Implied term
Landlord proposing to construct additional flats on existing block of flats — Whether lease contained implied term against construction of additional flats — Whether construction of additional flats breach of landlord's repairing obligation — Held, although lease part of letting scheme no limit on additional flats — Construction of additional flats no breach of repairing covenant — *Devonshire Reid Properties Ltd* v *Trenaman* [1997] 1 EGLR 45 distinguished and not followed
***Hannon* v *169 Queen's Gate Ltd (Ch),* [2000]** 1 **40**

Landlord and Tenant (Covenants) Act 1995
Assignment — Refusal of consent — Development lease — Unregistered option to grant sublease in preliminary agreement — Agreement to pay commission — Landlord refused consent on ground that assignee not prepared to be bound by option and commission liability — Whether grounds of refusal concerned with relationship of landlord and tenant — Whether obligations to grant sublease and pay commission 'tenant covenants' under Landlord and Tenant (Covenants) Act 1995 and founded valid reason for refusal of consent to assignment as being within relationship of landlord and tenant — Held, unreasonable withholding of consent to assignment — Grounds for withholding outside relationship of landlord and tenant — Covenants for grant of sublease and payment of commission personal and not tenant covenants — If covenants tenant covenants then benefit passed with freehold reversion
***First Penthouse Ltd* v *Channel Hotels & Properties (UK) Ltd; Channel Hotels & Properties (UK) Ltd* v *Al Tamimi and others (Ch),* [2004]** 1 **16**

Meaning of "landlord" — Meaning of "landlord covenant" — Whether personal obligations in agreement to grant lease capable of being released by s8 notice — Whether original landlord released from obligations — Held, "landlord covenant" an obligation of the landlord for the time being — Original landlord not released from personal obligation by s8 notice
***BHP Petroleum Great Britain Ltd* v *Chesterfield Properties Ltd (CA),* [2002]** 2 **121**

Letting scheme
Parade of shops let under leases containing covenants imposing restrictions on specified uses — Whether constituting "letting scheme" — Whether one lessee entitled to enforce restrictive covenant directly against another lessee — Use of premises for sale of tobacco and confectionery — Whether breach of covenant — Competition laws — Whether letting scheme imposing restrictions on competition contrary to competition laws — Held, rules for letting scheme established — Lessee entitled to enforce restrictive covenant directly against another lessee in scheme — Breach of covenant
***Williams* v *Kiley (t/a CK Supermarkets Ltd) (CA),* [2003]** 1 **46**

Liability of surety — *See* **Landlord and tenant** (Assignment — Liability of surety)
Meaning of "Keep open"
Covenant by tenant to "keep open" retail premises — Whether landlord entitled to specific implement (performance) of covenant — Held, position in Scotland different from that in England — Party to contractual obligation entitled as of right to specific implement of contract obligation
***Highland & Universal Properties Ltd* v *Safeway Properties Ltd (Ct Sess: Inner House),* [2000]** 3 **110**

Personal
Side letter to lease allowing tenant to surrender on terms — Whether obligation to accept surrender binding on successor in title to reversion — Whether covenant personal — Held, covenant binding on successor in title by construction and under s142 of Law of Property Act 1925
***System Floors Ltd* v *Ruralpride Ltd (CA),* [1995]** 1 **48**

Principles of good estate management
Premises let for banking purposes — Whether landlord in breach of express covenant to abide by principles of good estate management and/or obligation not to derogate from grant in letting adjoining premises for same use — Held, express covenant did not cover letting policy — *Port* v *Griffith* remained good law — No breach of covenant or obligation
***Romulus Trading Co Ltd* v *Comet Properties Ltd (QB),* [1996]** 2 **70**

Right of entry
Landlord entitled to enter to carry out works — Whether landlord entitled to carry out improvements — Whether entry limited to remedying breaches of landlord or tenant covenants — Held, no right of entry to carry out improvements — Entry restricted to remedy breaches of landlord and tenant covenants
***Yeoman's Row Management Ltd* v *Bodentien-Meyrick (CA),* [2002]** 2 **39**

Surety — See **Landlord and tenant** *(Surety covenant)*
User
Lease imposing obligation on original tenant to construct buildings for certain specified use classes — Whether covenant implied restricting use to specified use classes — Held, by reason of interaction of clauses for building, insuring and rebuilding by reference to specified use classes, covenant restricting use to specified classes
***Ashworth Frazer Ltd* v *Gloucester City Council (CA),* [2000]** 1 **44**

Lease imposing obligation upon original tenant to construct buildings for certain specified use classes — Whether covenant implying restricted use to specified use classes — Held, no such covenant implied restricting use to specified classes
***Ashworth Frazer Ltd* v *Gloucester City Council (HL),* [2002]** 1 **15**

User covenant — See **Landlord and tenant** *(User covenant)*

Derogation from grant — See also **Derogation from grant** (Lease — Shopping mall); **Landlord and tenant** (Repudiation — Breach of covenant)
Breach
Lease of kiosk near exit to underground station — Exit closed for long hours — Whether derogation from grant — Whether damages payable on "no–transaction" basis — Whether set–off for increased profits from bureau on same station — Held, closure of exit derogation of grant — Damages not on "no–transaction" basis — Set–off for increased profits from bureau
***Platt* v *London Underground Ltd (Ch),* [2001]** 2 **121**

Repudiation
Shopping centre — Whether landlord derogated from grant — Implied obligations of landlord — Whether landlord in breach of implied obligations — Whether tenant entitled to treat lease as terminated — Whether tenant entitled to damages — Held, change of tenant mix and introduction of anchor tenant no breach
***Petra Investments Ltd* v *Jeffrey Rogers plc (Ch),* [2000]** 3 **120**

Covenant against alteration
Consent for alterations
Conditions — Tenant prohibited from carrying out alterations without landlord's consent — Such consent not to be unreasonably withheld — Grant of conditional consent — Whether condition unreasonable — Whether landlord entitled to impose condition to protect its business in adjoining property — Whether condition wider than necessary to protect landlord's business — Held, landlord entitled to impose condition to protect legitimate business interests in adjoining property — Condition imposing restriction on user wider than required to protect landlord's interest
***Sargeant* v *Macepark (Whittlebury) Ltd (Ch),* [2004]** 3 **26**

Tenant prohibited from carrying out alterations without landlord's consent — Such consent not to be unreasonably withheld — Consent withheld on grounds of structural stability and effect on car-parking area — Structural elements to be removed — Principles upon which requirement of reasonableness exercised — Whether consent unreasonably withheld — Whether landlord entitled to know tenant's solution to structural problems or should impose condition to ensure structural stability — Held, principles in *International Drilling Fluids Ltd* v *Louisville Investments (Uxbridge) Ltd* [1986] 1 EGLR 39 in relation to consent by a landlord to an assignment of a lease applicable with the necessary changes, together with tests in *Tollbench Ltd* v *Plymouth City Council* [1988] 1 EGLR 79 — Landlord entitled to withhold consent where tenant failing to provide technical solutions to structural stability problems — Landlord not required to impose condition to ensure structural stability
***Iqbal* v *Thakrar (CA),* [2004]** 3 **21**

LANDLORD AND TENANT — *continued*

Covenant for quiet enjoyment

Repairs

Landlord carrying out repairs — Carrying out of repairs causing loss and damage to tenant — Whether breach of covenant for quiet enjoyment — Construction of lease — Whether covenant for quiet enjoyment having priority over or parity with covenant to repair — Whether landlord required to take all reasonable steps or all possible steps to avoid damage to tenant — Held, covenant for quiet enjoyment had parity with, not priority over, covenant to repair — Landlord taking all reasonable steps to avoid losses — No obligation to take all possible steps

***Goldmile Properties Ltd* v *Lechouritis (CA),* [2003]** 1 **60**

Dilapidations

Damages

Terminal dilapidations — Whether air-conditioning and heating system required renewal or reconditioning — S18(1) of Landlord and Tenant Act 1927 — Whether diminution in value of reversion — Alternative valuation bases for determining dimimution in value — Held, landlord not entitled to renewal of systems — None of alternative valuations bases supported any diminution in reversionary value

***Ultraworth Ltd* v *General Accident Fire & Life (QB:TCC),* [2000]** 2 **115**

Measure of damages

Repairs — Damages — Appropriate method of repairing roof — Whether premises should be valued for industrial or storage purposes for purposes of s18(1) of Landlord and Tenant Act 1927 — Whether VAT should be added — Held, overcladding appropriate repair to roof — VAT should be added to cost of works — Industrial storage appropriate basis of valuation for s18(1) cap

***Sun Life Assurance plc* v *Racal Tracs (QB:TCC),* [2000]** 1 **138**

Valuation

S18(1) of Landlord and Tenant Act 1927 — Whether premises should be valued for storage purposes for purposes of s18(1) of Landlord and Tenant Act 1927 — Whether VAT should be added - Held, premises valued for storage purposes — VAT added

***Sun Life Assurance plc* v *Racal Tracs (QB:TCC),* [2000]** 1 **138**

Disclaimer of lease — See also **Insolvency** (Disclaimer of lease — Determination of loss); **Landlord and tenant** (Forfeiture — Possession) (Insolvency — Disclaimer of lease) (Rent arrears — Liability of original tenant)

Guarantor

Obligation to take new lease — Disclaimer of lease — Obligation on guarantor to take new lease — Whether landlord recovered possession after disclaimer — Whether s2 of Law of Property (Miscellaneous Provisions) Act 1989 applied to obligation to take up new lease — Held, landlord did not take possession after disclaimer — S2 of 1989 Act did not apply to put option to take new lease — S2 satisfied by lease containing put option

***Active Estates* v *Parness (Ch),* [2002]** 3 **13**

Liability of original lessee

Whether disclaimer of lease by liquidator of assignee company discharges liability of earlier assignee and surety for arrears of rent — Held, original tenant and earlier assignee and surety not discharged from liability

***Hindcastle Ltd* v *Barbara Attenborough Associates Ltd (HL),* [1996]** 1 **94**

Distress

Distraining goods of assignee

Whether landlord can distrain against goods of assignee in respect of rent due before assignment and owed by assignor — Held, landlord cannot distrain goods of assignee for rent owed by assignor

***Wharfland Ltd* v *South London Co-operative Building Co Ltd (QB),* [1995]** 2 **21**

Interim rent

Interim rent determined by county court — Tenant appealed order determining interim rent — Landlord sought to levy a distress in respect of arrears of interim rent — Whether tenant entitled to a stay pending hearing of appeal — Held, landlord not entitled to a distress as rent not certain until appeal — Stay granted

***Eren* v *Tranmac Ltd (CA),* [1997]** 2 **211**

Reputed ownership

Van under hire agreement to tenant sold by landlord following distress — Whether van in reputed ownership of tenant — Whether landlord entitled to sell van — Held, landlord should have known that van might be subject of hire agreement — Van not in reputed ownership of tenant — Wrongful sale

***Salford Van Hire (Contracts) Ltd* v *Bocholt Developments Ltd (CA),* [1995]** 2 **50**

Set-off

Legal and equitable set-off — Overpayment of rent — Damages — Whether tenant entitled to legal and/or equitable set-off against rent — Whether landlord must have regard to tenant's rights of set-off on levying distress — Whether distress interference with human rights — Held, no prior demand required in respect of overpayment of rent under a mistake of fact and/or law — Overpayment due without demand — Equitable but not legal set-off applies to rent due where distress levied — Distress interferes with Art 8 of European Convention on Human Rights and Art 1 of First Protocol

***Fuller* v *Happy Shopper Markets Ltd (Ch),* [2001]** 2 **32**

Walking possession agreement

Walking possession agreement — Whether bailiff and sheriff entitled forcibly to re-enter to remove goods — Held, bailiff not entitled to re-enter forcibly in absence of tenant's agreement — Sheriff not entitled to re-enter by force save in case of non-residential premises where initial entry forcible

***Khazanchi* v *Faircharm Investments Ltd; McLeod* v *Butterwick (CA),* [1998]** 3 **147**

Double rent

Holding over

Break clause — Holding over — Double rent — Distress for Rent Act 1737 — Appeal from [1998] 3 60 — Held, landlord did not treat tenant as trespasser and not entitled to double rent

***Ballard (Kent) Ltd* v *Oliver Ashworth (Holdings) Ltd (CA),* [1999]** 2 **23**

Enforcement of covenants

S141 of Law of Property Act 1925

Unregistered headlease — Whether landlord entitled to enforce covenant in respect of breaches prior to date of registration of title — Purported forfeiture — Whether purported forfeiture can be disregarded — Whether presumption of legality — Whether landlord taking advantage of own wrong — Whether landlord estopped from denying forfeiture — Whether estoppel fed — Whether feeding of estoppel required application of s141 to ante-date forfeiture — Held, landlord with unregistered reversionary interest not entitled to take advantage of own wrong — Presumption of legality applied — Landlord estopped from denying forfeiture — Estoppel fed on registration of title — Application of s141 not ante-dated to forfeiture

***Rother District Investments* v *Corke (Ch),* [2004]** 1 **47**

Equitable set-off

Assignment of right to recover rent

Claim by tenant for breach of covenant to repair — Claim by landlord for possession based upon rent arrears — Previous landlord breaching covenant for repair — Previous landlord assigning right to recover rent arrears — Present landlord claiming rent arrears — Whether tenant entitled to set off claim for damages for previous landlord's breach of covenant to repair against claim for rent by present landlord — Whether equitable set-off — Whether rule that assignee of chose of action takes subject to equities applying — Held, general equitable set-off not applying — Rule of equity applying — Assignee of chose of action (right to recover arrears of rent) taking subject to equities (right to damages against assignor) — Tenant entitled to set off claim for damages against liability to pay rent

***Muscat* v *Smith (CA),* [2003]** 3 **11**

Eviction — See **Landlord and tenant** (Lease — Surrender)

Forfeiture

Administration order

S10(1) (*b*) Insolvency Act 1986 — Whether landlord entitled to forfeit lease held by tenant company where petition for administration order presented — Held, right of forfeiture not security and landlord not secured creditor — Landlord entitled to forfeit lease

***Lomax Leisure Ltd, In re (Ch),* [1999]** 2 **37**

Compromise of proceedings

Forfeiture by service of writ — Whether lease remained forfeited following compromise of forfeiture proceedings — Held, upon compromise of proceedings, forfeiture became a nullity

***Hynes* v *Twinsectra Ltd (CA),* [1995]** 2 **69**

Costs — See **Landlord and tenant** (Service charge — Legal costs)

County Courts Act 1984 s138

Charging order — Whether holder of charging order entitled to relief from forfeiture — Whether holder of charging order owner of interest within meaning of s138(9C) — Whether possession order should be set aside — Held, owner of charging order holds interest within s138(9C) and entitled to relief from forfeiture

***Croydon Unique Ltd* v *Wright (CA),* [1999]** 3 **28**

Property forfeited by issue of proceedings for arrears of rent — Whether "all arrears of rent" in s138 includes rent due between date of forfeiture and date relief obtained by payment — Held, "all arrears of rent" means rent up to date of forfeiture only

***Maryland Estates Ltd* v *Joseph (CC),* [1997]** 2 **96**

Equitable charge

Relief — Registered equitable charge of lease — Whether equitable chargee entitled to relief from forfeiture — Held, equitable chargor owing obligations to protect chargee's security — Equitable chargee entitled to seek relief from forfeiture by joining lessee chargor to proceedings

***Bland* v *Ingrams Estates Ltd (CA),* [2001]** 2 **23**

Relief — Registered equitable charge of lease — Whether equitable chargee entitled to relief from forfeiture — Terms of relief — Whether landlord required to give credit against arrears of rent for rents actually received after entry from third party — Held, equitable chargee entitled to seek relief from forfeiture by joining lessee chargor to proceedings — Relief granted on terms — Landlord required to give credit for rents received from third party after entry

***Bland* v *Ingrams Estates Ltd (No 2) (CA),* [2001]** 3 **34**

Insolvency Act 1986

Peaceable re-entry — Whether leave of court required under s252 of Insolvency Act 1986 before peaceable re-entry — Held, leave of court not required for peaceable re-entry

***In re Debtors no 13A10 and 14A10 of 1994 (Ch),* [1995]** 2 **33**

Whether forfeiture by peaceable re-entry constitutes the enforcement by a secured creditor of his security for purposes of s285(3)-(4) of the Insolvency Act 1986 — Whether forfeiture a remedy requiring leave of court before exercise — Held, exercise of forfeiture neither security nor remedy under s285(3) — No leave required

***Razzaq* v *Pala (Ch),* [1997]** 2 **53**

Irremediable breach

Whether breaches of covenants other than against assignment capable of remedy — Wrongful carrying out of alterations and signage display without consent — Whether s146 notice required remedial action — Whether lease forfeited — Held, breach remediable if mischief can be removed by removing alterations and signage — S146 notice failed to require remedy — No forfeiture

***Savva* v *Houssein (CA),* [1996]** 2 **65**

Mistake

Landlord served writ claiming forfeiture upon mistake about breach of covenant — By its defence tenant accepted forfeiture — Whether landlord entitled to avoid forfeiture — Held, lease forfeited and forfeiture accepted by tenant — Forfeiture cannot be avoided

***GS Fashions Ltd* v *B&Q plc (Ch),* [1995]** 1 **62**

Possession

Rent arrears — Liability of original tenant — Disclaimer of lease — Payment of rent by related company — Whether leases determined by taking possession following disclaimer — Whether leases determined by forfeiture — Held, no termination of leases — Original tenants liable for rent arrears

***Cromwell Developments Ltd* v *Godfrey (CA),* [1998]** 2 **62**

Relief

County court — Arrears of rent — County Courts Act 1984 s138 — Whether "all rent in arrears" limited to arrears at date of forfeiture — Whether county court discretion extends to order payment of all arrears — Held, "all rent in arrears" includes rent accrued after commencement of proceedings — Liability to pay rent continues after summons seeking possession

***Maryland Estates Ltd* v *Joseph (CA),* [1998]** 2 **47**

Mortgagee of lease informed by landlord of intention to grant new lease — Whether mortgagee of lease entitled to relief from forfeiture — Whether relief to mortgagee subject to new lease — Held, relief granted by reversionary lease to new lease

***Bank of Ireland Home Mortgages* v *South Lodge Developments (Ch),* [1996]** 1 **91**

Refusal of consent to assign, change of use and make alterations — Whether refusal reasonable — Whether landlord entitled to consider estate management objectives of its estate — Whether relief from forfeiture for deliberate breach — Held, landlord entitled to consider estate management objectives — Relief refused for deliberate breach

***Crown Estate Commissioners* v *Signet Group plc (Ch),* [1996]** 2 **200**

Registered equitable charge — Whether equitable chargee of lease entitled to relief from forfeiture — Held, equitable chargee of lease not entitled to relief from forfeiture under s146(4) of Law of Property Act 1925 or under inherent jurisdiction of the court

***Bland* v *Ingrams Estates Ltd (Ch),* [1999]** 2 **49**

Relief — S38 of Supreme Court Act 1981 — Tenant commencing collateral claim against landlord for damages — Whether tenant entitled to relief from forfeiture — Whether relief can be granted on terms that permit tenant first to pursue collateral claim against landlord — Held, tenant not entitled to relief from forfeiture — Tenant cannot rely on possibility of collateral claim succeeding

***Inntrepreneur Pub Co* v *Langton (Ch),* [2000]** 1 **34**

S38 of Supreme Court Act 1981 — Whether mortgagee by subdemise entitled to retrospective relief by reinstatement of lease upon payment of arrears — Whether tenant must be informed — Held, such relief can be granted where tenant cannot reasonably be found

***Escalus Properties Ltd* v *Robinson; Sinclair Gardens Investments (Kensington) Ltd* v *Walsh; Escalus Properties Ltd* v *Cooper-Smith; Escalus Properties Ltd* v *Dennis (CA),* [1995]** 2 **23**

Whether application for relief can be made by originating summons — Whether equitable assignee entitled to relief from forfeiture — Held, relief can be applied for by originating summons — Equitable assignee a "lessee" under s146(5) (*b*) of the Law of Property Act 1925 and entitled to relief

High Street Investments Ltd* v *Bellshore Property Investments Ltd (CA),

S146 notice

Alterations — Creation of two flats without permission — S146 notice alleging the same — Whether notice invalid for failing to give particulars of the breach — Held, s146 complied with — Breach of covenant adequately specified

***Adagio Properties Ltd* v *Ansari (CA),* [1998]** 2 **69**

Surety covenants

Compromise of forfeiture action by acquisition of lease by landlord — Whether sureties liable for rent arrears up to merger of lease — Held, issue and service of forfeiture proceedings do not forfeit lease — Sureties remain liable

***Ivory Gate Ltd* v *Spetale (CA),* [1998]** 2 **43**

Tenant's fixtures

Bankruptcy — Whether tenant entitled to remove tenant's fixtures following peaceable re-entry — Held, no right for tenant to recover tenant's fixtures

***Palmiero: debtor 3666 of 1999, Re (Ch),* [1999]** 3 **27**

Undertenant

Mortgage — Lease and supplement lease — Disclaimer of lease — Forfeiture of lease — Whether mortgagee as undertenant entitled to relief under s146(4) Law of Property Act 1925 — Whether undertenant entitled to relief in respect of lease only without taking supplemental lease — Held, jurisdiction to grant undertenant relief under s146(4) — Undertenant not entitled to have relief only for original lease without taking supplemental lease

***Barclays Bank plc* v *Prudential Assurance Co Ltd (Ch),* [1998]** 1 **44**

Unjust enrichment

Scotland — Conventional irritancy (forfeiture) — Unjust enrichment — Irritancy clause invoked — Whether landlord unjustly enriched — Whether tenant entitled to recompense remedy against landlord — Held, no remedy of recompense — Advantage to landlord consequence of clear contractual agreement

***Dollar Land (Cumbernauld) Ltd* v *CIN Properties Ltd (Scotland) (HL),* [1998]** 3 **79**

Validity

Assignment of reversion — Assignment back to assignor of right to rent arrears — Whether assignee entitled to forfeit lease — Whether s141(3) of Law of Property Act 1925 requires assignee to have interest in arrears before effecting a re-entry — Held, assignee entitled to re-enter — Unnecessary for assignee to be owed arrears

***Kataria* v *Safeland plc (CA),* [1998]** 1 **39**

Waiver

Development lease — Tenant in breach of covenants for timeous completion of development — Whether continuing or once-and-for-all breach — Whether waiver of right to forfeit — Held, breach once and for all — Forfeiture right waived

***First Penthouse Ltd* v *Channel Hotels & Properties (UK) Ltd; Channel Hotels & Properties (UK) Ltd* v *Al Tamimi and others (Ch),* [2004]** 1 **16**

LANDLORD AND TENANT — *continued*
Forfeiture — *continued*
Waiver — *continued*

Whether landlord waives right to forfeit lease following demand for rent by action after period for payment — Held, landlord may re-enter once an instalment of rent is unpaid — Nothing inconsistent in prior demand for rent
In re Debtors no 13A10 and 14A10 of 1994 (Ch), **[1995]** 2 **33**

Guarantee in lease
Liability for rent
Claim for rent arrears against guarantor of original tenant — Whether limitation period of six or 12 years applies — Held, under s19 of Limitation Act 1980 limitation period of six years applies to claims against guarantors — Cause of action arises when rent demanded from guarantor
Romain* v *Scuba TV Ltd (CA), **[1996]** 1 **103**

Headline rent — See **Landlord and tenant** (Rent review clause — Construction: hypothetical lease)

Housing Act 1988
Assured shorthold tenancy
Statutory periodic tenancy — S21(4) notice requiring possession on day after last day of period of tenancy — Whether notice valid — Held, notice under s21(4) not notice to quit — Notice must specify last day of period of tenancy — Notice invalid
McDonald* v *Fernandez (CA), **[2003]** **3** **22**

Improvements — See **Landlord and tenant** (Rent review clause — Disregards)

Insolvency — See also **Insolvency**
Disclaimer of lease
Rent arrears — Liability of original tenant — Disclaimer of lease — Payment of rent by related company — Whether leases determined by taking possession following disclaimer — Whether leases determined by forfeiture — Held, no termination of leases — Original tenants liable for rent arrears
Cromwell Developments Ltd* v *Godfrey (CA), **[1998]** 2 **62**

Vesting order
Variation of lease — Liquidation of second assignee of term — Disclaimer of lease — Liability of original tenant and first assignee — Principles in *Friends Provident Life Office* v *British Railways Board* [1995] 2 EGLR 55 — Whether liable under rent review memorandum varying rents payable — Vesting order — Rent payable under lease created by vesting order — Held, original tenant and first assignee not bound by rent memorandum varying rents — On vesting order rent payable as subject to reviews under lease
Beegas Nominees Ltd* v *BHP Petroleum Ltd; Sevington Properties Ltd* v *Adams (CA), **[1998]** 2 **57**

Insurance against damage — See also **Insurance** (Landlord's covenant to insure)

Joint tenancy
Assignment
Housing Act 1985 — Non-assignable joint tenancy — Deed of release of beneficial interest by one joint tenant to another — Distinction between assignment and release of interest of joint tenant — Whether deed effective to turn non-assignable joint tenancy into sole secure tenancy — Held, deed of release ineffective to release one joint tenant from tenancy by reason of s91 of 1985 Act
Burton* v *Camden London Borough Council (HL), **[2000]** 1 **49**

Notice to quit
One joint tenant only serving notice to quit — Trusts of Land and Appointment of Trustees Act 1996 — Whether service of notice to quit a "function" under 1996 Act — Whether duty to consult and give effect to wishes of co-joint tenant applies to service of notice to quit — Whether 1996 Act changed law — Whether notice to quit valid — Held, law remained unchanged — One joint tenant entitled to serve valid notice to quit
Notting Hill Housing Trust* v *Brackley (CA), **[2001]** 3 **11**

Landlord and Tenant Act 1927
Improvements
Notice of intention to make improvements — Notice of objection on ground that landlord intending to carry out improvements — Tenant deciding not to proceed with improvements — Whether landlord entitled to carry out improvements — Held, landlord not entitled to do so
Norfolk Capital Group Ltd v Cadogan Estates Ltd (Ch), **[2004]** 2 **50**

Service of notices
Landlord and Tenant (Covenants) Act 1995 — Whether s17 notice validly served by recorded delivery at last known abode — Whether s17 notice that includes errors valid — Whether affidavit must identify person obtaining rent records from computer — Held, notice served by recorded delivery and undelivered validly served — S17 notice valid even when it claims more than landlord entitled to — Affidavit need not identify person obtaining rent records from computer
Commercial Union Life Assurance Co Ltd* v *Moustafa (QB), **[1999]** 2 **44**

Service — Service by recorded delivery under s23 of Landlord and Tenant Act 1927 — Whether s7 of Interpretation Act 1978 applying to s23 of 1927 Act — Whether notice served when posted — Whether on proper construction of s23 notice served when delivery attempted or actually made — Whether s23 required to be construed compatibly with European Convention on Human Rights — Whether notice valid — Held, s7 of 1978 Act had no application — S23 of 1927 Act cannot be construed that service effected when delivery attempted or made — No breach of s3 of Human Rights Act 1998 — Notice served under s23 when posted by recorded delivery
CA Webber (Transport) Ltd* v *Railtrack plc (CA), **[2004]** 1 **49**

Landlord and Tenant Act 1954
Application for new tenancy
Parties and time limits — Application to court for new tenancy — Incorrect parties named as landlord — Application to substitute correct landlord — Expiry of statutory time limit for making application for new tenancy — Civil Procedure Rules — Whether tenant entitled to substitute correct landlord as party to proceedings — Held, CPR 19.5(1)(c) permitting amendment and substitution of parties where statutory time limits not prohibiting the same after expiry of time limits — S29(3) of Landlord and Tenant Act 1954 not so prohibiting — Substitution of landlord permitted after expiry of s29(3) time limit
Parsons* v *George (CA), **[2004]** 3 **49**

Terms of new tenancy — Test for redevelopment break clause — Whether evidence can be admitted after trial — Whether appeal to High Court review or rehearing — Period of time before redevelopment break clause capable of being exercised — Held, redevelopment break clause exercisable after first five years of new term within court's discretion — New evidence admitted in appeal at discretion of court — Date of appeal relevant date for determining terms of tenancy — On facts period before break clause exercisable reduced
Davy's of London (Wine Merchants) Ltd* v *City of London Corporation (Ch), **[2004]** 3 **39**

Compensation
Entitlement to compensation — Failure of tenant to serve counternotice to s25 notice — S25 notice opposing grant of tenancy on ground (f) — Tenant holding over under tenancy at will — Lease excluding compensation where tenant in occupation for less than five years — Whether tenant entitled to s37 compensation — Held, lease effectively excluding compensation where occupation under protected tenancy less than five years — Tenant not entitled to compensation under s37
London Baggage Co (Charing Cross) Ltd* v *Railtrack plc (No 2) (Ch), **[2003]** 1 **141**

Contracting out
Validity — Application to court for contracting-out order — Applicants including proposed tenant and sureties — Lease granted to parties intended to be sureties — Whether agreement to contract out effective — Whether court order authorising contracting out valid — Held, order validly made — Order authorised grant of lease to three parties of whom two becoming lessees — Valid contracting out
Brighton & Hove City Council* v *Collinson (CA), **[2004]** 2 **65**

Counternotice
Tenant's counternotice indicating willingness to give up possession — Second counternotice denying willingness to give up possession — Whether positive counternotice irrevocable — Whether second counternotice valid — Held, second negative counternotice may be served within two-month time limit — Issue in each case whether actual tenant entitled to resile from earlier negative counternotice
Pennycook* v *Shaws (EAL) Ltd (Ch), **[2003]** 3 **28**

Tenant's counternotice indicating willingness to give up possession — Second counternotice denying willingness to give up possession — Whether positive counternotice irrevocable — Whether second counternotice valid — Whether statutory construction of legislation adopted in *Re 14 Grafton Street* [1971] 1 Ch 935 engaging and violating Art 1 of First Protocol to, and Art 6(1) of, European Convention on Human Rights — Held, service of first counternotice irrevocable — Second counternotice ineffective — Art 1 of First Protocol engaged but not violated — Art 6(1) not engaged
Pennycook* v *Shaws (EAL) Ltd (CA), **[2004]** 2 **55**

Notices

Service — Service by recorded delivery under s3 of Landlord and Tenant Act 1927 — Whether s7 of Interpretation Act 1978 applying to s23 of 1927 Act — Whether notice served when posted — Whether s23 required to be construed compatibly with European Convention on Human Rights — Whether notice valid — Held, notice served on date of posting — S7 of 1978 Act not applying — S23 of 1927 Act cannot be construed that service effected when delivery attempted or made — No breach of s3 of Human Rights Act 1998

***Beanby Estates Ltd v Egg Stores (Stamford Hill) Ltd (Ch)*, [2003]** 3 **85**

Occupation

Claim for new tenancy — Occupation for purposes of s23 — Tenant a company limited by guarantee — Tenant unable to distribute profits to shareholders or members — Whether tenant carrying on a business for profit — Whether activities of tenant within meaning of s23 — Whether tenant in occupation for purposes of business — Held, tenant company in occupation for purposes of business — Irrelevant that tenant unable to distribute profits to shareholders or members

***Hawkesbrook Leisure Ltd* v *The Reece-Jones Partnership (Ch)*, [2004]** 2 **61**

Request for new tenancy

Tenant deciding to relocate to other premises before serving s26 requests — Landlord opposing request on ground (*f*) of s30(1) of Landlord and Tenant Act 1954 — Statutory compensation — Whether s26 notices validly given — Whether tenant must have genuine intention of requiring new tenancy when serving s26 notice — Whether evidence of tenant's state of mind admissible — Whether tenant entitled to statutory compensation — Held, state of mind of tenant irrelevant — Request valid — Tenant entitled to statutory compensation

***Sun Life Assurance plc* v *Thales Tracs Ltd (CA)*, [2001]** 2 **57**

Termination of tenancy

Effect of s64(1)(c) of Landlord and Tenant Act 1954 — Landlord serving s25 notice — Tenant applying for new tenancy — Tenant ceasing occupation prior to term date — Whether tenancy continuing until expiry of three months after final disposal of proceedings — Whether tenant liable for rent after term date — Held, s25 notice having no effect where tenant vacates prior to term date — Tenancy not continuing where tenant vacates by term date — S64(1)(c) having no effect where tenancy not continued by s24 — Tenant not liable for rent after term date

***Single Horse Properties Ltd* v *Surrey County Council (CA)*, [2002]** 2 **43**

Landlord and Tenant Act 1954 Part I — See also **Landlord and tenant** (Leasehold enfranchisement — Price of freehold)

Joint tenants

Whether one of two joint tenants entitled to protection of 1954 Act — Whether continuing tenancy held by joint tenants or tenant in occupation — Whether interest in continuing tenancy part of bankrupt's estate — Held, one of two joint tenants entitled to protection of 1954 Act, Part I — Interest of tenant under continuing tenancy derived from Act and not mere continuation of contractual tenancy — Continuation tenancy is property capable of being disclaimed by trustee in bankruptcy of tenant

***de Rothschild* v *Bell (CA)*, [1999]** 1 **35**

Validity of s4 notice

Two self-contained maisonettes held under one long lease — Two s4 notices served on tenants — Whether validity of notices can be determined on preliminary point of law — Held, trial court must first determine whether one or more tenancies continue under Rent Act 1977 before determining validity of notices

***St Ermins Property Co Ltd* v *Patel (CA)*, [1997]** 2 **61**

Landlord and Tenant Act 1954 Part II

Application for new tenancy

New tenancy agreement — Landlord and Tenant (Covenants) Act 1995 — Construction — Terms of new tenancy — Whether terms should provide that on assignment landlord entitled to authorised guarantee agreement or only where reasonable — Held, requirement for authorised guarantee agreement under s16(3)(b) of the Landlord and Tenant (Covenants) Act 1995 only lawfully imposed where landlord's condition for requiring authorised guarantee agreement reasonable — Assignment provisions in new lease cannot make landlord's entitlement to authorised guarantee agreement automatic on assignment — Requirement of authorised guarantee agreement only where reasonable

***Wallis Fashion Group Ltd* v *CGU Life Assurance Ltd (Ch)*, [2000]** 2 **49**

Compensation

Twice rateable value — Determination of tenancy on grounds (*f*) and (*g*) — S25 notice expiring after contractual term date — Tenant giving up possession on term date — Whether tenant entitled to disturbance compensation based on twice rateable value — Whether 14-year occupational period must be satisfied up to date specified in s25 notice — Held, 14-year occupational period must be satisfied up to s25 termination date — Tenant not entitled to twice rateable value

***Sight & Sound Education Ltd* v *Books etc Ltd (Ch)*, [1999]** 3 **45**

Compensation for disturbance

Lease excluded statutory disturbance compensation — Whether occupation throughout five years immediately preceding termination of tenancy — S38(2) — Meaning of "occupied for the purposes of a business" — Whether tenant entitled to compensation following termination of his business tenancy — Held, claim to compensation unaffected by giving up possession prior to termination of tenancy — Tenant entitled to compensation

***Bacchiocchi* v *Academic Agency Ltd (CA)*, [1998]** 3 **157**

Contracted-out tenancy

County court jurisdiction — Whether order of county court outside jurisdiction valid — Held, order valid — Not open to co-applicant to challenge jurisdiction at later date

***St Giles Hotel Ltd* v *Microworld Technology Ltd (CA)*, [1997]** 2 **105**

Exclusion order

Court making s38 exclusion order authorising draft lease — Lease entered into differing in terms from lease authorised by court — Whether tenant held tenancy protected by Part II of 1954 Act — Lease containing break clauses — Whether lease for "term of years certain" within s38 — Held, difference in terms not sufficiently substantial — Lease therefore protected — Lease for "term of years certain"

***Receiver for the Metropolitan Police* v *Palacegate Properties Ltd (CA)*, [2000]** 1 **63**

Interim continuation under s64

Application for extension of time for appeal refused — Whether proceedings on appeal for purpose of s64 — Whether tenancy determined three months after county court decision — Held, procedural irregularities as to manner of lodging appeal — If extension of time for appealing granted, then there are proceedings on appeal and s64 applies — No extension granted

***Shotley Point Marina (1986) Ltd* v *Spalding (CA)*, [1997]** 1 **233**

Ground (*g*) — Landlord's chance of obtaining planning permission without personal occupancy condition tied to tenant company — Whether landlord established necessary intention — Burden of proof on landlord — Chance of obtaining planning permission — Whether Court of Appeal entitled to consider planning permission granted after date of county court hearing — Held, landlord established chance of obtaining planning permission — Burden of proof established — Court of Appeal entitled to consider permission granted after county court hearing

***Gatwick Parking Services Ltd* v *Sargent (CA)*, [2000]** 2 **45**

S25 notice terminating tenancy — Landlord and Tenant Act 1954, Part II (Notices) Regulations 1983 (as amended) — Notice differed from prescribed form — Warning notes omitted — Whether notice valid — Held, notices invalid — *Morris* v *Patel* [1987] 1 EGLR 75 wrong to suggest negligible matters irrelevant

***Sabella Ltd* v *Montgomery (CA)*, [1998]** 1 **65**

Landlord's opposition to new tenancy

S30(1)(b) — Persistent delay in paying rent — Whether landlord giving express notice requiring strict compliance with obligations of lease to pay rent timeously — Whether tenant entitled to new tenancy — S30(1)(a) — Breach of repair covenant — Held, landlord failing to give express notice requiring timeous compliance with lease payment obligations — No persistent delay until such notice given — Tenant entitled to new lease

***Hazel* v *Akhtar (CA)*, [2002] 1 45**

S30(1) (*f*) — Break clause — Whether notice under s25 of the 1954 Act and accompanying letter effective notice to exercise break clause for development purpose — Whether s30(1) (*f*) satisfied — Held, notice valid and s30(1) (*f*) satisfied

***Aberdeen Steak Houses Group Ltd* v *Crown Estate Commissioners (Ch)*, [1997]** 2 **107**

S30(1)(*f*) — "Eggshell tenancy" — Load-bearing walls not forming part of demised premises — Landlord proposing demolition and reconstruction — Whether demolition or reconstruction of "premises" — Whether tenant entitled to rely upon s31A — Whether landlord entitled to oppose grant of new tenancy — Held, tenant not entitled to rely upon s31A — Tenant's right to undo landlord's works under terms of new tenancy relevant — Demise under "eggshell tenancy" capable of being demolished or reconstructed having regard to rights of support from structure under s30(1)(f)

***Pumperninks of Piccadilly Ltd* v *Land Securities plc (CA)*, [2002]** 2 **147**

S30(1)(*g*) — Whether landlord established reasonable prospect of establishing proprietary golf club after termination of tenancy — Held, statutory test of intention applied — Determination of the reality of landlord's intention to start business, not probability of achieving a start or ultimate success
***Dolgellau Golf Club* v *Hett (CA),* [1998]** 2 **75**

Licence
S25 notice to terminate tenancy — Letter accepting tenancy — Whether licensor estopped from denying tenancy — Held, although limited representations, no detrimental reliance on representations — No estoppel
***Wroe* v *Exmos Cover Ltd (CA),* [2000]** 1 **66**

New tenancy application
Rent and rent review clause — Whether ransom payable for rent of electricity substation under lease with restriction for that purpose — Whether rent review clause should be included in 14-year term — Held, no ransom as rent must reflect restrictions — No rent review clause ordered having regard to low rent
***Northern Electric plc* v *Addison (CA),* [1997]** 2 **111**

S26 request — Whether s26 notice addressed to and served on landlord's agent valid — Whether addressee of notice had general agency — Whether impliedly authorised to accept notice — Held, by virtue of s23(1) of the Landlord and Tenant Act 1927 notice valid
***Railtrack plc* v *Gojra (CA),* [1998]** 1 **63**

New tenancy grant
Rent — Whether betting office premises "licensed premises" — Whether s34(1) (*d*) of Landlord and Tenant Act 1954 applied — Held, s34(1) (*d*) applied — Rent must disregard value attributable to licence
***Ganton House Investments* v *Crossman Investments (CC),* [1995]** 1 **239**

Notice to terminate tenancy
Early determination of lease by break clause — Whether request for new tenancy to commence earlier than date lease would determine by effluxion of time valid — Appeal from [1996] 1 113 — Held, request for new tenancy to commence earlier than expiration of the term of lease invalid
***Garston* v *Scottish Widows Fund & Life Assurance Society (CA),* [1998]** 2 **73**

Estoppel and representation — S25 notice served not in prescribed form — Tenant served counternotice and made application for new tenancy — Whether tenant entitled to dispute validity of s25 notice claiming freehold under Leasehold Reform Act 1967 — Held, tenant knew of defective s25 notice and estopped from disputing its validity — Claim under 1967 Act invalid
***Keepers and Governors of the Possessions Revenues and Goods of the Free* v *Mayhew (CA),* [1997]** 1 **88**

Tenant gave up occupation of premises before contractual term date — Whether notice under s27(2) required to determine tenancy — Held, tenancy continued after term date until determined by s27(1) notice
***Provident Mutual Life Assurance Association* v *Greater London Employer's Association Ltd (Ch),* [1996]** 1 **106**

Tenant giving up occupation before expiration of contractual term — Whether tenancy continues — Whether s27(2) notice required to determine tenancy — Appeal from [1995] 2 61 — Held, 1954 Act no longer applied after tenant gave up occupation — Tenancy expired by effluxion of time on contractual term date — S27(2) notice not required
***Esselte AB* v *Pearl Assurance plc (CA),* [1997]** 1 **73**

Notices
Landlord serving conflicting s25 notices — One notice incapable the other capable of being an effective s25 notice — Landlord serving third notice capable of being effective — Whether tenant entitled to rely upon one of the two conflicting notices — Whether third notice only valid notice — Held, as two conflicting notices served, both would be treated by recipient as invalid — Third notice only valid notice
***Barclays Bank plc* v *Bee (CA),* [2001]** 3 **41**

Occupation
S23 — Fire damage rendered premises unfit for occupation — Whether tenant occupied fire-damaged premises on expiry of lease within s23(1) — Whether tenant intended to reoccupy the premises — Held, on the evidence the tenant retained control and occupation on term date and later
***Flairline Properties Ltd* v *Hassan (QB),* [1999]** 1 **138**

Occupation for purposes of tenant's business
Tenant holding lease of market hall — Stall holders having exclusive occupation of stalls — Whether tenant occupied for purposes of its business of market operator — Held, subtenants occupied the stalls under business tenancies and therefore tenant did not so occupy
***Graysim Holdings Ltd* v *P&O Property Holdings Ltd (HL),* [1996]** 1 **109**

Opposition to grant of new tenancy
Ground (g) — Controlling interest of landlord in relevant business — Tenant raising failure to satisfy s30(3) in closing submissions — Landlord satisfying s30(3) during adjournment — Whether trial judge entitled to adjourn hearing — Whether tenant required by CPR to raise issue at earlier stage — Held, judge entitled to adjourn hearing — Overriding objective of CPR requiring issues to be raised at earlier stage — Ground (g) satisfied
***Ambrose* v *Kay (CA),* [2002]** 1 *49*

Ground (g) — Five-year period of ownership — Group of companies — Succession of interests — Whether principle in Artemiou v Procopiou [1966] 1 QB 878 applied where successive leasehold interests follow freehold interest — Held, principle applied — Five-year period of ownership satisfied by succession of leases preceded by freehold interest
***VCS Car Park Management Ltd* v *Regional Railways North East Ltd (CA),* [2000]** 1 **57**

Rent
New tenancy application — Determination of rent — Tenancy renewal — Landlocked holding — Whether landlocked characteristic required to be disregarded — Whether discount to open market rent to reflect landlocked characteristic — Held, landlocked characteristic cannot be ignored — Discount to open market rent to reflect landlocked characteristic
***J Murphy & Sons Ltd* v *Railtrack plc (CA),* [2002]** 2 **48**

Rent review
Lease with provisions for upwards-only rent review — Operation of break clause and service of s25 notice — Whether landlord entitled to operate rent review clause during continuation tenancy — Held, "term" did not continue for purposes of rent review — Landlord not entitled to operate rent review clause
***Willison* v *Cheverell Estates Ltd (CA),* [1996]** 1 **116**

Representation
Licence granted — S25 notice to terminate tenancy — Letter accepting tenancy — Whether licensor estopped from denying tenancy — Held, although limited representations, no detrimental reliance on representations — No estoppel
***Wroe* v *Exmos Cover Ltd (CA),* [2000]** 1 **66**

Request for new tenancy
Tenant finding new premises before service of s26 notice — Whether s26 notice validly given — Whether tenant must have genuine intention of requiring new tenancy when serving s26 notice — Whether s26 notices valid — Whether tenant entitled to statutory compensation under s37 — Held, although notices validly given they were invalid as tenant did not have genuine intention that new tenancy required — No entitlement to statutory compensation
***Sun Life Assurance plc* v *Racal Tracs (QB:TCC),* [2000]** 1 **138**

S43(3)
Occupation as tenant for two successive periods less than 12 months — Further occupation as tenant at will — Whether occupation for period exceeding 12 months for purposes of s43(3) (*b*) — Whether tenancy protected by 1954 Act — Held, tenancy at will following periods of occupation as tenant cannot be added to make continuous period of occupation as tenant for purposes of s43(3)(*b*)
***Cricket Ltd* v *Shaftesbury plc (Ch),* [1999]** 2 **57**

Statutory compensation
Tenant finding new premises before service of s26 notice — Whether s26 notices validly given — Whether tenant must have genuine intention of requiring new tenancy when serving s26 notice — Whether s26 notices valid — Whether tenant entitled to statutory compensation under s37 — Held, although notices validly given they were invalid as tenant did not have genuine intention that new tenancy required — No entitlement to statutory compensation
***Sun Life Assurance plc* v *Racal Tracs (QB:TCC),* [2000]** 1 **138**

Tenancy by estoppel
Interim rent application — Whether landlord under tenancy by estoppel entitled to interim rent — Whether tenancy by estoppel protected by 1954 Act — Held, tenancy by estoppel protected by 1954 Act — Landlord under tenancy by estoppel entitled to interim rent
***Bell* v *General Accident Fire & Life Assurance Corporation Ltd (CA),* [1998]** 1 **69**

LANDLORD AND TENANT — *continued*

Landlord and Tenant Act 1987

Determination of price

Transfer without notice to tenants — No consideration specified in transfer documents — Transfer part of larger transaction — Whether property to be transferred at nil consideration — Held, flats to be acquired at nil consideration

***Woodridge Ltd* v *Downie (LT),* [1997]** 2 **193**

Landlord's address for service of notices

Whether inclusion of landlord's registered address in tenancy agreement without reference to address for service of notices satisfied s48 — Whether notice must be in writing — Held, s48 can be satisfied by address in agreement where reasonable tenant would so understand address as place for service of notices — Notice must be in writing

***Rogan* v *Woodfield Building Services Ltd (CA),* [1995]** 1 **72**

Leasehold valuation tribunal

Purchase notice referred to the tribunal — Whether all matters in dispute between the parties within jurisdiction of tribunal — Held, only those matters specifically referred to in purchase notice within jurisdiction of tribunal

***134 Shirland Road Management Co Ltd* v *Lester (LVT),* [1997]** 2 **207**

Manager

Manager appointed under 1987 Act — Whether manager owing duties of landlord under lease to tenant — Whether tenant entitled to set off damages for breach of landlord's repairing covenant against manager's claim for service charges — Held, manger not owing duties of landlord under leases — Tenant not entitled to set off claim for damages for breach of landlord's repairing covenant against manager's claim for arrears of service charges

***Maunder Taylor* v *Blaquiere (CA),* [2003]** 1 **52; [2002]** 2 **35**

Pre-emption rights

Block of flats — Acquisition of freehold — Landlord paid vendor purchase price and sum for arrears of rents and service charges — Whether sum payable by nominee purchaser should include additional sum — Held, landlords entitled to additional sum as being part of consideration — Irrelevant that service charges might not be recoverable

***Linkproud Ltd* v *Ballard (LT),* [1997]** 1 **181**

Purchase notices

Whether landlord obliged to comply with purchase notice — Whether individual buildings or complex of buildings "premises" for purposes of determining majority of qualifying tenants — Whether purchase notice invalid by including buildings or parts of buildings tenants not entitled to acquire — Held, landlord must comply with purchase notice — "Premises" identified as individual building — Purchase notice requirements in s12(3) directory and not mandatory

***Kay-Green* v *Twinsectra Ltd (CA),* [1996]** 2 **43**

Rights of first refusal

Qualifying tenants' rights to information — S11A notice requiring information — Notice failing to specify tenants' addresses pursuant to s54 — Whether requirement to specify addresses substantive or secondary requirement — Whether notice valid — Held, requirement to specify addresses was secondary requirement — Notice valid

***Tudor* v *M25 Group Ltd (CA),* [2004]** 1 **23**

Sale of freehold

S12 purchase notice served following s18 notice — New landlord granted new interests — Statutory repair notices served — Whether tribunal entitled to consider identity of property going to validity of purchase notice — Determination of reduction in consideration attributable to the grant of lease and to costs of carrying works specified in statutory notices of repair — Held, tribunal would not consider identity of premises which founded validity of purchase notice already considered by county court — Consideration reduced by premium received for lease grant and estimated cost of complying with statutory repair notices

***Saga Properties Ltd* v *Palmeira Square Nos 2-6 Ltd (LT),* [1995]** 1 **199**

S48 notice

Housing Act 1988 — Grounds for possession — Whether s48 notice served by notice under s21 of 1988 Act and application for possession — Held, s48 notice served by s21 notice but not by application for possession

***Drew-Morgan* v *Hamid-Zadeh (CA),* [1999]** 2 **13**

Tenant's rights of first refusal

Landlords contracted to sell flats without serving s5 notices — Whether exchange of contracts a "relevant disposal" — Whether landlords proposed to make a relevant disposal — Whether omission to serve s5 notices a failure to comply with statutory duties — Held, exchange of contracts not a relevant disposal — Landlords had failed to serve s5 notices — S5 notices must be served as soon as landlord proposes to make a relevant disposal

***Mainwaring* v *Henry Smith's Charity Trustees (CA),* [1996]** 2 **25**

Purchase notice — Whether price to be increased by death of regulated tenant — Extent of property to be acquired — Determination of tenants' rights over gardens and amenity areas — Whether chargee or landlord entitled to the consideration — Held, price increased following death of tenant — Act applies to premises of mixed flats and houses — Rights to gardens granted without termination — Consideration to be paid to chargee

***Kay-Green* v *Twinsectra Ltd (LVT),* [1997]** 1 **219**

S5 notice — Landlord proposing disposition of estate of 55 flats — Flats contained in four buildings and sharing common appurtenances — Meaning of "building" — Whether four buildings constituting "building" — Whether s5 notice offering estate of 55 flats in four buildings valid — Held, "building" meaning either single building or one or more buildings where tenants sharing use of same appurtenant premises — Notice valid

***Long Acre Securities Ltd* v *Karet (Ch),* [2004]** 2 **121**

S11 notice — Landlord provided copy of contract but refused to confirm compliance with notice — Contract terms differed from terms actually applying to disposal — Whether s11 complied with — Held, notice not complied with as terms of actual transaction differed from contract

***Staszewski* v *Maribella Ltd (CA),* [1998]** 1 **34**

Sale of flats to defendant company and sale of shares in defendant to purchaser — Whether a disposal to an associated company — Whether a relevant disposal — Whether s11 notice valid — Estoppel — Held, no relevant disposal — S11 notice invalid — Tenants estopped by acquiescence

***Michaels* v *Harley House (Marylebone) Ltd (Ch),* [1997]** 2 **44**

Whether LVT entitled to order grant of perpetual rights of gardens — Whether disposal price to include increased value of flat following death of regulated tenant — Held, tenants entitled to acquire interest disposed by original landlord — Interest included rights to gardens — Change of circumstances up to date of contract or conveyance relevant for s12(6) of 1987 Act

***Twinsectra Ltd* v *Jones (LT),* [1998]** 2 **129**

Landlord and Tenant Act 1988

Breach of statutory duty

Causation — Refusal of consent to underletting at rack-rental value — Refusal of consent to change of user — Whether landlord owed statutory duty under 1988 Act — Whether breach of statutory duty — Whether causation of damages proved — Whether challenge to refusal of consent to change of user would have succeeded — Held, breach of statutory duty in unreasonably withholding consent to letting at rack-rental value — Tenant failing to show causation of damages — No evidence that refusal of consent to change of user unreasonably withheld

***Clinton Cards (Essex) Ltd* v *Sun Alliance & London Assurance Co Ltd (ChD),* [2002]** 3 **19**

S1(3) duty on landlord

Reasonable time" to perform s1(3) duty — Whether "reasonable time" ending when landlords giving notice withholding consent — Whether breach of statutory duty occurring when landlords giving notice withholding consent — Whether correspondence between parties extending period of "reasonable time" — Whether breach of statutory duty by landlords — Held, "reasonable time" expiring with service of notice withholding consent — Landlords in breach upon service of such notice — Post-notice correspondence not releasing landlords from notice withholding consent — No waiver or estoppel

***Go West Ltd* v *Spigarolo (CA),* [2003]** 1 **133**

Landlord and Tenant (Covenants) Act 1995

Guarantors

Whether s17(2) notice must be served on original tenant before guarantor liable following notice under s17(3) — Held, not precondition of guarantor's liability that s17(2) notice must first be served on tenant

***Cheverell Estates Ltd* v *Harris (QB),* [1998]** 1 **27**

Lease

Beer supply agreement

Whether beer supply agreement prohibited by and void under Arts 85(1) and (2) of European Union Treaty — Whether prohibition and voidness only temporaneous with relevant circumstances — Held, prohibition and voidness only temporaneous with circumstances making agreement contrary to Art 85(1) — Tie in the lease ceased to be void on reversion passing to new landlord

***Passmore* v *Morland plc (CA),* [1999]** 1 **51**

Beer tie

Licensee in occupation under agreement for lease — Arrears of licence fees — Whether agreement void under Art 85(1) of Treaty of Rome — Held, on assumption that beer tie a breach of Art 85(1), beer tie void and severable — Agreement otherwise valid

***Star Rider Ltd* v *Inntrepreneur Pub Co (Ch),* [1998]** 1 **53**

Disclaimer — See **Insolvency** *(Disclaimer of lease)*

LANDLORD AND TENANT — *continued*

Extent of demise
Roof terrace with french door access — Whether terrace part of "main structure" — Whether all components of terrace demised to tenants — Whether surface only demised to tenants — Held, only surface tiles of terrace demised to tenant
***Ibrahim* v *Dovecorn Reversions Ltd (Ch),* [2001]** 2 **46**

Notice
S23 of Landlord and Tenant Act 1927 — Service under express terms — Notice sent by recorded delivery and returned — Notice affixed to premises to which tenant had no access — Whether good service under express terms — Whether good service under s23 of 1927 Act — Whether good service by affixing notice to premises — Held, good service under express terms and s23
***Blunden* v *Frogmore Investments Ltd (CA),* [2002]** 2 **29**

Parking rights
Lease allowing landlord to specify parking areas — Landlord preventing car parking for three-year period — Landlord reducing parking spaces from 13 to four — Whether parking rights infringed for three-year period — Whether reduction in spaces an infringement — Award of damages — Held, substantial interference and infringement for three-year period — Reduction in spaces substantial interference — Award of damages of £6,300 affirmed
***Saeed* v *Plustrade Ltd (CA),* [2002]** 2 **19**

Premium — See **Stamp duty** *(Lease premium)*
Reversionary lease
Tenant not paying rent — Adverse possession — Whether "lease in writing" under para 5, Sched 1, Limitation Act 1980 — Whether tenant can be in adverse possession under Landlord and Tenant Act 1954 continuation tenancy — Held, reversionary lease granted by operation of law — Documents were not a lease in writing
***Long* v *Tower Hamlets London Borough Council (Ch),* [1997]** 1 **78**

Sealing of documents
Sealing of documents — Corporations — Grantor local authority retaining lease — S74(1) of Law of Property Act 1925 — Whether lease delivered upon sealing — Whether grant of lease — Held, deed not delivered upon sealing by corporation — Deed made when completion intended by the parties — No grant of lease
***Bolton Metropolitan Borough Council* v *Torkington (CA),* [2004]** 1 **28**

Surrender
Landlord evicted gypsies from premises — Whether surrender by operation of law — Held, intention to surrender never accepted by landlord — No surrender
***McDougalls Catering Foods Ltd* v *BSE Trading Ltd (CA),* [1997]** 2 **65**

Time
Term expressed to commence "from" specified date "until" specified date — Whether term commenced on day following first specified date — Held, term commenced on first specified day — Rule that lease expressed to commence "from" specified day commences on following day rebutted by indication to contrary
***Meadfield Properties Ltd* v *Secretary of State for the Environment (Ch),* [1995]** 1 **39**

Variation
Grant of long lease protected by Part II of Landlord and Tenant Act 1954 — Determination of tenancy — Whether variation of lease — Whether tenant held secure tenancy — Whether dwelling-house let as a separate dwelling — Whether s11 of Landlord and Tenant Act 1985 applied to tenancy — Held, no contractual variation of lease — Dwelling-house not let as a separate dwelling — S11 of 1985 Act did not apply — Tenancy not protected by Housing Act 1985
***Tomkins* v *Basildon District Council (CA),* [2002]** 3 **33**

Lease or licence
Car park
Agreement to operate shoppers' car park — Whether agreement tenancy or licence — Whether exclusive possession granted to grantee — Held, grantee not granted exclusive possession — Agreement a licence
***National Car Parks Ltd* v *Trinity Development Co (Banbury) Ltd (Ch),* [2001]** 2 **43**

Flat
Legal capacity to grant tenancies — Housing trust holding temporary housing accommodation as licensee — Whether housing trust capable of granting tenancy — Whether tenancy granted — Whether tenancy by estoppel relevant approach — Held, agreement granted exclusive possession — No grounds for agreement being other than a lease
***Bruton* v *London & Quadrant Housing Trust (HL),* [1999]** 2 **59**

Hotel room
Occupiers introducing cooking facilities — Whether occupiers of hotel rooms had assured tenancies — Whether occupiers licensees — Housing Act 1988 — Whether hotel room a dwelling-house — Held, introduction of cooking facilities by occupier may indicate hotel room became dwelling house if authorised by owner
***Uratemp Ventures Ltd* v *Collins (CA),* [2000]** 1 **156**

Refuse tip
Exclusive right to deposit refuse — Whether lease of licence — Whether payment to be made for capping materials as well as dry waste — Held, licence granted — Payments must include airspace occupied by capping materials
***Hunts Refuse Disposals Ltd* v *Norfolk Environmental Waste Services Ltd (CA),* [1997]** 1 **16**

Tenancy by estoppel
Housing trust holding property under licence from local housing authority — Trust granting purported licence to occupy to grantee upon terms reciting trust's lack of title — Whether grantee held tenancy by estoppel — Whether tenancy or licence — Held, terms of grant not inconsistent with intention — No estoppel — Grantee held licence
***Bruton* v *London Quadrant & Housing Trust (CA),* [1997]** 2 **91**

Leasehold enfranchisement
Acquisition of freehold
Leasehold Reform, Housing and Urban Development Act 1993 — Collective enfranchisement — Appeal against leasehold valuation tribunal — Function of Lands Tribunal on appeal — Whether valuation of freehold interest in "no 1993 Act world" — Held, appeal was dismissed as no point of law or valuation in error — Freehold interest must be valued in "no 1993 Act world"
***Daejan Properties Ltd* v *Weeks (LT),* [1998]** 3 **125**

Acquisition of freehold — continued
Leasehold Reform, Housing and Urban Devlopment Act 1993, s24 — Reversion to three flats — Rising ground rents — Whether open market value will reflect value of insurance commissions — Held, capitalisation rate of 11% — Value of insurance commissions to special purchasers not reflected in open market value
***Moore* v *Escalus Properties Ltd (LVT),* [1997]** 1 **200**

Covenants
Leasehold Reform Act 1967 — Restrictive covenants — Whether covenants against development can be included in conveyance under section 10(4) — Held, all restrictive covenants as fall within the Land Charges Act 1925 can be included
***Langevad* v *Chiswick Quay Freeholds Ltd (CA),* [1999]** 1 **61**

Determination of freehold price
Value of insurance commission — Acquisition of block of flats — Whether any value in right to receive insurance commission — Held, on the evidence an additional sum added to the price to reflect the value of insurance commission at 8.5 YP
***Blackstone Investments Ltd* v *Middleton-Dell Management Co Ltd (LT),* [1997]** 1 **185**

Enfranchisement price
Collective enfranchisement price — Whether price includes "hope value" in respect of flats of non-participating tenants — Appropriate yield and "uplift" — Held, yield of 9% and 4% uplift — "Hope value" can be included as part of freehold value
***Becker Properties Ltd* v *Garden Court NW8 Property Co Ltd (LT),* [1998]** 1 **121**

Extended lease
Leasehold Reform, Housing and Urban Development Act 1993 — Determination of premium payable on grant of extended lease — Proper capitalisation rate — Appropriate marriage value share — Held, appropriate capitalisation rate of 7.5% — 50% appropriate marriage value share in absence of evidence
***Black* v *Eton College (LVT),* [1995]** 1 **223**

"House"
Leasehold Reform Act 1967 — Basement of house under adjoining building — Whether the whole a "house" for purposes of s2(1) — Whether basement a material part for purposes of s2(2) — Held, s2(1) was satisfied — Basement was a material part under s2(2)
***Duke of Westminster* v *Birrane (CA),* [1995]** 1 **89**

"House and premises"

Landlord and tenant — Leasehold Reform Act 1967 — Enfranchisement — House and premises — Whether garden held under series of short leases enfranchisable with house and premises — Held, adjoining land held under short leases not enfranchisable
***Burr* v *Eyre (CC),* [1998]** 2 **92**

Lease of main house and mews house — Mews house containing flat and garage — Whether mews house enfranchisable with main house — Whether garage part of "house and premises" — Held, main house and garage only constituted "house and premises" — Flat not enfranchisable
***Dugan-Chapman* v *Grosvenor Estates (CC),* [1997]** 1 **96**

Leasehold Reform Act 1967

Determination of price payable under s9(1C) — Appeal from leasehold valuation tribunal — Whether appeal a rehearing — Determination of freehold and leasehold values — Appropriate yield rate — Whether yields derived from settlements reliable guide — Held, appeal is a rehearing and not review of evidence before LVT — Yields derived from settlement give reliable guide
***Trustees of the Eyre Estate* v *Saphir (LT),* [1999]** 2 **123**

Price payable — "*Delaforce*" effect on settlement evidence — Appropriate risk rates — Valuation method — Costs — Held, 5.5% and 7% appropriate risk rate for valuing rising ground rent — *Delaforce* effect of between 5% and 45% — Tenants pay 50% of landlords' costs
***Windsor Life Assurance Co Ltd* v *Austin (LT),* [1996]** 2 **169**

Leasehold Reform, Housing and Urban Development Act 1993

Acquisition of reversion to block of flats — Whether decisions of other tribunals admissible — Determination of yield — Whether compensation payable for hope value, loss of insurance commissions and loss of ability to enforce covenants on other properties — Held, other tribunal decisions admissible — Appropriate yield 8.5% — Compensation payable for marriage value, loss of insurance premiums and loss of right to enforce covenants
***Solent Pines Residents Association Ltd* v *Campbell (LVT),* [1996]** 1 **213**

Collective enfranchisement — Appeal to Lands Tribunal — Whether Lands Tribunal should disturb decision of LVT — Whether hope value compensatable — Whether 65% share of marriage value — Held, appeal will not necessarily be treated as a rehearing — No evidence to support hope value — 50% marriage value
***Verkan & Co Ltd* v *Byland Close (Winchmore Hill) Ltd (LT),* [1998]** 2 **139**

Collective enfranchisement — Forfeiture — Effect of forfeiture on s13 enfranchisement notice — Whether landlord's counternotice valid — Whether retrospective leave under para 3 of Sched 3 to the Act — Held, s13 notice invalid — Forfeiture proceedings were pending on date of notice — Retrospective leave of court cannot be given
***Martin* v *Maryland Estates Ltd (CA),* [1998]** 2 **8**

Collective enfranchisement — Purchase price — Value of leases in "non-1993 Act world" — Effect of lack of mutual enforceability clause — Whether price should include hope value attributable to non–participating tenant's flat — Reasonableness of legal fees — Held, leases in "no-1993 Act world" valued at 78% of long leasehold value — No discount for non mutual enforceability clauses — Yield rate was 10% — Non–participating flat had hope value of 15% of marriage value — Legal costs determined including counsel's fees
***In Shulem B Association Ltd's appeal (LT),* [2001]** 1 **105**

Collective enfranchisement — Whether s13 notice containing errors in plan and proposing one purchase price where headlease to be acquired valid — Whether headlessee capable of being qualifying tenant of caretaker's flat — Held, notice valid — Headlessee a qualifying tenant
***Crean Davidson Investments Ltd* v *Earl Cadogan (CC),* [1998]** 2 **96**

Determination of price for acquisition of freehold to two maisonettes — Appropriate risk rates — Marriage values — Held, 6% appropriate risk rate — 50% of marriage value payable
***Donath* v *Second Duke of Westminister's Trustees (LVT),* [1997]** 1 **203**

Determination of price of freehold under Sched 6 — Appeal from leasehold valuation tribunal — Nominee purchaser withdrawing notice of intention to appear — Valuation evidence — Insurance commission — Marriage value — Determination of costs — Held, insurance commission valued at 6 YP — Marriage value derived from analysis of comparable transactions to produce a relation between freehold investment value and marriage value rejected — Nominee purchaser to pay costs notwithstanding that it did not appear
***Maryland Estates Ltd* v *63 Perham Road Ltd (LT),* [1997]** 2 **198**

Lease extension — Whether attic box storeroom part of flat — Whether an appurtenance or an outhouse — Held, storeroom an appurtenance within s62(2) and tenant entitled to new lease of flat and storeroom under s7(6)(b)
***Cadogan* v *McGirk (CA),* [1996]** 2 **75**

New lease — S42 notice — Whether notice rendered invalid if signed by agents on behalf of tenant under s99(5) — Whether county court has jurisdiction to determine validity of notice — Held, notice invalid — Mandatory that the tenant personally sign a s42 notice — Court has jurisdiction to determine validity of notice
***Viscount Chelsea* v *Hirshorn (CC),* [1998]** 2 **90**

Premium for extended lease — Appellate jurisdiction of Lands Tribunal — Whether LVT erred in determining value of extended lease — Amount of discount for "sitting tenant" rights — Proportion of marriage value — Costs — Held, premium increased — 15% discount for Landlord and Tenant Act 1954 Part I rights — 72.5% marriage value share — No cost order
***Cadogan Estates Ltd* v *Shahgholi (LT),* [1999]** 1 **189**

Purchase price — Determination — Whether price should reflect value of insurance commission and marriage value — Held, insurance commission not separately valued but reflected in auction price — Element of marriage value payable
***63 Perham Road Ltd* v *Manyfield Ltd (LVT),* [1995]** 2 **206**

Leaseholder's notice of claim

Estoppel and representation — Notice under Landlord and Tenant Act 1954 s25 not served in prescribed form — Tenant served counternotice and made application for new tenancy — Whether tenant entitled to dispute validity of s25 notice in claiming freehold under Leasehold Reform Act 1967 — Held, tenant knew of defective s25 notice and estopped from disputing its validity — Claim under Leasehold Reform Act 1967 invalid
***Keepers and Governors of the Possessions Revenues and Goods of the Free* v *Mayhew (CA),* [1997]** 1 **88**

Management scheme

Application by landlord for approval of management scheme under s70 of Leasehold Reform, Housing and Urban Development Act 1993 — Held, scheme approved — Repairing and management fee provisions reasonable and lawful
***Grosvenor Estate (Mayfair) London's Application, Re (LVT),* [1995]** 2 **202**

New lease

Leasehold Reform, Housing and Urban Development Act 1993 — Determination of premium — Lease expired by date of hearing — Whether deduction for Landlord and Tenant Act 1954 Part I rights — Held, 25% deduction appropriate to reflect risk of security of tenure
***Horden* v *Viscount Chelsea (LVT),* [1997]** 1 **195**

Notices

Extended lease — Tenants' initial notices specifying date for counternotices less than two months after tenants' notice — Whether notices required to be served on third party to leases — Delay in service on third party until after expiration of specified counternotice date — Whether tenants' notices valid — Held, notices served on later of dates of service on more than one party — Notices invalid as not specifying not less than two months for counternotices — Para 9 of Sched 12 to 1993 Act does not apply to avoid errors in notices
***Keepers and Governors of John Lyon Grammar School* v *Secchi (CA),* [1999]** 3 **49**

S42 notice claiming new lease — Registration — Notice unregistered — Collective acquisition of freehold — Whether new freeholder bound by s42 notice — Held, unregistered notice not binding on purchaser of freehold
***Melbury Road Properties 1995 Ltd* v *Kreidi (CC),* [1999]** 3 **108**

Notice to acquire freehold

Lease held by joint tenants — Whether notice must be signed by or with authority of both joint tenants — Held, notice invalid if signed by one of two joint tenants or without authority of both
***Wax* v *Viscount Chelsea (CC),* [1996]** 2 **80**

Price of freehold

Acquisition of freehold — Price under s9(1C) of Leasehold Reform Act 1967 — Yield rate — Value of improvements — Decision, price of £36,000 — 7% yield rate — Value of improvements to be deducted
***Williams* v *Portman Family Settled Estates Trustees (LVT),* [1996]** 1 **218**

Determination of price under s9(1) (A) — Whether any deduction in freehold value to reflect risk of tenant claiming protection of Landlord and Tenant Act 1954, Part I — Held, Part I of 1954 Act now amended by Local Government and Housing Act 1989 — Tenant could only claim assured tenancy at market rent — No deduction from freehold value
***Eyre Estate Trustees* v *Shack (LVT),* [1995]** 1 **213**

LANDLORD AND TENANT — *continued*
Leasehold enfranchisement — *continued*
Price of freehold — continued

Leasehold Reform Act 1967 — Price of freehold under s9(1A) — Whether any deduction for Landlord and Tenant Act 1954, Part I rights — Whether works improvements — Whether unsuccessful tenant should pay landlords' costs — Held, no deduction from freehold price for improvements — No certainty that tenant would claim security of tenure under Part I of the Landlord and Tenant Act 1954 — Tenant ordered to pay costs
***Vignaud* v *Keepers and Governors of the Possessions Revenues and Goods of the Free Grammar School of John Lyon (LT),* [1996]** 2 **179**

Leasehold Reform, Housing and Urban Development Act 1993 — Acquisition of freehold — Determination of price — Capitalisation rates — Value of insurance commission — Decision, capitalisation rate of 13% — No value of insurance commission or administration charges — Marriage value of £7,200 supported by evidence
***Sked* v *Towngran Developments Ltd (LVT),* [1995]** 1 **216**

Leasehold Reform, Housing and Urban Development Act 1993 — Determination of price for acquisition of freehold of block of 10 flats — Whether loss of development value — Whether additional losses — Held, no loss of development value, marriage value or other losses — 10 YP applied to ground rents
***Scarlett* v *Halcrow Ltd (LVT),* [1995]** 1 **219**

S9(1) determination — Whether in absence of evidence from respondents tribunal bound to allow appeal — Reliability of sales to tenants — Yield rate — Held, appeal dismissed — Respondents not bound to call evidence although risky course — Tribunal disregards as unreliable and wrong landlord's reliance on sales and yield rate
***Swann* v *White (LT),* [1996]** 1 **199**

Purchase price
Acquisition of property portflolio without compliance with 1987 Act Part I — Purchase notice by tenants in respect of part of portfolio — Determination of price — Held, 11 years' purchase of ground rents alternatively seven years plus allowance for insurance commission.
***Groundpremium Property Management Ltd* v *Longmint Ltd (LVT),* [1998]** 1 **131**

Surveyor's fees
Landlord incurred valuation fee when instructing surveyor to set down hearing before LVT — Whether landlord entitled to recover fee from tenant — Held, fee irrecoverable as fee in connection with reference pursuant to para 5 of Sched 22 to Housing Act 1980
***Covent Garden Group Ltd* v *Naiva (CA),* [1995]** 1 **243**

Validity of initial notice
Leasehold Reform, Housing and Urban Development Act 1993 — Collective enfranchisement — Whether tenant's initial notice valid — Whether plan must be served with notice — Held, initial notice invalid — S13 assumes that service of plan not a technicality — Act uses mandatory language
***Mutual Place Property Management Ltd* v *Blaquiere (CC),* [1996]** 2 **78**

Validity of notice claim
Whether valid s6 Landlord and Tenant Act 1987 notice served — Whether tenant entitled to withdraw from acquisition process notwithstanding irrevocable agreement with other tenants to proceed— Held, no valid s6 notice — In context of 1987 Act tenant cannot by contract deprive himself of right to withdraw acceptance of landlords' offer — Appointment by tenants of one of their number as agent not irrevocable in this context
***Mainwaring* v *Henry Smith's Charity Trustees (No 2) (CA),* [1997]** 1 **93**

Lease held by joint proprietors — Notice of leaseholder's claim — Whether one of two joint tenants can serve notice — Leasehold Reform Act 1967 — Held, both joint tenants must join the leaseholder's notice under s8 — One may serve notice with the authority of the other
***Wax* v *Viscount Chelsea (CC),* [1996]** 2 **80**

Validity of s6 notice
Photostat copy of s6 notice served after one tenant had withdrawn support — Whether valid s6 notice served — Whether tenant entitled to withdraw from acquisition process notwithstanding irrevocable agreement with other tenants to proceed — Held, no valid s6 notice served — One tenant entitled to withdraw from acquisition process despite contract with other tenants
***Mainwaring* v *Henry Smith's Charity Trustees (No 2) (CA),* [1997]** 1 **93**

Leasehold Property (Repairs) Act 1938 s1
Mortgagee in possession
Counternotice — S146 of Law of Property Act 1925 — Mortgages — Mortgage of lease — Whether mortgagee in possession "lessee" for purposes of s146 — Whether mortgagee in possession "lessee" for purposes of s1 of 1938 Act — Whether mortgagee in possession entitled to serve 1938 Act counternotice — Held, mortgagee not "lessee" for purposes of s146(1) of 1925 Act or s1 of 1938 Act — Mortgagee not entitled to serve 1938 Act counternotice
***Smith* v *Spaul (CA),* [2003]** 1 **70**

Leasehold Reform Act 1967
Forfeited lease
Forfeiture by service of writ — Whether lease remained forfeited following compromise of forfeiture proceedings — Whether tenant entitled to enfranchise under Leasehold Reform Act 1967 — Held, upon compromise, forfeiture became a nullity — Tenant entitled to enfranchise
***Hynes* v *Twinsectra Ltd (CA),* [1995]** 2 **69**

Leasehold Reform, Housing and Urban Development Act 1993 — See also **Landlord and tenant** (Leasehold enfranchisement)
Assignment of tenant's notice to new lease
Contract to assign — Failure to complete assignment of notice — Whether Act requires legal assignment of notice of claim — Whether lease assigned without benefit of notice — Whether notice deemed withdrawn — Held, 1993 Act requires legal assignment of notice of claim at same time as assignment of lease — On failure to assign notice the latter is deemed withdrawn
***Aldavon Co Ltd* v *Deverill (CC),* [1999]** 2 **69**

Liability of original tenant — See **Landlord and tenant** (Assignment — Liability of original tenant) (Forfeiture — Possession) (Insolvency — Disclaimer of lease) (Rent arrears)

Long lease
Notice to terminate
Part I of Landlord and Tenant Act 1954 — Whether notice terminating tenancy addressed to ex-tenant valid — Whether tenant entitled to Rent Act protection and fair rent limit — Local Government and Housing Act 1989 — Notice terminating long residential tenancy and proposing assured tenancy with rent above £25,000 limit — Whether rent assessment committee entitled to determine rent above £25,000 limit — Whether tenancy an assured tenancy — Held, notice invalid as not addressed to tenant — Notice proposing rent in excess of £25,000 valid — Committee entitled to determine rent in excess of limit — Tenancy would not continue as an assured tenancy
***R (on the application of Morris)* v *London Rent Assessment Committee (CA),* [2002]** 2 **13**

Merger
Commutation covenant
Equity-sharing headlease — Merger of underlease with headlease — Capital sum payable — Whether breach of covenant against commutation of rent — Held, no breach — Arrangement not within standard commutation clause
***Associated Newspapers Property Ltd* v *Master and Wardens and Brethren and Sisters of the Guild or Fraternity of the Blessed Mary the Virgin of the Mystery of Drapers in the City of London (CA),* [1997]** 2 **88**

Mesne profits
Whether recoverable
Mortgage — Whether landlord entitled to recover *mesne* profits between forfeiture and date of relief — Held, where mortgagee by subdemise claims relief under s146(2) of Law of Property Act 1925 *mesne* profits irrecoverable
***Escalus Properties Ltd* v *Robinson; Sinclair Gardens Investments (Kensington) Ltd* v *Walsh; Escalus Properties Ltd* v *Cooper-Smith; Escalus Properties Ltd* v *Dennis (CA),* [1995]** 2 **23**

Misrepresentation
Damages
Premises represented as having planning permission and suitable for car repairs — Negligent misrepresentations — Whether landlord liable in damages for misrepresentation — Whether period of liability continued after expiration of initial six-month term — Whether misrepresentation causative of losses — Held, landlord liable for damages — Reasonable for tenants to remain in possession
***Bridgegrove Ltd* v *Smith (CA),* [1997]** 2 **40**

Negligence
Breach of statutory duty
Fire — Windows — Lockable window catches with removable keys — Inability to escape during fire through locked windows — Whether council landlords liable to occupiers for lockable windows with removable keys — Held, landlords not liable for using lockable windows with removable keys
***Adams* v *Rhymney Valley District Council (CA),* [2000]** 3 **25**

Notice to quit — See also **Agricultural holdings** (Notice to quit)

Service

Notice to quit served at premises — Residential tenant abroad — Whether valid service in accordance with s196(5) of Law of Property Act 1925 — Held, no valid service — Tenancy agreement did not "require" service of notice for purposes of s196(5)

***Wandsworth London Borough Council* v *Attwell (CA),* [1996]** 1 **57**

Subtenancy

Whether subtenancy determines when notice to quit served by headtenant on landlord — Held, subtenancy determines with tenancy

***Pennell* v *Payne (CA),* [1995]** 1 **6**

Validity — See **Landlord and tenant** *(Break clause — Validity of notice)*

Notices — See also **Landlord and tenant** (Break clause — Notice) (Forfeiture — S146 notice) (Leasehold enfranchisement — Leasehold Reform, Housing and Urban Development Act 1993)

Conflicting notices

Part II Landlord and Tenant Act 1954 — Notices —Landlord serving conflicting s25 notices — One notice incapable the other capable of being an effective s25 notice — Landlord serving third notice capable of being effective — Whether tenant entitled to rely upon one of the two conflicting notices — Whether third notice only valid notice — Held, as two conflicting notices served, both would be treated by recipient as invalid — Third notice only valid notice

***Barclays Bank plc* v *Bee (CA),* [2001]** 3 **41**

New lease

Leasehold Reform, Housing and Urban Development Act 1993 — S42 notice — Whether failure to specify genuine premium renders notice invalid— Whether county court has jurisdiction to determine validity of notice — Held, purchase notice valid

***Viscount Chelsea* v *Morris (CC),* [1997]** 2 **100**

Service — See **Landlord and tenant** (Landlord and Tenant Act 1987 — Landlord's address for service of notices)

Validity of notice

Notices determining fixed term lease one day earlier than date specified in break clause — Whether notices valid — Appeal from [1996] 1 69 — Held, notices valid on test of reasonable recipient — *Hankey* v *Clavering* overruled

***Mannai Investment Co Ltd* v *Eagle Star Life Assurance Co Ltd (HL),* [1997]** 1 **57**

Notice to terminate

Validity

Notice to terminate — Part I of Landlord and Tenant Act 1954 — Whether notice terminating tenancy addressed to ex-tenant valid notice — Whether tenant entitled to Rent Act protection and fair rent limit — Local Government and Housing Act 1989 — Notice terminating long residential tenancy and proposing assured tenancy with rent above £25,000 limit — Whether rent assessment committee entitled to determine rent above £25,000 limit — Whether tenancy an assured tenancy — Held, notice invalid as not addressed to tenant — Notice proposing rentin excess of £25,000 valid — Committee can determine rent in excess of limit — Tenancy would not continue as assured tenancy

***R (on the application of Morris)* v *London Rent Assessment Committee (QB),* [2001]** 2 **51**

Open market rent — See **Landlord and tenant** (Rent review clause — Construction: hypothetical lease)

Option

Enforceability

Assignment of part of lease — Grant of subleases with terms longer than headlease — Whether grant of option *ultra vires* local authority landlord — Whether option remained exercisable after grant of subleases with longer terms than headlease — Whether tenant entitled to exercise option — Held, grant of option not authorised by ss164 or 172(3) of Local Government Act 1933 — Assignment of part of lease did not disentitle exercise of option by headlessee

***Stretch* v *West Dorset District Council (CA),* [1998]** 3 **62**

Option to determine lease

Validity

Option ceasing to have effect after assignment of lease — Lease assigned but assignee never registered as proprietor under Land Registration Act 1925 — Whether exercise of option effective — Held, legal estate not assigned —*Gentle* v *Faulkner* applied — Option validly exercisable

***Brown & Root Technology Ltd* v *Sun Alliance & London Assurance Co Ltd (CA),* [1997]** 1 **39**

Option to renew

Local authority

Validity — Lease — Option for further term purportedly granted by local authority — Grant of option *ultra vires* local authority's powers — Tenant not entitled to acquire new term under English law — Whether breach of Art 1 of First Protocol to European Convention on Human Rights — Held, interference of deprivation of possessions — Not justified and disproportionate — Violation of Art 1 — Award of damages

***Stretch* v *United Kingdom (ECHR),* [2004]** 1 **11**

Peaceable re-entry — See **Landlord and tenant** (Forfeiture — Insolvency Act 1986)

Possession — See also **Adverse possession**; **Housing** (Secure tenancy — Possession claim) (Secure tenancy — Right to buy); **Landlord and tenant** (Assured tenancy) (Secure tenancy) (Housing Act 1988 — Damages for unlawful eviction)

European Convention on Human Rights

Secure tenancy — Tenancy held by wife only — Allegation of domestic violence — Tenant leaving dwelling — Council applying policy on domestic violence and requiring wife to terminate tenancy before offer of rehousing — Tenancy determined — Possession proceedings against husband — Whether breaches of Arts 6, 8 and 14 of Convention — Held, no breaches of Arts 6, 8 and 14 of Convention — Council's conduct lawful — *Harrow London Borough Council* v *Qazi* [2003] 3 EGLR 109 applied

***Newham London Borough Council* v *Kibata (CA),* [2004]** 1 **56**

Order of possession

Suspension of warrant — Tenant breaching terms of suspended possession order — Landlord obtaining warrant for possession — Tenant failing to apply to suspend warrant of possession before eviction — Whether warrant should be suspended under principle in *Leicester City Council* v *Aldwinckle* (1991) 24 HLR 40 — Whether warrant should be set aside under court's inherent jurisdiction — Held, in absence of any fault by either landlord or court principle in Aldwinckle not invoked — Court will not exercise jurisdiction to set aside in absence of oppression

***Jephson Homes Housing Association* v *Moisejevs (CA),* [2001]** 3 **14**

Procedure

Landlord obtaining judgment in default for arrears of rent — Order for possession executed — No express dispensation of service of summons under Ord 19, r 5 — Whether judgment liable to be set aside for irregularity — Held, judgment upheld — Preferable for dispensation to appear expressly in order

***Bass Holdings Ltd* v *Brodie (QB),* [1998]** 1 **51**

Suspended order

Application to set aside — Appeal — Tenant not appearing at possession order hearing — Tenant seeking to set aside possession order — Whether merits of defence or reason for non-appearance predominant consideration — Whether objection to long suspended possession order — Held, non-appearance at hearing predominant consideration — No objection to possession order likely to be suspended for long period

***Lambeth London Borough Council* v *Henry (CA),* [2000]** 1 **33**

Quiet enjoyment

Breach of covenant

Nuisance — Inadequate sound insulation between flats — Whether breach of landlord's covenant for quiet enjoyment — Whether landlord liable in nuisance — Appeal from [1998] 3 46 — Held, no breach of covenant for quiet enjoyment — Landlord not liable in nuisance

***Southwark London Borough Council* v *Mills (HL),* [1999]** 3 **35**

Nuisance — Inadequate sound insulation between flats — Whether breach of landlord's covenant — Whether liable in nuisance — Held, landlord liable for mischief from natural and necessary result of what he has authorised — Capable of liability in nuisance and breach of covenant

***Baxter* v *Camden London Borough Council (CA),* [1998]** 2 **29**

Rent — See **Insolvency** (Disclaimer of lease — Determination of loss); **Landlord and tenant** (Landlord and Tenant Act 1954 Part II — New tenancy grant)

Rack-rents — See **Landlord and tenant** (Rent review clause — Construction: general)

Recovery of overpayment

Mistake — Sums paid as rent under mistake — Whether sums repayable — Whether mistakes of fact or law — Whether defence to recovering of sums paid under mistake of fact — Held, no defence to recovery of sums paid under mistake or fact — Payments under mistake of law recoverable

***Nurdin & Peacock plc* v *DB Ramsden & Co Ltd (No 2) (Ch),* [1999]** 1 **15**

LANDLORD AND TENANT — *continued*

Rent — *continued*

Rent suspension clause

Insured risk — Premises unfit for occupation on happening of insured risk — Whether rent suspension applies to insurance and service charges reserved as rent — Held, on construction of lease rent suspension clause does not include insurance and service charges reserved as rent

***P&O Property Holdings Ltd* v *International Computers Ltd (Ch),* [1999]** 2 **17**

Set-off

Breaches of Art 85 EC Treaty — Tied public houses — Whether liability for sums payable for any breach of Art 85 or implied term in leases can be set off against rent and other sums payable by tenants — Held, no set-off of any claim for breach of Art 85 against rent — Insufficient connection between rent and cross-claim

***Courage Ltd* v *Crehan (CA),* [1999]** 2 **145**

Licensee in occupation under agreement for lease — Arrears of licence fees — Whether licensee entitled to set off claim for damages against licence fees — Held, no set-off of damages claim against licence fees or rent arrears

***Star Rider Ltd* v *Inntrepreneur Pub Co (Ch),* [1998]** 1 **53**

Rent Acts

Claim for damages

S102 of Rent Act 1977 — Order for possession of dwelling-house under Case 9 of Sched 15 to Rent Act 1977 — Landlord recovering possession — Landlord selling dwelling-house on recovering possession — Tenant claiming damages under s102 of Rent Act 1977 — Whether tenant entitled to damages — Measure of damages — Held, tenant entitled to claim damages — Landlord giving false representations of intentions at trial of possession proceedings — Damages assessed at £60,000

***Clements* v *Simmonds (QB),* [2002]** 3 **22**

Fair rent

Determination by rent assessment committee — Whether decision containing adequate reasons — Whether committee erring in respect of requirement to disregard of breaches of repair — Whether committee failing to give adequate reasons for choice of scarcity area and scarcity discount — Held, committee failing to give adequate reasons with regard to disregard of tenant's breach of repair obligations and in relation to scarcity

***R (on the application of Wolters (London) Ltd)* v *London Rent Assessment Committee (Admin),* [2003]** 3 **17**

Rent Acts (Maximum Fair Rent) Order 1999 — Capping order — Failure to register fair rent for 10 years — Whether capping order applies to original registered rent or to rent level that would have been registered two years previously — Assessment of fair rent — Comparables derived from assured tenancies — Voids — Whether correct to make deductions from rents to reflect voids — Held, capping order applied to registered rent — Committee wrong to make deductions for voids without evidence

***Spath Holme Ltd* v *North Western Rent Assessment (Admin),* [2001]** 3 **30**

Rent Acts (Maximum Fair Rent) Order 1999 — Whether 1999 Order *ultra vires* powers in Landlord and Tenant Act 1985 — Whether 1999 Order unlawful — Held, powers in 1985 Act restricted to counter-inflationary measures — 1999 Order outside the purposes of the order-making power

***R* v *Secretary of State for the Environment, Transport and the Regions, ex parte Spath Holme Ltd (CA),* [2000]** 1 **167**

Rent Acts (Maximum Fair Rent) Order 1999 — Whether 1999 Order *ultra vires* powers in s31 of Landlord and Tenant Act 1985 — Whether 1999 Order unlawful — Held, Secretary of State had power under s31 to make the 1999 order — S31 was not ambiguous or unclear — Not necessary to consider earlier legislation

***R* v *Secretary of State for the Environment, Transport and the Regions, ex parte Spath Holme Ltd (HL),* [2001]** 1 **129**

Rent registration — Repairs — Dwelling–house put into repair — Application for registration of fair rent less than two years after previous registration — S67(3)(a) of Rent Act 1977 — Whether change of condition — Whether carrying out repairs amounts to change of condition permitting mid–term application for fair rent application — Held, carrying out repairs can constitute change of condition — Mid–term application for registration of fair rent permitted

***R (on the application of Haysport Properties Ltd)* v *Rent Officer of the West Sussex Registration Area (CA),* [2001]** 2 **63**

Service charges — Services provided and charged for by separate management company — Landlords having no liability under the leases to provided services or repairs — Whether services provided by management company are "services provided by the landlord" for purposes of s71(4) of Rent Act 1977 — Whether service charges fell to be included as part of registered rent — Held, management company not agent of landlords — Service charges did not fall to be included as part of registered rent

***Eaton Square Properties Ltd* v *Ogilvie (RAC),* [2000]** 1 **73**

Fair rent determination

Rent assessment committee — Whether committee made proper determinations — Whether proper use made of market rent comparables — Held, committee improperly added increase to old fair rents — Guidance on use of market rent comparables

***Northumberland & Durham Property Trust Ltd* v *Chairman of the London Rent Assessment Committee (QB),* [1998]** 3 **85**

Possession — See also **Negligence** (Duty of care and liability — Letting agents)

Case 1 of Sched 15 to Rent Act 1977 — Statutory tenancy — Ground for Possession — Bankruptcy of tenant — Forfeiture clause — Whether tenant in breach of an "obligation" not to go bankrupt — Held, bankruptcy event under forfeiture clause amounted to breach of an obligation — Case 1 satisfied

***Cadogan Estates Ltd* v *McMahon (HL),* [2001]** 1 **47**

Landlord and tenant — Rent Act 1977 — Possession proceedings — Suspended possession order — Whether order should be suspended on terms that rent arrears payable for indefinite period of time — Held, possession order should not be suspended on terms that rent arrears payable over indefinite period of time

***Taj* v *Ali (CA),* [2000]** 3 **35**

Protected tenancy

Statutory successor — Same-sex partner — Para 2 of Sched 1 to Rent Act 1977 — Whether same-sex partner entitled to succeed to statutory tenancy — Whether *Fitzpatrick* v *Sterling Housing Association* [2001] 1 AC 27 survived coming into force of Human Rights Act 1998 — Whether Art 14 read with Art 8 of European Convention on Human Rights engaged — Whether para 2 of Sched 1 to 1977 Act can be interpreted to apply to homosexual couples — Held, Convention rights engaged — Para 2 of Sched 1 to 1977 Act interpreted as applying to homosexual couples

***Ghaidan* v *Godin-Mendoza (HL),* [2004]** 2 **132**

Rent assessment

Fair rent — Whether fair rents should be determined by reference to market rent comparables — Whether rent officers and rent assessment committees should give reasons and arithmetical workings — Held, market rent comparables preferred over comparable registered fair rents — Committees must give workings especially when rejecting market rent comparables

***Curtis* v *Chairman of the London Rent Assessment Committee (CA),* [1998]** 1 **79**

Fair rent — Whether rent assessment committee wrong to reject assured tenancy rents in preference to evidence of fair rents — Held, assured tenancy rent should be considered without adjustment for security but discounted for scarcity

***Spath Holme Ltd* v *Chairman of the Greater Manchester & Lancashire Rent Assessment Committee (CA),* [1995]** 2 **80**

Guidelines for determination of fair rents — Held, committee entitled to consider fair rents settlements as well as market evidence

***Northumberland & Durham Property Trust Ltd* v *London Rent Assessment Committee (No 2) (QB),* [1998]** 3 **85**

Whether rent assessment committee erred in considering evidence of market and fair rent assessments — Held, committee entitled to consider fair rent settlements as well as market evidence

***Northumberland & Durham Property Trust Ltd* v *London Rent Assessment Committee (QB),* [1997]** 1 **236**

Resident landlord

Landlord acquires new property and moves in one week after granting existing tenant of old home a new tenancy — Whether in residence for purposes of s12(1) (*b*) of Rent Act 1977 — Held, landlord in residence — Outward and visible signs of intention to reside present

***Barnett* v *O'Sullivan (CA),* [1995]** 1 **93**

Security of tenure — See **Housing** *(Secure tenancy)*

Statutory tenancy

Whether statutory tenancy can be held by joint tenants — Held, landlords estopped from denying mother and daughter to be treated as joint statutory tenants

***Daejan Properties Ltd* v *Mahoney (CA),* [1995]** 2 **75**

LANDLORD AND TENANT — *continued*
Rent Acts — *continued*

Structure
Bungalow erected on site — Annual licence fee for site — Whether tenancy of bungalow — Whether bungalow fixture or chattel — Held, bungalow part of the land — Protected tenancy of site and bungalow
***Elitestone Ltd* v *Morris (HL),* [1997]** 2 **115**

Subtenancy
Headlease of premises protected by Landlord and Tenant Act 1954 Part II — Part of premises sublet to tenants protected by Rent Act 1977 — Whether subtenants remain protected against superior landlord on termination of headlease — Held, *Pittalis* v *Grant* [1989] 2 EGLR 90 decided *per incuriam* — Subtenants remained protected by s137(3) of the Rent Act 1977
***Church Commissioners for England* v *Baines; Ebied* v *Hopkins; Wellcome Trust Ltd* v *Hammad (CA),* [1998]** 1 **73**

Rent arrears — See also **Insolvency** (Individual voluntary arrangement — Rent arrears); **Landlord and tenant** (Assignment — Liability of original tenant) (Forfeiture — Relief)

Arrears: liability of guarantor
Claim for rent arrears against guarantor of original tenant — Whether limitation period of six or 12 years applied — Held, under s19 of Limitation Act 1980 limitation period of six years applies to claims against guarantors — Cause of action arises when rent demanded from guarantor
***Romain* v *Scuba TV Ltd (CA),* [1996]** 1 **103**

Liability of original tenant
Disclaimer of lease — Payment of rent by related company — Whether leases determined by taking possession following disclaimer — Whether leases determined by forfeiture — Held, no termination of leases — Original tenants liable for rent arrears
***Cromwell Developments Ltd* v *Godfrey (CA),* [1998]** 2 **62**

Rent-free periods — See **Landlord and tenant** (Rent review clause — Construction: hypothetical lease)

Rent review — See also **Agricultural holdings** (Arbitration); **Landlord and tenant** (Insolvency — Vesting order) (Landlord and Tenant Act 1954 Part II — Notice to terminate tenancy)

Appointment of expert
Suitability of appointed expert — Appointment by president of RICS — Appointment made from president's panel of experts — Experts from Channel Islands not considered — President choosing expert from persons with experience of variety stores — Whether expert required to have experience of variety store or general retail — Relevance of zoning and GIA approaches to valuation — Whether appointment contrary to terms of lease — Validity of appointment — Role and liability of president in making appointments — Held, appointment valid — President correctly categorising premises and concluding no local expert had relevant experience — Guidance given on role and liability of president in relation to appointment of experts
***Epoch Properties Ltd* v *British Home Stores (Jersey) Ltd (Channel Islands CA) (CA: Ch Islands),* [2004]** 3 **34**

Assumptions
Tenant must be assumed to have complied with repairing covenants — Air conditioning plant kept going by patch repairs — Whether arbitrator applied correct test in holding that tenant complied with repairing covenants by patch repairing air conditioning plant in place of plant's complete replacement — Held, no error of law by arbitrator — Tenant entitled to less onerous method of complying with covenant
***Land Securities plc* v *Westminster City Council (No 2) (Ch),* [1995]** 1 **245**

Construction
Hypothesis that comparable premises in specified areas in same location as actual premises — Market rents in specified areas greater than actual location of premises — Whether rents of comparable premises requiring adjustment or discount to allow for location of actual premises — Held, rent review clause containing clear express words — No adjustment required to rents of comparables to allow for actual location of subject premises
***Beegas Nominees Ltd* v *Decco Ltd (Ch),* [2003]** 3 **25**

Rent geared to 8% of rack-rents receivable from sublessees — Whether any increase where premises not sublet — Held, rents receivable include rents which could be receivable — Rent liable to increase
***Ashworth Frazer Ltd* v *Gloucester City Council (CA),* [1997]** 1 **104**

Originating summons issued to determine questions arising in rent review — Whether questions to be determined by expert under lease — Whether originating summons should be struck out — Appeal from [1998] 2 85 — Held, role of expert determined by construction of lease — Lease did not give to expert role of construing lease
***National Grid Co plc* v *M25 Group Ltd (CA),* [1999]** 1 **65**

Time of the essence — Specified period for service of trigger notice — Expert to determine rent review before review date — Late service of trigger notice — Whether time of the essence — Whether contra-indications — Held, time not of the essence — No contra-indications
***McDonald's Property Co Ltd* v *HSBC Bank plc (Ch),* [2001]** 3 **19**

Construction: general
Assumptions — Tenant's repairing obligations — Whether it must be assumed that tenant carried out extensive repairs to poorly built building or carried out more limited repairs which had regard to building's "commercial life" — Held, no authority for "commercial life" concept — Assumption that tenant carried out more extensive repairs
***Ladbroke Hotels Ltd* v *Sandhu (Ch),* [1995]** 2 **92**

Meaning of Tenant with capital "T"— Casino — Tenant assumed to have necessary licences — Whether assumption referred to actual tenant or hypothetical tenant — Held, on true construction "Tenant" reference only to actual tenant
***Parkside Clubs (Nottingham) Ltd* v *Armgrade Ltd (CA),* [1995]** 2 **96**

Provision for expert to be appointed only upon application by landlord — Whether rent reviewable for each review period — Whether landlord could be compelled to seek appointment and rent reviewed — Appeal from [1995] 2 87 — Held, rent must be reviewed each rental period — Provisions for appointment of expert machinery — Where machinery breaks down court can intervene
***Royal Bank of Scotland plc* v *Jennings (CA),* [1997]** 1 **101**

Rent geared to rack-rents "receivable" on rent review date — Premises occupied by tenant — Whether rack-rents receivable means whatever would be the market rack-rents — Where no rack-rents whether court should devise valuation machinery — Held, as premises not sublet, rent review machinery had broken down — Court could not imply term or machinery to determine rent
***Fraser Pipestock Ltd* v *Gloucester City Council (Ch),* [1995]** 2 **90**

Rent to be determined of notional premises in defined radius — Whether notional premises at any location within radius — Held, notional premises at location comparable to actual location
***Dukeminster (Ebbgate House One) Ltd* v *Somerfield Properties Co Ltd (CA),* [1997]** 2 **125**

Upwards/downwards review provision — In default of agreement landlord entitled to request appointment of surveyor — Whether in default of such appointment tenant entitled to request court to determine rent — Held, rent payable under previous review payable in default of further review — No room for court to determine rent
***Harben Style Ltd* v *Rhodes Trust (Ch),* [1995]** 1 **118**

Whether rent at review to be upwards-only — Obligation to pay current rent or an increased rent — General approach to construction of documents — Held, rent review upwards-only — Guidance on construction of documents
***Melanesian Mission Trust Board* v *Australian Mutual Provident Society (PC),* [1997]** 2 **128**

Whether rent to be reviewed upwards-only — Held, by reason of references to "increased" and absence of right for tenant to initiate review, rent "upwards-only"
***Standard Life Assurance* v *Unipath Ltd (CA),* [1997]** 2 **121**

Construction: hypothetical lease
Hypothetical lease on same terms and conditions as actual lease — Whether later deed containing options varied lease for purposes of hypothesis — Held, later deed binding on successors in title — Same as if lease actually varied — Hypothetical lease includes provisions of later deed
***Commercial Union Life Assurance Co Ltd* v *Woolworths plc (Ch),* [1996]** 1 **237**

Hypothetical letting on same terms of actual lease — Residue of actual lease 120 years — Whether term of hypothetical letting residue of hypothetical lease — Held, term of hypothetical letting of such length as would be reasonably agreed in the open market
***Brown* v *Gloucester City Council (CA),* [1998]** 1 **95**

Open market rent — On assumption that no reduction or allowance made of any rent-free period or other rent concession which in new letting might be granted to an incoming tenant — Whether headline rent payable at review — Held, headline rent discounted — Assume rent fixed from day one of tenancy
***Prudential Nominees Ltd* v *Greenham Trading Ltd (CA),* [1995]** 1 **97**

Open market rent after expiration of rent-free or concessionary rent or other inducement which may be offered in case of new letting — Whether headline rent payable at review — Held, headline rents from comparables must be discounted to allow for rent-free periods
***Co-operative Wholesale Society Ltd* v *National Westminster Bank plc (CA),* [1995]** 1 **97**

LANDLORD AND TENANT — *continued*
Rent review — *continued*
Construction: hypothetical lease — continued

Open market rent after expiration of rent-free period of such length as would be negotiated in the open market between willing landlord and willing tenant upon a letting of the premises or other inducement that may be offered in case of new letting — Whether headline rent payable at review — Held, rent at review is headline rent
***Broadgate Square plc* v *Lehman Brothers Ltd (CA),* [1995]** 1 **97**

Open market rent after expiration of rent-free period or periods at concessionary rents that might be granted on a new letting — Whether headline rent payable at review — Held, open market rent is to be headline rent
***Scottish Amicable Life Assurance Society* v *Middleton Potts & Co (CA),* [1995]** 1 **97**

Construction: valuation
Expert determination — Whether assumption can be made that hypothetical tenant would subdivide — Whether wrong to use overall rather than zoning method — Whether *Iceland Frozen Foods plc* v *Starlight Investments Ltd* [1992] 1 EGLR 126 correctly decided — Held, overall method not inappropriate — *Iceland* case wrongly decided
***Lewisham Investment Partnership Ltd* v *Morgan (Ch),* [1997]** 2 **150**

Date of review
Rent reviewable in certain specified years "of the said term" — Meaning of "the said term" — Whether date of grant or date from which term expressed to run — Arbitration — Appeal — Whether arbitrator deciding correct date for rent review — Held, "specified term" meaning term from date of grant of lease and first rent payment — Appeal from arbitrator's decision allowed
***Bisichi Mining Ltd* v *Bass Holdings Ltd (Ch),* [2002]** 2 **4**

Disregard of improvements
Rent review clause incorporating disregards in s34(2) of Landlord and Tenant Act 1954 — Improvements carried out by third party under agreement with tenant — Whether tenant "carried out" improvements — Whether tenant must pay for improvements — Whether arbitrator erred in deciding that improvements not to be disregarded — Held, improvements need not be directly carried out or paid for by tenant — Sufficient if tenant had some involvement in identifying, supervising and/or financing improvements
***Durley House* v *Cadogan (Ch),* [2000]** 1 **60**

Disregards
Improvements — Licence permitting improvements deemed premises improved from commencement of term — Whether disregard of tenant's improvements avoided at review — Held, disregard of tenant's improvements still applied
***Historic Houses Hotels Ltd* v *Cadogan Estates (CA),* [1995]** 1 **117**

Fixtures
Assumption that premises fully equipped and fitted out — Whether tenant's fixtures taken into account and rentalised at rent review — Held, tenant's fixtures not rentalised at rent review
***Ocean Accident & Guarantee Corporation* v *Next plc (Ch),* [1996]** 2 **84**

Hypothetical term
Commencement of hypothetical term — Actual term of 25 years commencing seven years prior to grant of lease — Whether hypothetical term of 25 years commenced on date of actual term — Whether hypothetical term of 25 years commenced on relevant rent review date — Held, hypothetical term of 25 years commenced on date of actual term
***Chancebutton Ltd* v *Compass Services UK & Ireland Ltd (Ch),* [2004]** 2 **47**

Term of actual lease 25 years — Hypothetical term for rent review purposes 25 years — Whether hypothetical term running from date of grant of actual lease or relevant rent review date — Whether presumption of reality applying — Held, hypothetical term running 25 years from relevant rent review date — Presumption of reality not applying
***Canary Wharf Investments (Three)* v *Telegraph Group Ltd (Ch),* [2003[** 3 **31**

Issue estoppel
Arbitration — Construction of rent review clause — Whether leave should be given to appeal arbitrator's award — Whether issue estoppel would arise at future reviews by reason of arbitrator's award on construction of clause — Held, leave refused — Unjust if landlords would be entitled to object to construction summons on point of construction decided by arbitrator throughout rest of 125-year term
***British Railways Board* v *Ringbest Ltd (Ch),* [1996]** 2 **82**

Notice and counternotice
Landlord's trigger notice specifying rent required at review — Whether rent figure in trigger notice must be determined in accordance with definition of review rent — Held, landlord entitled to specify rent in trigger notice without a determination in accordance with rent definition — Trigger notice valid
***Fox & Widley* v *Gurham (QB),* [1998]** 1 **91**

Rent review
Rent review clause with deeming provision — Whether time of the essence —Whether tenant's counternotice timeously served — Whether counternotice valid — Whether *Mecca Leisure Ltd* v *Renown Holdings Ltd* or *Trustees of Henry Smith's Charity* v *AWADA Trading & Promotion Services Ltd* should be followed — Held, *Mecca* followed — Counternotice valid — Time not of the essence
***Starmark Enterprises Ltd* v *CPL Distribution Ltd (Ch),* [2000]** 3 **37**

Tenants' rights of first refusal
Sale of two reversionary leases — Grant of individual leases of flats by new landlord — Whether new leases incumbrances — Whether new landlord under continuing duty to comply with s16 — Held, new leases were not incumbrances — Tenants entitled to enforce acquisition of reversion lease and compel new landlord to serve notices under s16
***Belvedere Court Management Ltd* v *Frogmore Developments Ltd (CA),* [1996]** 1 **59**

Time of the essence
Construction — Whether time limit to be implied — Whether time of the essence — Whether landlord "neglecting" to apply for appointment of surveyor within reasonable time — Whether permission for late estoppel claim should be given — Held, time limit could not be implied — Time not of the essence — Trigger notice valid — Late application to advance estoppel argument dismissed as no real prospect of success
***Iceland Foods plc* v *Dangoor (Ch),* [2002]** 2 **5**

Rent review determinable by independent expert upon landlord's application — Rent review clause containing no express time limits — Tenant serving notice making time of the essence — Whether time limit to be implied — Whether notice effective in making time of the essence — Held, time limit implied to give business efficacy — Time limit a reasonable period — Tenant's notice to make time of the essence effective
***Barclays Bank plc* v *Savile Estates Ltd (CA),* [2002]** 2 **16**

Time limits
Counternotice to "trigger" notice — Rent payable from rent review date to be rent specified by trigger notice unless counternotice served — Whether presumption that time not of the essence rebutted by deeming provision — Held, deeming provision not sufficient contra-indication — Presumption applied
***Bickenhall Engineering Co Ltd* v *Grandmet Restaurants Ltd (CA),* [1995]** 1 **110**

Deeming provision — Time limit for application to appoint expert — On expiration of time limit rent deemed to be rent in tenant's notice — Whether time of the essence — Held, time limit of the essence — Visionhire Ltd v Britel Fund Trustees Ltd [1992] 1 128 followed
***Banks* v *Kokkinos (Ch),* [1999]** 3 **133**

Trigger notice — Counternotice to be served within three months — Time of the essence — Whether extension of time for trigger notice under Arbitration Act 1996 s12 — Held, no extension of time — Tenant failed to satisfy requirements of s12
***Fox & Widley* v *Gurham (QB),* [1998]** 1 **91**

Trigger notice — Interrelationship with break clause — Whether time of the essence for service of trigger notice — Whether landlord entitled to rent review — Held, mutuality of rights under rent review and break clause — Time of the essence — No rent review
***Central Estates Ltd* v *Secretary of State for the Environment (CA),* [1997]** 1 **239**

Trigger notices
Rent to be higher of current market rental value or initial rent — Current market rent below initial rent — Landlord only entitled to serve trigger notice — Whether rent must be reviewed — Whether landlord must serve trigger notice — Held, rent must be reviewed — Landlord must serve trigger notice
***Addin* v *Secretary of State for the Environment (Ch),* [1997]** 1 **99**

Service — S196(4) of Law of Property Act 1925 — "Ordinary course of post" — Whether recipient required for deemed service — Whether *Stephenson & Son* v *Orca Properties Ltd* [1989] 2 EGLR 129 correctly decided — Whether implied term that recipient of notice precluded from avoiding service — Held, no requirement in deeming provision in s196(4) that no service without available recipient — Question of fact when service would be effected in "ordinary course of post" — No implied term relating to the avoidance of service
***WX Investments Ltd* v *Begg (Ch),* [2002]** 3 **47**

Trigger notice to be served not more than 12 months prior to expiry of five-year period of term — Whether lease permitting service of trigger notice after expiry of five-year period — Whether trigger notice required to be served during 12-month period prior to expiry of relevant five-year period — Whether time of the essence — Held, upon true construction of lease, review notice could be served only during relevant 12-month period — Time was of the essence of the service of the review notice
***First Property Growth Partnership LP* v *Royal & Sun Alliance Property Services Ltd (Ch),* [2002]** 2 **11; *(CA),* [2003]** 1 **39**

Valuation hypothesis

Open market rental value — Large retail unit — Hypothesis directs overall psf basis rather than zoning basis of rental valuation — Whether arbitrator wrongly excluding all zoning basis of rental valuation — Whether rent review hypothesis allowing landlord to threaten alternative option of dividing up large unit and assuming tenant would offer rent reflecting value to landlord of subdivision option — Whether serious irregluarity under s68 of Arbitration Act 1996 — Whether error of law within meaning of s69 — Held, no error by arbitrator in relation to application of zoning comparables — Landlord not entitled to contend that in valuation hypothesis it could threaten alternative transaction such as subdivision at higher aggregate rates
***Marklands v Virgin Retail Ltd (Ch),* [2004]** 2 **43**

Repairs — See also **Landlord and tenant** (Roof space — Creation of flat) (Service charges — Set-off) (Subletting — Enforcement of covenants)

Breach

Terminal dilapidations — Damages — S18(1) of Landlord and Tenant Act 1927 — Valuation date in recession period — Unhelpful comparables — Landlord not having carried out repairs — Whether any diminution in reversionary value — Whether cost of repairs relevant — Factors that would influence hypothetical purchaser — Held, hypothetical purchaser would be speculator in recession period — Several factors would influence purchaser — Some diminution in value less than cost of repairs appropriate
***Craven (Builders) Ltd* v *Secretary of State for Health (Ch),* [2000]** 1 **128**

Landlord proposing to construct additional flats on existing block of flats — Whether lease contained implied term against construction of additional flats — Whether construction of additional flats breach of landlord's repairing obligation — Held, although lease part of letting scheme no limit on additional flats — Construction of additional flats no breach of repairing covenant — *Devonshire Reid Properties Ltd* v *Trenaman* [1997] 1 EGLR 45 distinguished and not followed
***Hannon* v *169 Queen's Gate Ltd (Ch),* [2000]** 1 **40**

Breach of implied covenant

Dampness — Whether existence of damp a cause of action — Whether tenant proving breach of covenant to repair by landlords — Whether landlords owing any duty to warn tenant of consequences of creating damp conditions — Whether any disrepair of structure or exterior — Whether landlords' omission to investigate dampness relevant — Held, landlords owing no duty to advise tenant to avoid moisture-creating activity — Tenant failing to prove breach of covenant by failure of repair of structure or exterior — Existence of damp giving rise to no cause of action in itself
***Southwark London Borough Council* v *McIntosh (Ch),*[2002]** 1 **25**

Condensation

Defect in design — Whether local authority landlords owing statutory duty under s6 of Human Rights Act 1998 to do works to comply with Art 8 of ECHR — Whether *Quick* v *Taff-Ely Borough Council* [1985] 2 EGLR 50 wrongly decided — Whether s3 of Human Rights Act 1998 requiring wider interpretation of s11 of Landlord and Tenant Act 1985 — Whether implied term in tenancies granted by local authority landlords — Whether any duty under s6 of Defective Premises Act 1972 — Held, no statutory duty owed under s6 of Human Rights Act 1998 — *Quick* not wrongly decided — S11 of 1985 Act not requiring wider interpretation — No implied covenant to remedy defects in design in tenancies granted by local authority landlords — No duty arising under s6 of 1972 Act
***Lee* v *Leeds City Council; Ratcliffe* v *Sandwell Metropolitan Borough Council (CA),* [2002]** 1 **103**

Covenant

Damages for breach — Landlord and Tenant Act 1985 — Defective Premises Act 1972 — Breach of implied obligation to repair dwelling — Assessment of damages — Held, four relevant propositions — Damages for discomfort and inconvenience assessed on global basis but should be cross-checked to rents payable
***Wallace* v *Manchester City Council (CA),* [1998]** 3 **38**

Damages for terminal dilapidation claim — Effect of tenant demand for new specifications — Negative reversionary value — Whether cost of repairs measure of damages — Proper application of limit applied by s18(1) of Landlord and Tenant Act 1927 to negative reversionary value — Held, cost of repairs appropriate measure — S18(1) can apply to negative values
***Shortlands Investments Ltd* v *Cargill plc (Off Ref),* [1995]** 1 **51**

Defective electric underfloor storage heating — Whether tenant discharged covenant to repair by installing space heaters rather than restoring and maintaining underfloor heating — Held, underfloor system capable of repair — Tenant not entitled to provide space heaters in performance of repairing covenant
***Creska Ltd* v *Hammersmith and Fulham London Borough Council (CA),* [1998]** 3 **35**

Landlord's covenant to repair — Whether replacement of single-glazed window frames with double-glazed window frames repair — Whether landlord entitled to interlocutory mandatory injunction allowing access for repairs — Whether undue delay in seeking interlocutory relief — Held, replacement by double-glazed window frames a repair — Landlord entitled to injunction
***Minja Properties Ltd* v *Cussins Property Group plc (Ch),* [1998]** 2 **52**

Lease containing repairing and yielding up clause — Whether clause limited recoverable damages — Whether repairing clause so ambiguous whole clause should be struck down — Waste — Whether tenant for years liable for permissive waste — Held, clause contained unintelligible words that should be disregarded — Tenant for years liable for permissive waste
***Dayani* v *Bromley London Borough Council (QB:TCC),* [1999]** 3 **144**

Mould growth — External walls of 9" solid brick — Landlord's obligation to maintain in good condition and repair — Whether obligation limited to structure and exterior — Whether obligation to maintain condition included obligation to provide thermal insulation — Held, obligation to maintain in good condition wider than obligation to repair — Dwelling suffering from mould not in good condition — Landlord liable for failure to provide thermal insulation
***Welsh* v *Greenwich London Borough Council (CA),* [2000]** 3 **41**

Tenants' repairing covenant — Whether specific performance will be ordered — Held, no ground for withholding specific performance in appropriatecases
***Rainbow Estates Ltd* v *Tokenhold Ltd (Ch),* [1998]** 2 **34**

Damages

Breach of covenant to provide services — Assessment of damages — Commercial tenant in occupation — Whether tenant entitled to damages assessed by reference to diminution in rental value — Whether general damages approach relevant — Held, damages assessed as diminution in value of premises to tenant — Evidence of rental value may be relevant
***Electricity Supply Nominees Ltd* v *National Magazine Co Ltd (CC),* [1999]** 1 **130**

Covenant to repair — Disrepair of front door — Whether landlord liable for burglary loss following entry of burglar — Held, burglary loss caused by breach of covenant to repair front door — Landlord liable in damages
***Marshall* v *Rubypoint Ltd (CA),* [1997]** 1 **69**

Defective Premises Act 1972

S4 of Defective Premises Act 1972 — Duty of care — Defective gas fire — Whether landlord liable for carbon monoxide poisoning — Whether breach of duty of care — Whether necessity for notice of defect — Whether tenant contributorily negligent — Held, liability under s4 of 1972 Act does not depend upon notice of defects — Liability depends upon general tortious obligations concerning standard of care — Tenant 80% contributorily negligent
***Sykes* v *Harry (CA),* [2001]** 1 **53**

Dilapidations

Payment of moneys into court in two tranches — Costs — Landlord beating first payment in but failing against second payment — Whether landlord entitled to costs of establishing claim based upon costs of repair up to date when tenant admitted that cost of repairs bound to exceed any diminution in value — Whether landlord entitled to costs upon diminution in value issue up to date of second payment in — Held, landlord required to prove repair costs until issue formally admitted — Landlord entitled to costs upon diminution in value issue up to second payment in
***Johnsey Estates (1990) Ltd* v *Secretary of State for the Environment (CA),* [2001]** 2 **128**

Insurance

Covenant to insure — Landlord and Tenant Act 1987 — Whether implied term that sum charged by insurer should not be unreasonable — Whether cost of insurance exceeds what is reasonable under Landlord and Tenant Act 1985 — Held, no term implied — Insurance premiums not unreasonable

***Berrycroft Management Co Ltd* v *Sinclair Gardens Investments (Kensington) Ltd (CA),* [1997]** 1 **47**

Interest on late payments

Interest payable on late payments described as rent — Whether interest impressed with statutory trust under section 42 of Landlord and Tenant Act 1987 — Whether landlord entitled to retain interest on late payment of rents and service charges — Held, interest on rent and service charge arrears described as rent and belonging to landlord

***St Mary's Mansions Ltd* v *Limegate Investment Co Ltd (CA),* [2003]** 1 41

Landlord and Tenant Act 1985

Failure of landlord to obtain estimates and consult tenant — Whether repair costs reasonably incurred at reasonable expense — Landlord and Tenant Act 1985 ss19-20 —Whether landlord entitled to dispensation under s20(9) — Held, dispensation granted under s20(9) — Costs reasonably incurred and not excessive

***Wilson* v *Stone (LVT),* [1998]** 2 **155**

S20 consultation provisions — Lease of several flats and other properties — Tenant subletting flats — Whether tenant under lease a tenant of a flat — Whether s20 consultation provisions applied where tenant held and sublet several flats — Held, s20 applied to lease of premises containing several flats underlet to occupying tenants

***Heron Maple House Ltd* v *Central Estates Ltd (CC),* [2002]** 1 **35**

Landlord and Tenanct Act 1985 s19

Replacement of flat roof with pitched — Replacement of metal-framed windows with uPVC double—glazed units — Whether costs reasonable — Whether cost-in-use calculations relevant — Whether replacements were repairs — Whether costs reasonably incurred on matters as a result of past neglect — Held, replacements were repairs — Cost-in-use calculations relevant — Decision to effect replacements reasonable — Costs reasonable — Additional costs incurred as a result of past neglect not reasonably incurred

***Wandsworth London Borough Council* v *Griffin (LT),* [2000]** 2 **105**

Landlord and Tenant Act 1987

Jurisdiction of leasehold valuation tribunal (LVT) — Whether service charge dispute including matters outside its jurisdiction should be referred to LVT — Whether LVT entitled to consider matters relating to interpretation of leases or application of s20 of 1987 Act — Whether proceedings should be transferred to LVT — Held, LVT having limited jurisdiction — Disputes concerning interpretation of leases or compliance with s20 outside LVT jurisdiction — Not desirable to fragment dispute containing matters within and outside jurisdiction of LVT

***Wandsworth London Borough Council* v *Manuel (Ch),* [2002]** 2 **128**

Leasehold valuation tribunal

S19(2A) of Landlord and Tenant Act 1985 — Jurisdiction of LVT — Whether LVT has jurisdiction where service charges paid — Meaning of service charges "alleged to be payable" — Limitation period — Whether limitation period of 12 years applies to applications to LVT — Appeal from interim decision of LVT — Whether Lands Tribunal has jurisdiction to consider reasonableness of service charge costs notwithstanding payment of applicant lessees — Limitation Act 1980 has no direct application — Correctness of decisions of Lands Tribunal concerning lack of jurisdiction to hear interim determinations of LVTs not decided

***R* v *London Leasehold Valuation Tribunal, ex p Daejan Properties Ltd (QVB),* [2000]** 3 **44**

Legal costs

Legal costs incurring on legal advice, recovery of service charges and administration of landlord company — Lessees shareholders in company — Whether landlord entitled to claim as service charges costs of legal advice and proceedings — Held, legal advice costs not recoverable as service charges

***St Mary's Mansions Ltd* v *Limegate Investment Co Ltd (CA),* [2003]** 1 41

Whether litigation costs incurred in claims for forfeiture and relief recoverable as landlord's costs by way of service charges — Whether direction under s20C of Landlord and Tenant Act 1985 that such costs not relevant costs payable by tenant under service charge — Held, litigation costs were recoverable under service charge — However direction pursuant to s20C made that such costs were not payable by plaintiff tenant.

***Iperion Investments Corporation* v *Broadwalk House Residents Ltd (CA),* [1995]** 2 **47**

Payment of depreciation allowance

Tenant covenanted to pay depreciation allowance — Air-conditioning plant not replaced by term date — Whether tenant entitled to return of depreciation allowance fund — Held, landlord entitled to retain fund

***Secretary of State for the Environment* v *Possfund (North West) Ltd (Ch),* [1997]** 2 **56**

Reasonableness

Management fees at 15% of costs — Whether LVT having jurisdiction to construe lease — Whether lease providing for management fees at 15% of costs — Whether management fees service charges under s18 of Landlord and Tenant Act 1985 — Whether provision in lease for 15% management fees void by virtue of s19(3) — Held, LVT having jurisdiction to construe lease — Management fees service charges under s18 — Agreement in lease to pay 15% management fees void under s19(3)

***Longmint Ltd* v *Marcus (LT),* [2004]** 3 **171**

Repairs

Additional works arising in course of repairs — Landlord and Tenant Act 1985 — Whether discretion under s20(9) should have been exercised — Whether s20 £1,000 limit applies to course of work or separately available for additional works — Held, discretion under s20(9) arises only if landlord acted reasonably — On the facts £1,000 limit properly applied to course of works — £1,000 limit must be applied by common-sense test

***Martin* v *Maryland Estates Ltd (CA),* [1999]** 2 **53**

Whether proposals by landlord to recover roofs and repair windows of large shopping mall within covenant to keep premises in good and substantial repair and condition — Whether recoverable under service charge — Held, landlord not premature in commencing roof repairs and acted reasonably in works carried out

***Postel Properties Ltd* v *Boots the Chemist (QB),* [1996]** 2 **60**

Reserve fund

Surplus on year-end service charge accounts — Whether year-end surplus can be added to reserve fund — Held, surplus year-end service charges only added to reserve fund in limited circumstances provided by lease

***St Mary's Mansions Ltd* v *Limegate Investment Co Ltd (CA),* [2003]** 1 **41**

Section 20B Landlord and Tenant Act 1985

Service charges payable by interim payments in advance — Service charge accounts and certificates required at conclusion of each service charge year — Difference between interim payments and actual expenditure payable (or repayable) — Whether 18-month limit in s20B of Landlord and Tenant Act 1985 applying where actual expenditure less than interim payments — Held, s20B not applying where interim service charges exceeding actual expenditure — No demand for payment within s20B where no further payments required

***Gilje* v *Charlegrove Securities Ltd (No 2) (Ch),* [2003]** 3 **9**

Set-off

Tenant's claim for damages for disrepair — Whether landlord entitled to set-off unpaid service charges — Whether equitable set-off — Whether set-off should be struck out — Held, set-off true equitable set-off and a defence — No striking out of set-off as true defence not claim

***Filross Securities Ltd* v *Midgeley (CA),* [1998]** 3 **43**

Statutory trust

S42 of Landlord and Tenant Act 1987 — Manager appointed under 1987 Act — Whether statutory trust applying to arrears of service charges — Whether tenant seeking to set off counterclaim for damages against trust fund — Held, statutory trust applying only to sums contributed by tenants — Tenant not seeking to set off counterclaim against trust fund

***Maunder Taylor* v *Blaquiere (CA),* [2003]** 3 **52**

Whether "rent"

Whether service charge rent where lease so provided — Held, parties entitled to give service charge all characteristics of rent by express provision

***Escalus Properties Ltd* v *Robinson; Sinclair Gardens Investments (Kensington) Ltd* v *Walsh; Escalus Properties Ltd* v *Cooper-Smith; Escalus Properties Ltd* v *Dennis (CA),* [1995]** 2 **23**

Service tenancy

Housing Act 1985

Sched 1 para 2(1) — Accommodation for better performance of employee's duties — Whether tenant required to occupy premises for better performance of his duties — Whether essential for employee to occupy precise premises let — Held, term implied in contract of employment that house occupied for better performance of tenant's duties — Requirements of para 2(1) of Sched 1 to Housing Act 1985 satisfied

***Surrey County Council* v *Lamond (CA),* [1999]** 1 **32**

Services

Breach

Air-conditioning — Repeated failure of air-conditioning over six-year period — Whether breach of covenant — Assessment of damages — Held, covenant to provide services not construed as repair covenant — Landlord in breach of covenant to provide air-conditioning — Damages assessed on basis of one-third of rent for proportion of relevant period

***Clarke* v *Lloyds TSB Bank plc (QB: TCC),* [2002]** 3 **93**

Side agreement or letter

Turnover rent

Construction — Commission payable based on percentage of sales — Whether "turnover rent" payable after expiry of contractual term — Whether side agreement determinable by notice — Held, commission payable so long as lease held that substituted for first lease — Side agreement not determinable by notice

***Berthon Boat Co Ltd* v *Hood Sailmakers Ltd (QB),* [2000]** 1 **39**

Subletting — See also **Landlord and tenant** (Assignment — Consent, refusal of)

Consent, refusal of

Application to assign and sublet — Whether consent unreasonably withheld — Whether landlord can rely at trial on reasons first put in affidavits — Whether landlord can withhold consent to subletting where proposed rent less than market value — Held, consent unreasonably withheld — Expiration of reasonable time relevant date for considering reasons for withholding consent — Subletting at less than market rent not reasonable ground for withholding consent to subletting — Landlord and Tenant Act 1988

***Norwich Union Life Insurance Society* v *Shopmoor Ltd (Ch),* [1998]** 2 **167**

Covenant against subletting without landlord's consent — Whether landlord entitled to withhold consent where reasonable belief held that subletting at rent less than open market value and subletting would detrimentally harm reversionary interest — Held, landlord did not hold belief at time of withholding consent — Not reasonable to withhold consent on ground that market will improve and increase value of landlord's interest

***Blockbuster Entertainment Ltd* v *Leakcliff Properties Ltd (Ch),* [1997]** 1 **28**

Landlord's opposition to subletting premises — Whether landlord entitled to rely on reasons not put in writing within reasonable time of request — Whether reasons satisfactory — Whether test of profit greater than three times' rent — Held, landlord cannot adduce reasons before court not put in writing within reasonable time or request — No test of three time' rent — Landlord and Tenant Act 1988

***Footwear Corporation Ltd* v *Amplight Properties Ltd (Ch),* [1998]** 2 **38**

Enforcement of covenants

Headlandlord purportedly enforcing covenants in underlease to repair on notice and reimburse repair costs in default — Whether covenants enforceable by assignee of reversion to headlease against underlesseeunder s56 of the Law of Property Act 1925 — Held, notwithstanding references to superior landlord, relevant covenants not enforceable against underlessees

***Amsprop Trading Ltd* v *Harris Distribution Ltd (Ch),* [1997]** 2 **78**

Surety — See **Insolvency; Landlord and tenant** (Assignment — Liability of original tenant) (Assignment — Liability of surety)

Surety covenant — See also **Landlord and tenant** (Assignment — Liability of original tenant) (Assignment — Liability of surety) (Forfeiture — Surety covenant)

Contribution

One of three sureties securing deed of surrender in consideration of a permium — Whether surety entitled to recover from co-surety contribution to premium — Held, that in absence of agreement or representation co-surety not liable to contribute to premium

***BSE Trading Ltd* v *Hands (CA),* [1996]** 2 **214**

Release of obligation

Lease containing convenant of guarantee — Licence to assign permitted lessor to enter and carry out external improvements — Whether covenant of guarantee disentitled guarantor from being released from obligation to pay rent — Held, covenant of guarantee did not prevent guarantor from being released from obligation

***West Horndon Industrial Park Ltd* v *Phoenix Timber Group plc (Ch),* [1995]** 1 **77**

Surrender — See **Agricultural holdings** (Notice to quit — head tenancy)

Surrender of lease — See also **Adverse possession** (Lease — Registered land); **Agricultural holdings** (Notice to quit — Subtenancy); **Insolvency** (Surety — Duty to mitigate)

Implied surrender

Surety of lease — Company associated with tenant company in possession — Rent accepted from company in occupation — Tenant in liquidation — Lease disclaimed — Whether implied acceptance of surrender of lease — Whether surety liable for arrears of rent — Held, no surrender by operation of law

***Bhogal* v *Cheema (Ch),* [1998]** 2 **50**

Surrender by operation of law

Acceptance of rent from occupier in possession — No intention to permit assignment of lease — Held, no surrender by operation of law — No intention to create legal relations

***Mattey Securities Ltd* v *Ervin (CA),* [1998]** 2 **66**

Whether tenancy protected by Landlord and Tenant Act 1954 surrendered upon grant of tenancy of will — Held, tenant estopped from denying surrender of lease by acceptance of tenancy at will

***Gibbs Mew plc* v *Gemmell (CA),* [1999]** 1 **43**

Tenancy

Unfair terms in consumer contracts

Local authority letting accommodation to homeless persons — Procedure governed by local policy — Whether policy unlawful — Whether allocation of accommodation lawful — Whether Unfair Terms in Consumer Contract Regulations 1999 and Council Directive 93/13/EEC applying to contracts for the transfer of an interest in land — Whether the 1999 Regulations and directive applying to tenancies offered by public authorities — Held, local policy unlawful — 1999 Regulations and directive applying to contracts for transfer of an interest in land — 1999 Regulations and directive applying to grant of tenancies by public authorities

***R (on the application of Khatun)* v *Newham London Borough Council; R (on the application of Zeb)* v *Newham London Borough Council (Admin),* [2004]** 1 **34**

Tenancy at will

Calculation of rent

Tenant holding over — Whether landlord entitled to market rental value or previously reserved rent — Held, tenant liable to pay suitably adjusted market rent having regard to actual circumstances

***Dean and Chapter of the Cathedral and Metropolitan Church of Christ Canterbury* v *Whitbread plc (Ch),* [1995]** 1 **82**

Use and occupation

Compensation — Valuation — Failure of negotiations to grant lease of commercial premises — Value of use and occupation — Whether rent agreed for proposed lease best evidence of value of use and occupation — Whether value tenant at will actually obtained relevant — Held, unless basis of agreement substantially different from tenancy actually held, rent agreed for proposed lease good evidence of value of use and occupation

***Lewisham London Borough Council* v *Masterson (CA),* [2000]** 1 **134**

Tenancy by estoppel — See also **Landlord and tenant** (Landlord and Tenant Act 1954 Part II — Tenancy by estoppel) (Lease or licence)

Creation of

Housing trust holding property under licence from local housing authority — Trust granting purported licence to occupy to grantee upon terms reciting trust's lack of title — Whether grantee held tenancy by estoppel — Whether tenancy or licence — Held, terms of grant not inconsistent with intention — No estoppel — Grantee held licence

***Bruton* v *London & Quadrant Housing Trust (CA),* [1997]** 2 **91**

Tenant holding over

Basis of holding over

Whether tenant holding over as tenant at will or as trespasser — Held, tenant holding over as tenant at will

***Dean and Chapter of the Cathedral and Metropolitan Church of Christ Canterbury* v *Whitbread plc (Ch),* [1995]** 1 **82**

Termination of lease

Implied surrender

Discussions with surety to take new lease — Keys returned by tenant accepted by landlord — Whether surrender by operation of law and release of surety from obligation to take new lease — Held, in context of discussions for new lease, no surrender of lease by return of keys

***Proudreed Ltd* v *Microgen Holdings plc (CA),* [1996]** 1 **89**

Termination of tenancy

Notice — See **Landlord and tenant** (Landlord and Tenant Act 1954 Part II — Notice to terminate tenancy)

Tied public houses

Art 85

Whether beer ties in leases and beer-supply agreements prohibited by Art 85 of EC Treaty — Whether community law provides directly enforceable rights — Whether implied term in leases as to beer prices payable by tenants — Whether liability for sums payable for any breach of Art 85 or implied term can be set off against rent and other sums payable by tenants — Meaning and effect of block exemption — Held, question whether breach of Article creates directly enforceable rights referred to European Court of Justice — No implied term against beer prices — No set-off of any claim for breach of Art 85 against rent — Insufficient connection between rent and cross-claim

***Courage Ltd* v *Crehan (CA),* [1999]** 2 **145**

European legislation

Tenant of pub restricted to purchasing specified beers from landlord or nominated supplier — Whether breach of Art 81 of European Community Treaty — Whether court entitled to have regard to commission decisions on ties — Whether tied lease within block exemption — Whether damages claimed of type Art 81 intended to prevent — Whether damages assessible at date of loss or date of trial — Held, leases containing beer ties breach of Art 81 — Beer ties not within block exemption— Damages suffered type of damages Art 81 intended to prevent — Damages assessible at date of loss — Measure of loss

***Crehan* v *Inntrepreneur Pub Co (CPC) (CA),* [2004]** 3 **128**

Time of the essence — See **Landlord and tenant** (Break clause — Estoppel) (Break clause — Time-limit) (Rent review clause — Time-limits)

Trespass — See also **Election** (Waiver — Landlord and tenant)

Mesne profits

Specified apartments in hotel — Whether wholesale prices payable by tour operators less expenses represented reasonable rent — Held, wholesale prices appropriate basis for calculating reasonable rent — Irrelevant that plaintiff not able to let apartments for every day out of possession

***Inverugie Investments Ltd* v *Hackett (PC),* [1996]** 1 **149**

Underlease

Determination of headlease

Break clause — Contractual provision for continuation of underleases notwithstanding principle in *Pennell* v *Payne* [1995] 1 EGLR 6 that underleases determining with headlease — Whether underleases continuing on exercise of headlease break clause — Whether effective contracting out of principle in *Pennell* — Whether covenants between underlessee and freeholder enforceable — Whether covenants enforceable pursuant to s3 of Human Rights Act 1998 — Whether surrender of headlease — Whether estoppel by deed or convention — Whether estoppel between successors in title — Held, underleases determining with headlease — If underleases continuing, covenants enforceable pursuant to s3 of Human Rights Act 1998 — No surrender — Underleases not continuing by estoppel by deed or convention — Penalty provision invoked

***PW & Co* v *Milton Gate Investments Ltd (Ch),* [2004]** 3 **103**

User covenant

Breach

"Keep open" covenant — Anchor store — Whether tenant can be ordered to comply with covenant — Whether mandatory injunction appropriate remedy to enjoin person to carry on business — Appeal from [1996] 1 71 — Held, in accordance with settled practice the courts should not order a person to carry on a business

Co-operative Insurance Society Ltd* v *Argyll Stores (Holdings) Ltd (HL), [1997] 1 **52**

Consent for change of use

Premises restricted to specified use — Landlord's consent to change of use not to be unreasonably withheld — Whether consent to change of use unreasonably withheld — Whether landlord entitled to consider only its interests — Whether property and uses coming into existence after date oflease relevant — Whether consequences of refusal to tenant relevant — Held, consent reasonably refused — Landlord entitled to consider only its interests — Property and uses coming into existence after lease date can be relevant — Effect on tenant irrelevant

***Sportoffer Ltd* v *Erewash Borough Council (Ch),* [1999]** 3 **136**

Landlord and Tenant Act 1927

Landlord minded to consent to change of user on terms which included insertion of landlord's break clause on payment of specified sum — Whether "fine" within s19(3) — Whether structural alterations involved — Held, requirement of break clause a "fine" — However tenant not relieved of need for landlord's consent as change of use involved structural alterations and outside s19(3)

***Barclays Bank plc* v *Daejan Investments (Grove Hall) Ltd (Ch),* [1995]** 1 **68**

User covenant

Withholding of landlord's consent to change of use

Application for consent to underlet and change of use refused — Premises used as nightclub and underletting to fitness club proposed — Consent would cause alleged £300,000 damage to reversionary interest — Whether refusal reasonable — Whether valuer's approach reasonable and informed — Held, valuations supporting reasons on wrong basis and ill informed — Consent unreasonably refused

***Luminar Leisure Ltd* v *Apostole (Ch),* [2001]** 3 **23**

Waste, liability for

Tenant for years

Permissive waste — Whether tenant for years liable for permissive waste — Held, tenant for years liable for permissive waste

***Dayani* v *Bromley London Borough Council (QB:TCC),* [1999]** 3 **144**

Wrongful eviction — See **Landlord and tenant** (Housing Act 1988 — Damages for unlawful eviction)

LANDLORD AND TENANT ACT 1927 — See **Landlord and tenant** (Dilapidations) (Landlord and Tenant Act 1927) (Landlord and Tenant Act 1954 Part II — New tenancy application) (Repairs — Breach) (Repairs — Covenant) (User covenant)

LANDLORD AND TENANT ACT 1954 — See **Adverse possession** (Lease in writing — Tenancy); **Landlord and tenant** (Break clause) (Landlord and Tenant Act 1954 Part I) (Landlord and Tenant Act 1954 Part II) (Landlord and Tenant Act 1954 Part II — Application for new tenancy) (Landlord and Tenant Act 1954 Part II — Landlord's opposition to new tenancy) (Landlord and Tenant Act 1954 Part II — Opposition to grant of new tenancy) (Lease — Reversionary lease) (Leasehold enfranchisement — New lease) (Leasehold enfranchisement — Price of freehold) (Rent Acts — Subtenancy) (Repairs); **Negligence** (Solicitors — Damages)

Exclusion order — See **Landlord and tenant** (Landlord and Tenant Act 1954 Part II — Exclusion order)

Licence — See **Landlord and tenant** (Landlord and Tenant Act 1954 Part II — Licence)

New tenancy — See **Landlord and tenant** (Landlord and Tenant Act 1954 Part II — Request for new tenancy) (Landlord and Tenant Act 1954 Part II — Opposition to grant of new tenancy)

Rent review — See **Landlord and tenant** (Rent review)

Statutory compensation — See **Landlord and tenant** (Landlord and Tenant Act 1954 Part II — Statutory compensation)

LANDLORD AND TENANT ACT 1985 — See **Landlord and tenant** (Repairs — Covenant) (Repairs — Implied covenant) (Service charges — Caretaker's flat) (Service charges — Insurance) (Service charges — Landlord and Tenant Act 1985) (Service charges — Landlord and Tenant Act 1985 s20B) (Service charges — Legal costs); **Leasehold enfranchisement** (Leasehold Reform Act 1967 — Meaning of "house")

LANDLORD AND TENANT ACT 1987 — See **Landlord and tenant** (Landlord and Tenant Act 1987) (Leasehold enfranchisement — Validity of notice claim) (Service charges — Landlord and Tenant Act 1987) (Service charges — Leasehold valuation tribunal); **Leasehold Valuation Tribunal** (Jurisdiction); **Mistake** (Rectification — Form of relief)

LANDLORD AND TENANT ACT 1988 — See **Landlord and tenant** (Assignment — Collateral deed) (Assignment — Consent, refusal of) (Assignment — Landlord and Tenant Act 1988) (Subletting — Consent, refusal of)

LANDS TRIBUNAL — See also **Compensation for acquisition of land** (Disturbance compensation) (Limitation period — Lands Tribunal) **Leasehold enfranchisement** (Leasehold Reform Act 1967 — S9(1)(A) valuation)

Appeal — See also **Compensation for acquisition of land** (Disturbance — Riding school business); **Compulsory purchase** (Compensation — Limitation period); **Landlord and tenant** (Leasehold enfranchisement — Acquisition of land) (Leasehold enfranchisement — Leasehold Reform, Housing and Urban Development Act 1993); **Leasehold enfranchisement** (Leasehold Reform Act 1967 — Appeal); **Leasehold valuation tribunal** (Service charge — Appeal)

From leasehold valuation tribunal

Leasehold Reform Act 1967 — Price of freehold — S9(1) determination by leasehold valuation tribunal — Whether in absence of evidence from respondents tribunal bound to allow appeal — Held, respondents not bound to call evidence although risky course

***Swann* v *White (LT),* [1996]** 1 **199**

LANDS TRIBUNAL — *continued*

Appeal — *continued*

Judicial review

Appeal from LVT — Lands Tribunal refusing permission to appeal decision of LVT —Whether decision appealable to Court of Appeal — Application for judicial review of decision of Lands Tribunal — Whether judicial review correct procedure — Held, decision of Lands Tribunal refusing permission to appeal not appealable to Court of Appeal — Save in exceptional circumstances not present, decision of Lands Tribunal not judicially reviewable

***R (on the application of Sinclair Gardens Investments (Kensington) Ltd)* v *Lands Tribunal (Admin),* [2004]** 3 **15**

Point of law

Alleged double-counting in residual valuation — Whether allegation capable of founding ground of appeal on point of law— Meaning of point of law — Held, allegation of double-counting could found ground of appeal

***Railtrack plc* v *Guinness Ltd (CA),* [2003]** 1 **124**

Award

Reopening award

Compulsory purchase — Reference to Lands Tribunal — Pre-reference costs — Failure to adduce pre-reference costs evidence at hearing — Compensation determined by tribunal — Whether tribunal entitled to reopen hearing — Whether tribunal functus officio — Held, where tribunal determines all issues save for reference costs, tribunal's decision not final — Hearing may be reopened

***Shraff Tip Ltd* v *Highways Agency (LT),* [1999]** 2 **205**

Compensation

Limitation period — See **Compulsory purchase** *(Lands Tribunal — Limitation period)*

Compromise

Costs

Compensation claim — Settlement offer — Offer to settle received two working days before hearing accepted — Whether offer defective — Whether claimant entitled to costs up to date of hearing — Held, offer defective — Offer too close to hearing and failed to include claimant's costs

***Tague* v *Lancaster City Council (LT),* [1999]** 2 **103**

Costs

Calderbank offer to settle

Lands Tribunal allowing landlord's appeal from leasehold valuation tribunal — Award of tribunal exceeding determination of leasehold valuation tribunal — Nominee purchaser's Calderbank offer exceeding award — *Calderbank* offer omitting reference to costs — Tribunal awarding costs to landlord — Whether Calderbank offer should deal with costs — Whether tribunalentitled to award costs to landlord — Held, offer was capable of acceptance — Offer was on basis that each side pay own costs — Alternatively tribunal should have considered whether landlord acted reasonably in failing to seek clarification of offer

***Phyllis Trading Ltd* v *86 Lordship Road Ltd (CA),* [2001]** 2 **85**

Compensation reference

Extinguishment of business — Goodwill — Loss on forced sale — Accountancy fees — Effect of sealed offer two days prior to hearing — Held, s4(1) of Land Compensation Act 1961 not followed where sealed offer served lated and fails to deal with costs of reference

***Shevlin* v *Trafford Park Development Corporation (LT),* [1998]** 1 **115**

Discretion

Compensation for deemed revocation of planning permission — Reference to Lands Tribunal compromised after delays and procedural negotiations — Whether claimant entitled to costs — Held, claimant entitled to costs — S4 of Land Compensation Act 1961 had no direct application — Compensating authority under duty to make offer of compensation

***Barrett* v *Mole Valley District Council (LT),* [1997]** 1 **167**

Exercise of discretion

Acquiring authority raising issue of validity of purchase notice three days before hearing — Expert evidence already exchanged — Whether acquiring authority succeeding in challenge to validity of notice entitled on costs — Held, acquiring authority should have raised validity at earlier opportunity and before exchange of expert evidence — No order as to costs

***Cook* v *Winchester City Council (LT),* [1995]** 1 **179**

Equivalence principle

Relevance of CPR guidance on award of costs — Award substantially below claimant's valuation — Whether tribunal entitled to reduce costs recoverable by claimant — Whether "special reasons" for departure from principle of equivalence — Whether compliance with Lands Tribunal Practice Direction requiring proper regard to equivalence principle — Held, tribunal should reduce costs of successful claimant only where "special reasons" existing — Special reason might include exaggerated claim reasonably known to be based upon unreliable expert evidence — Application of CPR guidance must have regard to equivalence principle

***Purfleet Farms Ltd* v *Secretary of State for Transport, Local Government and the Regions (CA),* [2003]** 1 **9**

Interest

Compensation for the acquisition of land — Lands Tribunal — Costs — Interest on costs — Award of tribunal set aside — Whether acquiring authority liable to pay interest on costs — Date from which interest payable on costs where award set aside — Held, s20 of Arbitration Act 1950 applied — Interest payable on costs — Interest payable from date of effective award

***Barclays Bank plc* v *Kent County Council (CA),* [1998]** 2 **14**

Sealed offer

Compensation — Sealed offer sent 11 days before hearing — Sealed offer accepted at hearing — Liability for costs — Held, claimants should have accepted offer within five working days — Entitled to costs up to expiration of such five days — Liable for costs thereafter

***Stanford Marsh Ltd* v *Secretary of State for the Environment (LT),* [1997]** 1 **178**

Two hearings — Second hearing six years later after appeal — Withdrawal of sealed offers by first hearing — Effect of withdrawn sealed offers — Whether tribunal entitled to have regard to offers made prior to first hearing in relation to costs of second hearing — Held, withdrawn sealed offers have relevance as if *Calderbank* offers — Tribunal wrong to have considered sealed offers made in relation to earlier hearing six years previously

***Hepworth Building Projects* v *Coal Authority (CA),* [1999]** 3 **99**

Simplified procedure

Lands Tribunal Rules simplified procedure — Unconditional offer above award — Whether acquiring authority entitled to costs under simplified procedure — Held, acquiring authority entitled to costs from date of unconditional offer under s4(1)(*a*) of Land Compensation Act 1961

***Nicholls* v *Highways Authority (LT),* [2000]** 2 **81**

Evidence

Transactional evidence held by stamp duty office

Capital gains tax — Valuation — Comparables — Anonymised comparables derived from stamp duty office returns — Whether anonymised comparables admissible — Guidance on admissibility of transactional evidence within Inland Revenue stamp duty office — Held, tribunal should normally order disclosure of evidence held by stamp duty office — Guidance on admissibility of such evidence

***Newman (HMIT)* v *Hatt (LT),* [2002]** 1 **89**

Jurisdiction — See also **Compensation for acquisition of land** (Mortgage — Date of redemption); **Compensation for electricity lines** (Jurisdiction — Lands Tribunal); **Compulsory purchase** (Compensation — Mortgage); **Leasehold enfranchisement** (Leasehold Reform, Housing and Urban Development Act 1993 — Lease extension)

Abuse of process

Agreed statement of facts — Finding of fact in preliminary issue — Compensation — Total extinguishment of claimant's business — Late application by acquiring authority that subsidiary company conducting business — Subsidiary company referring no claim — Whether acquiring authority entitled to raise new issue — Whether issue estoppel — Whether abuse of process — Whether discretion exercisable in favour of acquiring authority — Held, no issue estoppel because issues different — Tribunal having jurisdiction to control abuse of process — Discretion exercisable in favour of acquiring authority to raise late issue on terms

***Mean Fiddler Holdings Ltd v Islington London Borough Council (No 2) (LT),* [2003]** 3 **61**

Appeal

Purchase price — Jurisdiction of Lands Tribunal on an appeal — Whether a rehearing — Held, Lands Tribunal hears appeal on evidence adduced before it — Must allow appeal if lower tribunal wrong on the evidence before the Lands Tribunal

***Wellcome Trust Ltd* v *Romines (LT),* [1999]** 3 **229**

LANDS TRIBUNAL — *continued*
Jurisdiction — *continued*

Interest
Sewer laid under s15 of Public Health Act 1936 — Compensation agreed — Whether power in 1961 Act to award interest — Whether tribunal had jurisdiction to hear claim on disputed interest only — Held, interest payable under s11 of 1961 Act — Tribunal has jurisdiction to determine interest only dispute
***Taylor* v *North West Water Ltd (LT),* [1995]** 1 **266**

Leasehold valuation tribunal
Appeal to Lands Tribunal
Whether rehearing or review — Burden of proof — Whether claimant before LVT or appellant to Lands Tribunal having burden of proof — Held, appeal from LVT treated as rehearing — Burden of proof on appellant
***Veena SA* v *Cheong (LT),* [2003]** 1 **175**

Procedure
Evidence
Decisions of other tribunals and the Lands Tribunal on yields — Whether decisions admissible — Held, decisions of other tribunals on fact and opinion inadmissible
***Wellcome Trust Ltd* v *Romines (LT),* [1999]** 3 **229**

Points raised by member
Mitigation — Compensation assessed for loss of business — Lands Tribunal making deduction for claimant's freed-up time following loss of business — Whether tribunal entitled to make deduction — Whether procedural unfairness — Held, tribunal not entitled to make deduction on its own volition without putting point to the parties and hearing submissions and evidence — Freed-up time aspect of mitigation and otherwise not necessarily exceptional
***Faraday* v *Carmarthenshire County Council (CA),* [2004]** 2 **5**

Right to light
Obstruction notice registered under Rights of Light Act 1959 — See **Easement** *(Right to light)*

Rules
Amendment of application
Application to discharge or modify restriction against erection of more than one house — Whether applicants entitled to amend application to add further restriction to be modified — Lands Tribunal Rules 1996 — Held, the rules do not permit amendment of substantive application to add additional restriction to be modified. Only the grounds in paras (*a*), (*aa*), (*b*) and (*c*) of section 84(1) of the 1925 Act can be amended under r 36 of the rules
***In Diggens' Application (LT),* [2000]** 3 **87**

Time-limits — See **Compensation for acquisition of land** (General vesting declaration)

LAW OF PROPERTY ACT 1925 — See **Easements** (Acquisition — Prescription); **Landlord and tenant** (Enforcement of covenants); **Leasehold enfranchisement** (Leasehold Reform Act 1967 — Lease for lives); **Restrictive covenant** (Application to modify — Grounds (*a*), (*aa*) and (*c*)) (Burden — Successors in title) (Modification — Application to suspend proceedings); **Vendor and purchaser** (Condition of sale — Implied easements)

LAW OF PROPERTY (MISCELLANEOUS PROVISIONS) ACT 1989 — See also **Contract** (Exchange of contract); **Conveyancing** (Contract — Law of Property (Miscellaneous Provisions) Act 1989, s2); **Landlord and tenant** (Assignment — Lease) (Covenants — Collateral agreement) (Disclaimer of lease — Guarantor); **Pre-emption right**; **Proprietary estoppel** (Agreement — Negotiation "subject to contract"); **Vendor and purchaser** (Sale of land)
Estoppel — See **Compulsory purchase** (Statutory contract — S2(10 Law of Property (Miscellaneous Provisions) Act 1989)
Proprietary estoppel
Constructive trust
Oral agreement — Whether doctrine of proprietary estoppel can operate to give effect to an agreement rendered void by s2 of the 1989 Act — Held, where proprietary estoppel can operate through constructive trust, s2(5) of the 1989 Act saves oral agreement
***Yaxley* v *Gotts (CA),* [1999]** 2 **181**

LEASE — See **Landlord and tenant** (Lease); **Rectification** (Lease — Bona fide purchaser)
Agreement to grant — See **Landlord and tenant** (Agreement to grant lease — Misrepresentation)
Beer tie — See **Landlord and tenant** (Lease — Beer supply agreement) (Lease — Beer tie) (Surrender of lease — Surrender by operation of law)
Disclaimer — See **Insolvency** (Disclaimer of lease); **Landlord and tenant** (Assignment — Liability of original tenant) (Insolvency — Disclaimer of lease) (Rent arrears — Liability of original tenant)
Forfeiture — See **Landlord and tenant** (Forfeiture) (Leasehold enfranchisement — Leasehold Reform, Housing and Urban Development Act 1993)
Implied term — See **Landlord and tenant** (Roof space — Creation of flat)
Premium — See **Landlord and tenant** (Leasehold enfranchisement — Leasehold Reform, Housing and Urban Development Act 1993); **Stamp Duty** (Lease premium)
Reversionary — See **Landlord and tenant** (Lease)
Surety — See **Insolvency** (Surety); **Landlord and tenant** (Surrender of lease — Implied surrender)

LEASE OR LICENCE — See **Landlord and tenant** (Lease or licence — Hotel room)

LEASEHOLD ENFRANCHISEMENT — See also **Landlord and tenant** (Leasehold enfranchisement)
Leasehold enfranchisement procedure
High Court jurisdiction
Specific performance sought against landlord — Whether High Court had jurisdiction under 1967 Act — Held, proceedings should have been commenced in county court — High Court may transfer to, and retransfer proceedings from, county court and obtain jurisdiction — S41 of County Courts Act 1984
***South* v *Chamberlayne (CA),* [2001]** 3 **54**

Leasehold Reform Act 1967
Appeal
Validity of appeal — Determination of landlord's s9(4) costs — Leasehold valuation tribunal — Appeal — Whether valid appeal where landlord did not appear at substantive hearing — Held, Lands Tribunal has no jurisdiction to hear appeal on interlocutory decision or where putative appellant did not appear before lower tribunal
***Speedwell Estates Ltd, Re (LT),* [1999]** 2 **121**

Assignment of statutory right of enfranchisement
Assignment of statutory right independent of any assignment of tenancy — Whether assignment ineffective —Whether statutory right of enfranchisement extinguished by assignment independent of, or prior to, assignment of tenancy — Whether High Court had jurisdiction under 1967 Act — Held, assignment of statutory right without tenancy ineffective — Statutory right to enfranchise not extinguished — Proceedings should have been commenced in county court — High Court may transfer to and retransfer proceedings from county court and obtain jurisdiction
***South* v *Chamberlayne (CA),* [2001]** 3 **54**

Desire notice
Validity — Low rent test — Notice failing to mention earlier lease — Whether two leases constituting single tenancy under s3(3) of 1967 Act — Whether failure to refer to earlier lease an inaccuracy in particulars saved by para 6(3) of Sched 3 to the Act — Whether desire notice invalid — Whether tenancy at a low rent under s4A(1) — Whether rent under later lease of garden relevant to rent payable during first year of term of single hypothetical tenancy — Held, single hypothetical tenancy commencing in 1972 — Failure to refer to 1972 lease an inaccuracy saved by para 6(3) of Sched 3 to 1967 Act — Desire notice valid — Rent payable in first year of tenancy more than two-thirds of rateable value on relevant date — Rent under 1983 lease of garden cannot be added to rent payable in 1972
***Earl of Cadogan* v *Strauss (CA),* [2004]** 2 **69**

Determination of enfranchisement price
S9(1) valuation — Effect of rent review provision — Meaning of "current open market ground rent levels" — Held, nominal rent payable under rent review provision
***Jarrett* v *Burford Estates & Property Co Ltd (LT),* [1999]** 1 **181**

S9(1A) valuation — Improvements — Landlord's enfranchisement costs — Whether interest payable — Held, small global sum on improvements — Price determined at £323,000 — Interest cannot be awarded under s49 of the Arbitration Act 1996
***Shina* v *Elghanian (LT),* [1999]** 3 **103**

Determination of price
Disregard of improvements — Property initially converted into flats — Conversion depressing price on valuation date — Property reconverted to single family house — Whether reconversion works an improvement for purposes of s9(1A)(d) — Whether disregard to be construed as a whole — Whether disregard falling to be applied in relation to totality of works affecting value on valuation date — Held, s9(1A)(d) to be construed as applying to improvements as a whole — Effect of works as a whole as at valuation date — Disregard not applying to works reconverting property to single family house following conversion to flats where latter works depressed value
***John Lyon's Charity* v *Shalson (CA),* [2002]** 2 **55**

Enfranchisement notice

Validity — Failure to provide required particulars — Failure to identify lease, provide rateable value information and details of occupation — Whether failure to provide particulars an "inaccuracy" not invalidating notices — Whether sched 3, para 6(3), saved enfranchisement notices — Held, failure to provide correct details of leases an "inaccuracy" that did not invalidate notices — Failure to provide occupational details and financial information required by boxes 6, 7 and 8 of prescribed form invalidated notices

***Speedwell Estates Ltd* v *Dalziel (CA: Cr),* [2002]** 1 **55**

Enfranchisement price

Enfranchisement price — Lease expired — Whether house to be valued with vacant possession — Development value of mews house and fifth floor — Prospects of planning permission — Location and adjustment of comparable — Weight to be given to offer to purchase — Held, house valued with vacant possession —Price determined to reflect prospect of planning permission — Offer made three years after valuation date disregarded — Respondent landlord awarded costs

***Dyer* v *Cadogan (LT),* [2001]** 3 **149**

Inalienable property of landlord

Landlords holding Wimbledon Common under Wimbledon and Putney Commons Act 1871 — Land inalienable pursuant to s35 — Claim to acquire freehold of house under 1967 Act — Whether powers of enfranchisement under 1967 Act prevailing over bar against alienation in 1871 Act — Held, 1967 Act permitting and obliging conservators to sell freehold — Provision against alienation applying to voluntary dispositions — *R (on the application of O'Byrne)* v *Secretary of State for the Environment* [2002] UKHL 45; [2002] 1 WLR 3250 applying

***Blacker* v *Wimbledon and Putney Commons Conservators (CC),* [2004]** 3 **67**

Lease for lives

Claim to enfranchisement — Whether lease a long tenancy under 1967 Act — Whether lease within proviso to s3(1) of 1967 Act — Whether notice of determination required to be more than three months — Whether tenant entitled to enfranchise — Held, lease requiring notice of not less than one month within proviso to s3(1) — Lease not a long tenancy — Tenant not entitled to enfranchise

***Skinns* v *Greenwood (CA),* [2002]** 2 **53**

Lease for 70 years subject to cesser on death of survivor of tenant and wife — Effect of s 149(6) of Law of Property Act 1925 — Whether lease determined on death of tenant — Whether by proviso (*c*) in s149(6) lease determines on death of survivor — Whether Parry's life expectancy tables applicable — Whether lease for lives had any value — Whether freeholder entitled to 75% share of marriage value — Held, lease determines on death of survivor — Lease for life had no value — Freeholder entitled to 75% of marriage value

***Bistern Estate Trust's Appeal (LT),* [2000]** 2 **91**

Meaning of a "house"

Enfranchisement of two adjoining mews houses — Whether premises constituting "house... reasonably so called" — Whether premises excluded from constituting house by s2(1)(b) — Whether claimant entitled to enfranchise — Held, premises constituting "house... reasonably so called" — Premises not excluded from constituting house by s2(1)(b)

***Collins* v *Howard de Walden Estates Ltd (CA),* [2003]** 3 **35**

Enfranchisement of two adjoining mews houses — Whether premises "house... reasonably so called" — Whether tenant entitled to enfranchise — Held, premises let as one dwelling–house — Premises a "house... reasonably so called" — Tenant entitled to enfranchise

***Collins* v *Howard de Walden Estates Ltd (CC),* [2002]** 2 **61**

House reasonably so called — Material part underlying house — Whether house and mews cottage a "house" — Whether "material" part exclusion discriminatory — Whether breach of European Convention on Human Rights — Held, house and mews cottage a "house" — No breach of Convention in meaning given by the courts to "material" in s2(2) of 1967 Act

***Malekshad* v *Howard de Walden Estates Ltd (HL),* [2003]** 1 **151; *(CA),* [2001]** 3 **47**

Notice in reply

Validity — Tenant claiming entitlement to enfranchise — Tenant holding lease of house and qualifying tenant of flats under Leasehold Reform, Housing and Urban Development Act 1993 — Tenant a company — Whether tenant required to satisfy occupancy test under s1(1ZB) of 1967 Act — Whether landlord's notice in reply denying that tenant entitled to enfranchise since occupancy test not satisfied valid — Held, s1(1ZB) having no application as tenant of flat not tenant of house — Amendments made to 1967 Act permitting companies to enfranchise

***Earl Cadogan* v *Search Guarantees plc (CA),* [2004]** 3 **55**

S9(1A) valuation

Function of Lands Tribunal on an appeal — Whether LVT decision evidence before Lands Tribunal — Assessment of rent at review — Freehold capital value — Leasehold value — Compensation for severance of reversion — Held, wrong for appellant to show that LVT decision flawed — Appellant must adduce valuation evidence to support figures spoken to — Determination of ground rent on review, freehold and leasehold values — Determination of compensation for severance

***Carl* v *Grosvenor Estate Belgravia (LT),* [2000]** 3 **79**

Improvement disregard — Whether erection of house improvement — Whether improvements made by predecessor in title — Whether building agreement a "tenancy in equity" — Held, erection of house not capable of being an improvement — Building agreement not tenancy in equity under 1967 Act

***Rosen* v *Trustees of the Campden Charities (LT),* [1999]** 2 **213**

Improvement disregard — Whether erection of house improvement — Whether improvements made by predecessor in title — Whether building agreement a "tenancy in equity" — Held, erection of house on bare site not an improvement — Expenditure on erection of house in return for grant of lease not improvement within subsection (9A)(*d*) — Construction of house under building agreement did not arise under tenancy at law or in equity

***Rosen* v *Trustees of the Camden Charities (CA),* [2001]** 1 **59**

S9(4) costs

In-house costs — Landlord using in-house valuer to prepare valuations — Whether such costs recoverable — Whether costs excluded by para 5 of Schedule 22 to the Housing Act 1980 — Held, in-house costs recoverable — Costs not excluded under Sched 22 to the 1980 Act — Costs reduced

***Cressingham Properties Ltd, Re (LT),* [1999]** 2 **117**

Leasehold Reform, Housing and Urban Development Act 1993

Assessment of marriage value

Price payable — Collective enfranchisement — Assessment of marriage value — Factors relevant to assessment of marriage value — Held, para 4 of Sched 6 to 1993 Act does not preclude consideration of factors relevant to the control of freehold by tenants assessing marriage value

***Maryland Estates Ltd* v *Abbathure Flat Management Co Ltd (LT),* [1999]** 1 **100**

Claim for new extended lease

Validity of counternotice — Counternotice failing to admit claim — Whether counternotice valid — Whether tenant entitled to extended lease at premium specified in tenant's claim notice — Held, reasonable tenant would not have understood counternotice as other than one admitting claim — Counternotice specifying premium a valid s45(2)(a) notice

***Burman* v *Mount Cook Land Ltd (CC),* [2001]** 1 **62**

Validity of counternotice — Counternotice failing to admit claim — Whether counternotice valid — Whether tenant entitled to extended lease at premium specified in tenant's claim notice — Held, counternotice must state whether claim admitted or not admitted — S45(2) — If admitted, counternotice must state which of tenant's proposals are admitted

***Burman* v *Mount Cook Land Ltd (CA),* [2002]** 1 **61**

Collective enfranchisement

Acquisition of land and rights — Tenants having rights in common over access road and grounds surrounding blocks of flats — Whether right to acquire freehold over common use land satisfied by grant of permanent rights — Whether LVT having discretion to order acquisition of freehold or grant of permanent rights — Adopted road — Whether nominee purchaser entitled to acquire freehold to subsoil — Held, right to acquire commonly used land satisfied by grant of permanent rights — Where permanent rights over common use land offered by landlord, LVT having no discretion under s1(4)(a)(i) of 1993 Act — Discretion arising only where adequate rights not offered

***Shortdean Place (Eastbourne) Residents' Association Ltd* v *Lynari Properties Ltd (LT),* [2003]** 3 **147**

Determination of price — Valuation date — Uplift for control of freehold — Discount for risk of assured tenancies — Determination of marriage value share — Held, valuation date was date of determination of issue of grant of leasebacks — 3% uplift for control of freehold — 72.5% share of marriage value to landlord — Discount for risk of assured tenancies

***West Hampstead Management Co Ltd* v *Pearl Property Ltd (LT),* [2002]** 1 **115**

LEASE — *continued*
Leasehold Reform, Housing and Urban Development Act 1993 — *continued*
Collective enfranchisement — continued

Determination of price — Valuation date — Whether date of counternotice or date upon which terms of contract and transfer agreed — Determination of yield — Value of garden and parking areas — Hope value — Whether absence of any indication that non-participating tenants wanting lease extensions negating any addition for hope value — Valuation fees — Whether s33 valuation fees payable on percentage of value or hourly rate basis — Held, valuation date was date of landlords' counternotice when quality and extent of freehold agreed — Yield determined at 7% — Value of garden and parking areas included in virtual freehold values — Hope value payable where hypothetical purchaser likely to pay additional value — Unexpired terms and relatively large number of non-participating flats justified hope value — Reasonable s33 valuation fee based on hourly rate
***Blendcrown Ltd* v *Church Commissioners for England (LT),* [2004]** 1 **143**

Determination of price — Valuation date — Consent order by county court as to acquisition of freehold — Whether valuation date date of consent order or date of LVT hearing — Held, valuation date date of agreement of terms of grant of leasebacks at LVT hearing
***West Hampstead Management Co Ltd* v *Pearl Property Ltd (CA),* [2002]** 3 **55**

Landlord's counternotice — Failure to serve counternotice — Whether court has discretion as to terms of any order made under s25 — Held, in absence of any counternotice court must make s25 order if satisfied tenants entitled to enfranchise
***Willingale* v *Globalgrange Ltd (CA),* [2000]** 2 **55**

Extended lease
Valuation — Intermediate leasehold interest — Whether gross rent or profit rent taken for purposes of valuation of intermediate leaseholder's interest under Sched 13 — Held, gross rent should be used for valuation of intermediate leaseholder's interest
***Visible Information Packaged Systems Ltd* v *Squarepoint (London) Ltd (LT),* [2000]** 2 **93**

Initial notice
Claim for new long lease — Nominal premium of £100 specified by tenant — Whether initial notice invalid — Whether obligation to specify realistic premium figure in initial and counternotices applies to tenants and landlords respectively — Held, insertion of nominal unrealistic premium in tenant's initial notice or landlord's counternotice will invalidate respective notices
***Cadogan* v *Morris (CA),* [1999]** 1 **59**

Initial notice under s42
Validity — Notice specifying £100,000 as premium for extended lease — Valuation evidence that premium likely to exceed £200,000 — Effect of Cadogan v Morris [1999] 1 EGLR 59 — Whether initial notice specifying realistic premium —Whether s42 initial notice valid — Held, notice invalid — Premium specified was not realistic
Mount Cook Land Ltd* v *Rosen (CC), 75

Intermediate lease
Determination of proportion of premium payable to intermediate leaseholder — Schedule 13 of 1993 Act — Determination of diminution in value of intermediate leasehold interest — Ground rent geared to rents of occupational underleases — Whether "before valuation" requiring statutory right to acquire a new lease at peppercorn to be disregarded — Whether "before valuation" gearing applied to peppercorn rents — Held, "before valuation" to be made before grant of any lease under 1993 Act — For "before valuation" there is no extended lease at a peppercorn
***Grosvenor West End Properties* v *Whiston Properties Ltd (LT),* [2002]** 2 **87**

Lease extension
Notice of claim under s42 of 1993 Act — Failure to serve notice on third party to lease — Failure to serve notice on "other landlord" — Whether notice valid — Held, notice valid — No purpose served in giving notice to third party with continuing interest in lease — Failure to serve "other landlord" not invalidating notice — Para 4(2) of Sched 11 to Act rendering tenant liable for loss only
***Wellcome Trust Ltd* v *Bellhurst Ltd (CA),* [2002]** 2 **57**

Purchase price — Jurisdiction of Lands Tribunal on an appeal — Whether a rehearing — Whether potential for improvement — Determination of purchase price — Held, Lands Tribunal hears appeal on evidence adduced before it — Decisions of other tribunals on fact and opinion inadmissible — Yield of 6% — Purchase price increased
***Wellcome Trust Ltd* v *Romines (LT),* [1999]** 3 **229**

New extended lease
Business tenancy exclusion — Live/work unit — Application for new extended lease — Whether unit occupied under business lease — Whether permitted use under leases determined — Whether unit a flat — Whether unit occupied for business purposes — Whether residential use breach of lease — Whether tenant entitled to benefit from alleged breach of covenant — Held, unit a flat — Use for residential purposes without business use not a breach — Tenant entitled to new extended lease
***The Bishopsgate Foundation* v *Curtis (CC),* [2004]** 3 **57**

New extended lease application
S42 notice — Waiver — Competent landlord failing to serve counternotice — Parties proceeding on basis that sublessor granting new lease on terms to be determined by LVT — Tenant later relying upon Willingale v Globalgrange Ltd [2000] 2 EGLR 55 — Tenant claiming new lease on terms in original s42 notice — Whether doctrines of estoppel and waiver preventing tenant from relying upon failure to serve counternotice — Whether waiver of procedural or substantive requirements — Whether waiver going to jurisdiction of court — Held, failure to serve counternotice capable of being waived —Position on waiver of notices under Landlord and Tenant Act 1954 applying to 1993 Act — Tenant waiving failure to serve counternotice by conduct
***Latifi* v *Colherne Court Freehold Ltd (QB),* [2003]** 1 **78**

New lease
Determination of price — Discount for 1954 Act rights — Apportionment of marriage value — Held, apportionment rules assume sale by landlord and tenant of statutory 90-year lease — Haggle would have resulted in agreement of 72.5% marriage value share
***Goldstein* v *Conley (LT),* [1999]** 1 **95**

Non-residential element
Non-residential (business) element — Building — Whether non-residential element exceeding 25% — Whether non-residential areas and common parts can be aggregated for purposes of 25% proportion — Categorisation of converted areas — Held, common parts deducted from gross internal areas to arrive at aggregate of non-residential and residential areas — Non-residential proportion of aggregate having to exceed 25% to defeat collective enfranchisement
***Indiana Investments Ltd* v *Taylor (CC),* [2004]** 3 **63**

Notice claiming new lease
Validity — Notice signed by attorney under power of attorney — Whether notice valid — Whether s42 notice required personal signature of tenant — Held, notice invalid — S99(5)(a) of Leasehold Reform, Housing and Urban Development Act 1993 requires tenant's personal signature on s42 notice claiming new lease
***St Ermins Property Co* v *Tingay (Ch),* [2002]** 3 **53**

Section 20 notice
Obligation to deduce title — Nominee purchaser providing office copy entries dated prior to relevant date — Whether s20 requiring complete and conclusive proof as at relevant date and within 21-day time limit — Whether qualifying tenants failing to comply with s20 notice — Whether notice deemed to have been withdrawn — Held, no obligation to provide complete and conclusive proof of title within 21-day time limit — Landlord entitled to serve s21(2)(b) counternotice not admitting claim where titles not adequately proved
***Raymere Ltd* v *Belle Vue Gardens Ltd (CA),* [2003]** 3 **40**

Vesting order
Change of circumstances
Agreement as to valuation date and acquisition prices — County court making vesting order — Landlord applying under section 24(4)(b) for reassessment of price on ground of change of circumstances — Change of property values in three-year period — Preliminary issue — Whether LVT having jurisdiction to entertain application — Whether LVT having jurisdiction to entertain application to resolve service charge disputes — Held, LVT not having jurisdiction to consider application under section 24(4)(b) or in relation to service charges
***Castlegroom Ltd* v *Enoch (No 2) (LVT),* [2003]** 3 **46**

LEASEHOLD PROPERTY (REPAIRS) ACT 1938 — See **Landlord and tenant** (Repairs)

LEASEHOLD REFORM — See **Landlord and tenant** (Leasehold enfranchisement) (Leasehold Reform, Housing and Urban Development Act 1993)

LEASEHOLD REFORM ACT 1967 — See **Landlord and tenant** (Leasehold enfranchisement — "House") (Leasehold enfranchisement — "House and premises") (Leasehold enfranchisement — Leasehold Reform Act 1967); **Leasehold enfranchisement** (Leasehold Reform Act 1967 — Determination of price) (Leasehold Reform Act 1967 — Enfranchisement notice) (Leasehold Reform Act 1967 — Lease for lives) (Leasehold Reform Act 1967 — Meaning of "house") (Leasehold Reform Act 1967 — S9(1)(A) valuation); **Negligence** (Limitation Act — Date of damage)

LEASEHOLD REFORM, HOUSING AND URBAN DEVELOPMENT ACT 1993 — See **Landlord and tenant** (Leasehold enfranchisement); **Leasehold enfranchisement** (Leasehold Reform, Housing and Urban Development Act 1993 — S20 notice) (Leasehold Reform, Housing and Urban Development Act 1993 — Collective enfranchisement)

LEASEHOLD VALUATION TRIBUNAL — See **Landlord and tenant** (Service charges — Caretaker's flat) (Service charges — Leasehold valuation tribunal); **Leasehold enfranchisement** (Vesting order — Change of circumstance)

Appeal

Application for leave to appeal

Costs — Leave refused by Lands Tribunal — Costs incurred by respondent — Rules 2 and 52 of Lands Tribunal Rules 1996 — Service charges — S20C of Landlord and Tenant Act 1985 — Leasehold valuation tribunal refusing to order that landlord's legal costs not recoverable as service charge expenditure — Whether Lands Tribunal has jurisdiction to make costs order — Whether respondent landlord entitled to costs of responding to unsuccessful application for leave to appeal — Held, Lands Tribunal has jurisdiction under r52 to make award of costs to respondent for an unsuccessful application for leave to appeal — Award made upon robust lump sum approach

***Barrington Court Developments Ltd* v *Barrington Court Residents' Association (LT),* [2001]** 2 **99**

Appeal from — See **Lands Tribunal** (Appeal)

Jurisdiction

S19(2A) Landlord and Tenant Act 1987

Service charges challenged for years when lessees paid — Whether LVT has jurisdiction where service charges paid — Meaning of service charges "alleged to be payable" — Held, LVT's jurisdiction limited to reasonableness of unpaid services charge costs

***R (on the application of Daejan Properties Ltd)* v *London Leasehold Valuation Tribunal (CA),* [2001]** 3 **28**

Service charge

Appeal

Landlord and Tenant Act 1985 — Appeal to Lands Tribunal — Whether interlocutory decision of leasehold valuation tribunal appealable — Whether assignor of lease remains a "tenant" for the purposes of s30 of Landlord and Tenant Act 1985 — Held, interlocutory decisions not appealable — Assignor of lease may remain "tenant" for purposes of s30 of the Act if contractually bound to pay service charges for period of interest

***Sarum Properties Ltd's application (LT),* [1999]** 2 **131**

LICENCE — See also **Insolvency** (Disclaimer of licence — Waste-management licence); **Possession** (Rights to possession – Contractual rights) (RSC Ord 113 — Licence)

Eviction

Damages

S28 Housing Act 1988 — Whether occupier held lease or licence — Determination of reasonable period of notice to determine licence — Liability for receiver — Assessment of statutory damages — Trespass —Whether aggravated and exemplary damages payable — Assessment of damages — Held, agreement was a licence — Damages under s28 Housing Act 1988 assessed at £45,000 — In respect of trespass heads of claim for compensatory and aggravated damages assessed at £10,000 each — Exemplary damages of £7,500

***Mehta* v *Royal Bank of Scotland (QB),* [1999]** 3 **153**

Refuse tip — See **Refuse tip** (Licence — Rent formula)

LICENSED PREMISES — See **Rates and rating** (Public house)

LIMITATION — See also **Adverse possession** (*Animus possidendi* — Limitation Act 1980) (Limitation Act 1980) (Tenant — *Animus possidendi*); **Arbitration** (Appointment of arbitrator — Time limite extension); **Compensation** (Highways Act 1980 s28 — Footpath creation order) (Land Compensation Act 1973 Part I); (Limitation period — Waiver); **Compensation for acquisition of land** (Equivalent reinstatement — Interest) (General vesting declaration — Time-limits) (Limitation period — Lands Tribunal); **Compulsory purchase** (Compensation) (Lands Tribunal — Limitation period); **Landlord and tenant** (Service charges — Leasehold valuation tribunal); **Mortgage** (Limitation period — Arrears); (Possession proceedings); **Negligence** (Limitation Act) (Solicitors — Conveyancing) (Limitation period) (Surveyors — Limitation) (Valuers — Mortgage valuation) **Telecommunications Act 1984** (Telecommunications Code — Objection to apparatus)

Limitation Act 1980

Intestate estate

Failure to obtain administration — Possession by one of four beneficiaries for period exceeding 12 years — Whether beneficiaries time-barred — Held, interest of beneficiaries where no trust for sale or settlement sufficient interest for purposes of para 9 of Sched 1 to Limitation Act 1980 — Beneficiaries not time-barred

***Earnshaw* v *Hartley (CA),* [1999]** 2 **100**

LISTED BUILDING — See **Town and Country Planning** (Demolition — Listed building consent); **Value Added Tax** (Alterations)

LIVESTOCK QUOTAS — See **Agriculture** (Common market)

LOCAL GOVERNMENT

Sale of land

S123 Local Government Act 1972

Land capable of residential and/or employment-creating development — Higher bid received for land for residential development — Local authority proposing to sell land for employment-creating development at lower price — Economic benefit in job creation — Whether job creation benefits relevant in determining whether consideration best that could reasonably be obtained — Held, job creation benefits not relevant and not within the word "consideration" in s123 of 1972 Act

***R (on the application of Lemon Land Ltd)* v *Hackney London Borough Council (QB: Admin),* [2002]** 1 **81**

Surplus land — Cancellation of auction — Land sold to tenants without being offered in open market — Whether s128(2) prevented invalidity of contract of sale — Whether council in breach of law outside protection of s128 — Whether legitimate expectation that sale would be at auction to highest bidder — Whether local authority required to offer land in open market — Whether sale to tenants valid — Held, s128(2) only applies to protect purchaser after and not before completion — s128 does not protect purchaser where council acts unlawfully apart from failing to apply for ministerial consent — No legitimate expectation of sale by auction

***R* v *Hackney London Borough Council, ex parte Structadene Ltd (QB),* [2001]** 1 **15**

LOCAL GOVERNMENT AND HOUSING ACT 1989 — See **Landlord and tenant** (Long lease — Notice to terminate)

LOCAL LAND CHARGE

Local Land Charges Act 1975

Compensation

Valuation date — Planning condition — Agricultural occupancy condition — Failure of registration authority to register and/or give notice of local land charge — Whether valuation date for compensation purposes date of discovery of planning condition — Whether date of court hearing valuation date — Whether application of "breach date" would give rise to injustices — Held, unjust to use "breach date" derived from analogous breach of tortious duty — Claimants not in position to obtain alternative property until compensation paid or assessed — Appropriate and just valuation date was date of hearing of compensation assessment

***Smith and another* v *South Gloucestershire District Council (CA),* [2002]** 3 **1**

Protection

Insolvency vesting order

Local authority registered local land charges to protect expenses of dangerous structure notices — Whether local authority with interest in local land charges entitled to vesting order in priority to freehold and registered charge — Held, expenses of local authority protected by local land charges exceeding value of property — Local authority entitled to vesting order

***Hackney London Borough Council* v *Crown Estate Commissioners (Ch),* [1996]** 1 **151**

LONDON BUILDING ACTS (AMENDMENT) ACT 1939

Owners

Statutory tenants

Party wall award — Adjoining owner — Statutory tenants — Whether statutory tenants capable of being owners entitled to protection of 1939 Act — Held, owner must have interest greater than a year — Statutory tenants not capable of being "owners" for purposes of the Act

***Frances Holland School* v *Wassef (CC),* [2001]** 2 **88**

Party wall

Award of third surveyor

Appeal from surveyor's award — Whether appellate court entitled to receive additional evidence not before surveyor — Whether appellate court's jurisdiction limited — Held, award not an award of an arbitrator but a decision sui generis — Both High Court and county court have wide powers to hear additional evidence and remake award

***Chartered Society of Physiotherapy* v *Simmonds Church Smiles (Off Ref),* [1995]** 1 **155**

Claim for damages

Works commenced and damage caused prior to party wall award — Whether building owner liable in nuisance to adjoining owner for damages continuing after Act complied with — Held, damages caused by unauthorised works — Building owner liable for continuing nuisance

***Louis* v *Sadiq (CA),* [1997]** 1 **136**

LONDON BUILDING ACTS (AMENDMENT) ACT 1939 — *continued*

Party wall — *continued*

Enforcement of award

Surveyor's party wall award — Damage to adjoining owner's property to be made good at cost of building owner — Whether adjoining owner entitled to register caution against title of building owner — Held, neither benefit of party wall award nor right to enter building owner's land interests capable of being registered by adjoining owner

***Observatory Hill Ltd* v *Camtel Investments SA (Ch),* [1997]** 1 **140**

Party wall award

Ex parte award

Adjoining owner's surveyor making ex parte award — Whether surveyor entitled to make ex parte award — Adjoining owner — Statutory tenants — Whether statutory tenants capable of being owners entitled to protection of 1939 Act — Held, ex parte award can only be made on strict compliance with requirements of s55(e) of 1939 Act

***Frances Holland School* v *Wassef (CC),* [2001]** 2 **88**

LORD CAIRNS' ACT — See **Damages** (Lord Cairns' Act); **Restrictive covenant** (Breach — Damages)

MATERIAL DISPOSAL — See **Capital gains tax** (Retirement relief — Agricultural land)

MESNE PROFITS — See **Landlord and tenant** (Forfeiture — Relief) (Trespass)

MILK QUOTA

Disposal — See also **Capital gains tax** (Retirement relief — Sale of milk quota)

Mortgage of land

Bank taking possession of land and leasing quota pending sale of holding — Whether rents and part of proceeds of sale attributable to quota attached to land — Whether recoverable by mortgagor — Held, rents and value of quota attached to land for benefit of mortgagee

***Harries* v *Barclays Bank plc (CA),* [1997]** 2 **15**

Partnership asset

Dissolution of partnership

Arbitration — Termination of farming partnership — Whether milk quota partnership asset — Whether any claim to enhanced value of holding attributable to milk quota on partnership dissolution — Held, milk quota not partnership asset — No claim to enhanced value of holding

***Davies* v *H&R Ecroyd Ltd (Ch),* [1996]** 2 **5**

Security

Relief

Milk quotas used as security for loans by lender to farming borrowers — Quota transferred to third-party agricultural tenant — Third party's landlord seeking to obtain quota by terminating tenancy — Liquidators of lender seeking protection of quota as against third party and landlord — Whether lender entitled to direct that quota be retransferred to borrowers on repayment of loans — Whether equity in land sufficient to found a retransfer — Held, borrowers had equitable right to require retransfer of quota — Court entitled to order interim protection of quota

***Swift* v *Dairywise Farms Ltd (CA),* [2001]** 3 **101**

Slom II

Eligibility

Partnership — Slom II quota and compensation withdrawn — Whether partnerships between applicants and others sufficient merging of assets for purposes of EC regulations — Whether ministry acted unreasonably — Held, insufficient merging of assets — Ministry did not act unreasonably

***R* v *Ministry of Agriculture, Fisheries and Food, ex parte St Clere's Hall Farm (QB),* [1996]** 1 **15**

MINERALS — See **Rates and rating** (Rateability — Plant and machinery)

MISREPRESENTATION — See **Landlord and tenant** (Agreement for lease — Misrepresentation); **Mortgage** (Defence — Undue influence); **Property Misdescriptions Act 1991** (False and misleading statements — Pictures and show house)

Negligent — See **Landlord and tenant** (Misrepresentation — Damages)

Negligent misrepresentation

Assessment of damages

Acquisition of development properties at auction — Misrepresentation by vendors as to status of occupier — Occupier protected under Part II of Landlord and Tenant Act 1967 — Whether normal measure of damages applying (transaction date rule) — Whether application of overriding compensatory rule displacing normal measure in favour of consequential losses — Whether residual valuations appropriate to assess value of properties in actual state — Whether 50% division of marriage value appropriate price to obtain possession against occupier — Held, normal measure of damages applying — Difference in value — Residual valuations appropriate — 50% division of marriage value

***Pankhania* v *Hackney London Borough Council (Ch),* [2004]** 1 **135**

Public house lease — See **Landlord and tenant** (Agreement to grant lease)

MISTAKE — See also **Contract** (Rectification — Unconscionable conduct); **Landlord and tenant** (Rent — Recovery of overpayment); **Quasi-contract** (Restitution — Remedy of account and payment)

Rectification

Form of relief

Sale of land — Vendor intending to retain long leases of two flats — Purchaser understanding vendor's intention — Transfer failing to reserve rights in flats to vendor — Whether continuing common intention of parties — Whether rectification of transfer appropriate — Whether vendor making inconsistent pleading of common intention — Whether common intention as to form of transfer necessary — Whether requirements of Part I of Landlord and Tenant Act 1987 preventing rectification — Held, common intention that transfer would exclude flats — Inconsistent pleading of relief not inconsistent with common intention — Form of transfer requiring rectification to give effect to equity — 1987 Act not applying to rectification circumstance

***Swainland Builders Ltd* v *Freehold Properties Ltd (CA),* [2002]** 2 **71**

MOBILE HOMES

Mobile Homes Act 1983

Loss of protection

Porch extension to mobile home — Whether mobile home losing protection — Whether protection depending upon factual issue — Whether protection depending upon construction of agreement — Whether agreement terminated in accordance with terms of agreement — Held, agreement not terminable on ground that home no longer capable of being moved

***Howard* v *Charlton (CA),* [2002]** 3 **65**

Pitch fee

Determination

Owner incurred costs on trees on mobile homes site — Whether pitch fees can be increased under standard form agreement to recover expenses — Held, actual expenses a factor which can justify fee increase.

***Walker* v *Badcock (CA),* [1997]** 2 **163**

MORTGAGE — See also **Compulsory purchase** (Compensation); **Landlord and tenant** (Leasehold Property (Repairs) Act 1938, s1 — Mortgagee in possession); **Negligence** (Mortgage loan report — Valuation) (Solicitor — Measure of damages) (Surveyors — Mortgage valuation report) (Valuers); **Receiver** (Power of sale — Duties)

Agricultural holdings — See **Agricultural holdings;**

***Walker* v *Badcock (CA),* [1997]** 2 **163**

Clog

Enforcement

Legal charge — Later agreement requiring chargor to transfer interest — Whether duress — Whether unconscionable bargain — Whether clog on equity of redemption — Held, second agreement not invalidated by duress and not unconscionable bargain — Second agreement related to, or part of, legal charge and clog on equity of redemption

***Jones* v *Morgan (CA),* [2002]** 1 **125**

Compensation — See **Compensation for acquisition of land** (Mortgage — Date of redemption)

Contributory negligence — See **Negligence** (Valuers)

Defence

Undue influence

Pleading — Joint owners — Joint charge over matrimonial home — "All moneys" charge — Undue influence — Whether adequate pleading that mortgagee had notice of undue influence — Whether burden of proof in relation to constructive notice falls on mortgagor or mortgagee — Held, pleadings alleged all facts that would, as a matter of law, give rise to constructive notice of the alleged undue influence and misrepresentation — Burden of proof initially on mortgagor

***Barclays Bank plc* v *Boulter (HL),* [1999]** 3 **88**

MORTGAGE — continued

Forfeiture
Relief
Mortgagee of lease informed by landlord of intention to grant new lease — Whether mortgagee of lease entitled to relief from forfeiture — Whether relief to mortgage subject to new lease — Held, relief granted by reversionary lease to new lease.
***Bank of Ireland Home Mortgages* v *South Lodge Developments (Ch),* [1996]** 1 **91**

Interest
Simple or compound
Deed of priority — Capital sum and interest thereon — Whether deed giving priority to sum that included compound interest — Held, "interest" represented whatever interest payable by borrower — Compound interest therefore "interest"
***Whitbread plc* v *UCB Corporate Services Ltd (CA),* [2000]** 3 **60**

Limitation period
Arrears
Mortgaged property sold under power of sale — Subsequent claim for arrears of principal and interest — S20 of Limitation Act 1980 — Whether claim lying in simple contract under loan agreement — Whether claim lying by deed under mortgage — Whether 12-year limitation period applying to recovery of principal and interest — Whether six-year limitation period applying to recovery of arrears of interest — Whether limitation period in s20 applying to recovery of principal and interest — Held, claims for mortgage debt governed by s20 of Limitation Act 1980 — S20 applying where power of sale exercised and mortgagee bringing later proceedings seeking recovery of arrears of principal and interest — 12-year limitation period to sue for principal of debt — Six-year limitation period to sue for interest
***Bristol & West plc* v *Bartlett; Halifax plc* v *Grant; Paragon Finance plc* v *Banks (CA),* [2003]** 2 **85**

Mortgage indemnity insurance
Lending agreement
Whether proviso in policy a condition precedent — Whether lender's agent strictly liable for breach of lending criteria — Whether lender contributorily negligent in breach of criteria — Whether failed to mitigate losses — Held, proviso was a strict condition precedent — Damages payable were liquidated, so no room for mitigation of loss
***Den Danske Bank A/S* v *Skipton Building Society (QB),* [1998]** 1 **155**

Mortgage loan report — See **Negligence** (Limitation Act — Knowledge required to bring action) (Mortgage loan report) (Solicitors — Mortgage loan report) (Valuers — Mortgage loan report)**Mortgage of leasehold interest**
Mortgagee in possession
Forfeiture — Relief — Terms of relief — Whether mortgagee in possession entitled to serve counternotice under s1 of Leasehold Property (Repairs) Act 1938 — Whether landlord estopped from relying on time-limit for relief from forfeiture under s139 of County Courts Act 1984 — Held, mortgagee in possession entitled to protection of 1938 Act — Landlord estopped from relying on time-limits under s139 of County Courts Act 1984
***Target Home Loans Ltd* v *Iza Ltd (CC),* [2000]** 1 **23**

Mortgagee's power of leasing
Possession
First mortgagee granting lease — Second mortgagee seeking possession — Whether lease by first mortgagee binds second morgagee — Whether second mortgagee entitled to order for possession — Proper terms of possession order — Held, lease granted by first mortgagee binds second mortgage — Second mortgagee entitled to possession order subject to rights of first morgagee and its tenant
***Berkshire Capital Funding Ltd* v *Street (CA),* [1999]** 2 **92**

Mortgagees' possession
Milk quotas
Bank taking possession of land and leasing quota pending sale of holding — Whether rents and part of proceeds of sale attributable to quota attached to land — Whether recoverable by mortgagor — Held, rents and value of quota attached to land for benefit of mortgagee
***Harries* v *Barclays Bank plc (CA),* [1997]** 2 **15**

Non-transaction mortgage loan — See **Negligence** (Limitation Act)
Order for possession — See also **Trusts of Land and Appointment of Trustees Act 1996**
Warrant for possession
Whether order for possession can be stayed following execution of warrant for possession — Whether rehearing can be ordered of application for possession — Whether mortgagor can impugn mortgaged interest — Held, powers under s36 of Administration of Justice Act 1970 cannot be exercised after execution of warrant for possession — Mortgagor not entitled to impugn his security
***National Provincial Building Society* v *Ahmed (CA),* [1995]** 2 **127**

Possession
Tenancy
Agricultural holding mortgaged to bank — Mortgagors granting agricultural tenancy to company — Mortgagors defaulting — Bank entitled to possession — Whether tenancy a transaction at undervalue — Whether bank entitled to set aside tenancy under s423 of Insolvency Act 1986 — Whether bank entitled to possession as against agricultural tenant — Held, scheme for review of rent leaving tenant with greater consideration provided to it by mortgagors than tenant provided mortgagors — Transaction at undervalue — Tenancy set aside under s423 of 1986 Act
***Barclays Bank plc* v *Bean (Ch),* [2004]** 3 **71**

Possession proceedings
Limitation period
Claim for possession — Expiration of limitation period — Application to amend pleadings to add claim for money judgment — Whether mortgagee entitled to add money claim to possession proceedings after expiration of limitation period — Whether new remedy a new claim under s35(5)(a) of Limitation Act 1980 — Whether new claim raised new factual issues — Discretion — Held, additional remedy not a new claim under s35(5)(a) — If a new claim, based on substantially the same facts as pleaded — Discretion exercised under RSC Ord 20 r 5(5)
***Lloyds Bank plc* v *Rogers (CA),* [1999]** 3 **83**

Power to grant leases
Consent to lease
Duty of good faith — Modification of statutory power of leasing — Mortgagor applying for consent to lease — Whether mortgagee owes duty to mortgagor properly to consider application for consent — Whether duty of good faith owed — Held, no such duties by mortgagee on pleaded case — Alleged duties unlikely to be owed except in extreme case — Pleadings struck out
***Starling* v *Lloyds TSB Bank Ltd (CA),* [2000]** 1 **101**

Power of sale
Duty of mortgagee
Limitation period — Claim that mortgagee's expenses on sale excessive and sale made at an undervalue — Whether claim statute–barred by expiration of limitation period — Whether six–year limitation appropriate — Whether proper distribution of proceeds of sale — Held, mortgagee owes duty in equity to mortgagor in relation to exercise of powers of sale — Six–year limitation period applicable — Claim statute–barred — Mortgagee entitled to pay proceeds to subsequent chargee
***Raja* v *Lloyds TSB Bank plc (Ch),* [2001]** 2 **78**

Possession
Whether jurisdiction at common law to postpone execution of warrant for possession until completion of sale — Whether jurisdiction properly exercised — Held, there is limited jurisdiction to postpone possession at common law in rare cases — Postponement should not have been ordered
***Cheltenham & Gloucester plc* v *Booker (CA),* [1997]** 1 **142**

Quasi-contract — See **Quasi-contract** (Restitution — Remedy of account and payment)
Receiver
Duty of care
Pig farm — Duties owed to mortgagor — Whether receiver owes duty of care to mortgagor — Extent of duties owed to mortgagor by receiver appointed by mortgagee — Held, receiver owes equitable duty of care to mortgagor
***Medforth* v *Blake (CA),* [1999]** 2 **75**

Sale
Power of sale
Mortgagee's duty to obtain best price reasonably obtainable — Whether breach of duty — Whether use of bracket of valuations relevant in establishing breach of duty — Whether valuer negligent in failing to advise additional marketing — Held, use of bracket of valuations must be available to court in determining whether mortgagee in breach of his duty
***Michael* v *Miller (CA),* [2004]** 2 **151**

NATIONAL HEALTH TRUST HOSPITAL — See **Restrictive covenant** (Enforcement — Statutory powers)
NATIONAL TRUST — See Rates and rating (Assessment — Historic houses)

NEGLIGENCE

See **Estate agents** (Duties — Protection of vendor's interest)

Architects

Tortious liability

Chain of causation — Proximity — Patent defect — Valley gutter — Shortfall in design of valley gutter and lack of overflows — Surveyors inspecting prior to lease — Whether cause of action against architects — Whether lack of overflows patent defect — Whether chain of causation broken by discoverablity of patent defect — Whether absence of overflows of causative significance —Held, discoverability of patent defect broke chain of causation — Lack of overflows of causative significance — Lack of overflows patent defect discoverable by surveyors.

***Baxall Securities Ltd* v *Sheard Walshaw Partnership (CA),* [2002]** 1 **73**

Damages: measure and quantification — See also **Nuisance** (Sewers — Human Rights Act 1998); **Negligence** (Mortgage loan report — Damages)

Mitigation

Negligent structural survey — Purchase of long lease of maisonette — Landlord remedying defect at plaintiffs' instigation — Whether damages payable on Philips v Ward principles — Whether plaintiffs suffered any loss — Held, damages payable based on difference in value with and without defects at date of purchase — Mitigation principles did not apply

***Gardner* v *Marsh & Parsons (CA),* [1997]** 1 **111**

Duty of care

Local authority enquiries

Sale of land — Local authority providing reply in relation to public highway and highways maintainable at public expense — Local authority replying that public highway was a highway maintainable at public expense — After purchaser acquiring property, local authority correcting reply and stating highway not part of maintained public highway — Held, no duty of care owing where purpose of transaction not disclosed — Council unaware of purchaser's intention to develop — If duty owed, there was no negligence

***Gooden* v *Northamptonshire County Council (CA),* [2002]** 1 **137**

Duty of care and liability — See also **Negligence** (Mortgage loan report — Damages) (Mortgage loan report — Duty of care)

Disclaimer

Valuation report — Whether disclaimer effective against unknown lender — Whether duty of care negatived — Whether disclaimer satisfied reasonableness test under Unfair Contract Terms Act 1977 — Held duty negatived by disclaimer — Reasonableness test satisfied

***Omega Trust Co Ltd* v *Wright Son & Pepper (CA),* [1997]** 1 **120**

Estate agents

Exercise of mortgagee's power of sale — Whether estate agents instructed by mortgagee and receivers carried out negligent marketing — Whether opinion of expert on ultimate question of negligence admissible — Held, no negligent marketing — Expert's opinion on ultimate question of negligence

***Routestone Ltd* v *Minories Finance Ltd; Routestone Ltd* v *Bird (Ch),* [1997]** 1 **123**

Expert determination at rent review

Whether wrong to use overall rather than zoning method — Whether wrong to reject assumption that hypothetical tenant would subdivide — Whether negligent in relation to meaning and effect of Iceland Frozen Foods plc v Starlight Investments Ltd [1992] 1 EGLR 126 — Held, expert not negligent — Not negligent to use overall method or reject assumption that hypothetical tenant would subdivide — Not negligent in relation to Iceland

***Lewisham Investment Partnership Ltd* v *Morgan (Ch),* [1997]** 2 **150**

Landslip

Landowner claiming damages against adjoining owner — Whether defendant owner aware of hazard — Whether defendant in breach of duty of care — Held, defendant knew or ought to have been aware of landslip hazard — Breach of duty of care — Liable to landowner

***Holbeck Hall Hotel Ltd* v *Scarborough Borough Council; Scarborough Borough Council* v *Geotechnical Engineering (Northern) Ltd (Off Ref),* [1997]** 2 **213**

Letting agents

Owner instructing agents to find tenants on basis that possession could be recovered at term date — Letting under Rent Act 1977, Sched 15, Case 11 — Whether agents negligent — Held, agents negligent in failing to warn owner that possession under Case 11 depended on owner requiring premises as his residence

***Hellings* v *Parker Breslin Estates (Ch),* [1995]** 2 **99**

Reference for guarantor

Loan secured by charge and personal guarantee — Net worth of guarantor misstated — Whether duty of care owed — Whether breach of duty of care in giving advice or information — Whether contributory negligence — Held, duty of care owed — Breach of duty

***HIT Finance Ltd* v *Cohen Arnold & Co (QB),* [1998]** 1 **140**

Solicitors

Solicitor acting for purchaser and lender — Solicitors becoming aware of facts common to both clients — Whether solicitor under any duty to inform lender of simultaneous sale to vendor at price considerably below mortgage valuation — Held, solicitor owed duty to inform lender of such fact

***Mortgage Express Ltd* v *Bowerman & Partners (CA),* [1996]** 1 **126**

Expert determination of rent of restaurant premises at review — Whether valuer negligent in relying on retail comparables — Held, landlord chooses to suffer discount in letting for restaurant purposes but rents of comparable retail premises relevant

***Zubaida* v *Hargreaves (CA),* [1995]** 1 **127**

Turnover valuation of restaurant premises for mortgage purposes — Use of figures for turnover provided by borrowers — Whether valuer negligent in use of chosen figures and variables — Held, no negligence found

***Craneheath Securities Ltd* v *York Montague Ltd (CA),* [1996]** 1 **130**

Valuation — 46-year leasehold interest — Rental values — Yield rate — Whether advice on yield rate negligent without qualification — Whether rental values negligent — Appeal from [1998] 2 195 — Held, no negligence on rental values — Negligent to give yield rate without qualification

***Merivale Moore plc* v *Strutt & Parker (CA),* [1999]** 2 **171**

Limitation Act

Accrual of cause of action

Limitation Act 1980 — Survey report — Later discovery of subsidence — Whether plaintiffs entitled to rely on extended period of limitation under s14A — Whether plaintiffs had requisite knowledge — Held, plaintiffs had requisite knowledge and failed to commence proceedings within three years — Claim statute-barred

***Higgins* v *Hatch & Fielding (CA),* [1996]** 1 **133**

Compromised claim

Housebuyer's report — Claim for dry rot compromised — Later claim for structural defects — Whether period of limitation under s14 of Limitation Act 1980 ran from date of knowledge of surveyors' in relation to dry rot — Whether terms of compromise precluded further claim — Held, there was one single and indivisible cause of action arising out of one negligent act, the making of a single report — Present proceedings outside limitation period — Terms of compromise precluded all further claims

***Hamlin* v *Edwin Evans (CA),* [1996]** 2 **106**

Date of damage

Solicitors retained for purchase of lease already extended under Leasehold Reform Act 1967 — Whether damage suffered on acquisition or later date when entitlement to enfranchise arose — Held, damages suffered on acquisition — Claim statute-barred

***Sullivan* v *Layton Lougher & Co (CA),* [1995]** 2 **111**

Knowledge required to bring action

Mortgage valuation report omitted reference to serious defect — Writ issued after principal limitation period — Limitation Act 1980, s14A — Whether plaintiffs had knowledge to commence proceedings — Whether knowledge of plaintiffs included that of solicitors — Held, limitation period expired — Plaintiff bound by knowledge of solicitors

***Heathcote* v *David Marks & Co (QB),* [1996]** 1 **123**

Latent damage

Claim in negligence against valuers for failing to report Woolaway construction of house — Whether plaintiffs entitled to extension of limitation period under s14A of Limitation Act 1980 — Held, letter written by plaintiffs one day earlier than expiration of three-year period before issue of writ made clear plaintiffs had sufficient knowledge of facts for purpose of s14A

***Spencer-Ward* v *Humberts (CA),* [1995]** 1 **123**

Non-transaction mortgage loan

Lender would not have made loan had correct valuation been made — Writ issued more than six years after alleged negligent valuation — Whether statute-barred — Held, losses recoverable under principles in Swingcastle v Gibson — No losses incurred until after commencement of limitation period and not statute-barred

***First National Commercial Bank plc* v *Humberts (CA),* [1995]** 1 **142**

NEGLIGENCE — *continued*

Limitation period

Extension of period under s32(1)(b) Limitation Act 1980

Professional negligence — Whether limitation period capable of being extended by s32(1)(b) of Limitation Act 1980 — Whether concealment of error and breach of duty must be deliberate — Held, concealment means only that commission of act or omission must be intentional and involve breach of duty

***Brocklesby* v *Armitage & Guest (CA),* [2001]** 1 **67**

Local authority

Breach of statutory duty

Fire — Windows — Lockable window catches with removable keys — Inability to escape during fire through locked windows — Whether council landlord liable to occupiers for lockable windows with removable keys — Held, landlord not liable for using key-removable window locks

***Adams* v *Rhymney Valley District Council (CA),* [2000]** 3 **25**

Mortgage loan report

Contributory negligence

Non-status loan — Self-certification by borrowers — Negligent valuation — Whether contributory negligence by building society — Whether damages should be reduced by contributory negligence — Held, building society negligent in making non-status loan with inadequate loan to value cushion of 75% — Damages for negligent valuation not reduced by contributory negligence

***Coventry Building Society* v *William Martin & Partners (QB),* [1997]** 2 **146**

Restaurant premises — Profits method valuation — Whether valuers negligent — Whether responsible for plan of land valued — Whether lenders contributorily negligent — Held, valuers liable on net profit to turnover percentage — Lender contributorily negligent for not obtaining accounts and bank statements

***UCB Bank plc* v *David J Pinder plc (Off Ref),* [1998]** 2 **203**

Whether valuation an open market value — Whether mortgagee bank contributorily negligent in failing to reassess risk and in giving unclear instructions — Held, open market valuation provided — Bank contributorily negligent and damages reduced by 25%.

***South Australia Asset Management Corporation* v *York Montague Ltd (QB),* [1995]** 2 **219**

Damages

Negligent valuation — Damages — Whether interest on damages to run from date of valuation or date when property sold and security realised — Held, interest runs from date loss crystallised being date property sold

***Coventry Building Society* v *William Martin & Partners (QB),* [1997]** 2 **146**

Valuer failed to observe defect in concrete purlin in Bison-system terrace house otherwise observable in roof space — Plaintiffs lost sale, suffered job dismissal, temporary housing accommodation before obtaining council tenancy — Substantial arrears owing to building society and loss of equity — Held, damages awarded included worry, distress and physical inconvenience

***Ezekiel* v *McDade (CA),* [1995]** 2 **107**

Valuer negligent in failing to report missing chimney breast but not negligent in failing to observe other defects — Whether damages based on diminution in value should assume all defects known to hypothetical purchaser — Held, all defects would have been discovered by comprehensive survey on hypothetical purchaser being alerted by missing chimney breast — Damages should be assessed accordingly

***Sneesby* v *Goldings (CA),* [1995]** 2 **102**

Whether defendant's negligent valuation caused damage — Whether plaintiff contributorily negligent — Held, reliance and causation established at trial — Lender contributorily negligent in failing to question difference in valuations received

***Cavendish Funding Ltd* v *Henry Spencer & Sons Ltd (CA),* [1998]** 1 **104**

Duty of care

Valuation report to mortgage lender — Proper range of values or bracket — Liability — Held, no liability — 15% acceptable range for reasonably competent surveyor — True value within 13% of reported value

***BNP Mortgages Ltd* v *Barton Cook & Sams (Off Ref),* [1996]** 1 **239**

Whether valuer negligent in failing to report missing chimney breast — Whether general duty on making limited inspection for mortgage report to look into cupboards — Held, issue of fact not involving any general principle — Valuer negligent

***Sneesby* v *Goldings (CA),* [1995]** 2 **102**

Residual valuation

Valuation report of development site — Whether negligent if valuation within acceptable range of values — Held, valuers negligent — Failure to implement instructions to inspect — Bank would not have loaned if report had not been negligent

***Nykredit Mortgage Bank plc* v *Edward Erdman Group Ltd (CA),* [1996]** 1 **119**

Valuation

Value on 90-day sale by mortgagee in possession — Whether presumption that discount applies to sale on such basis — Held, no presumption — Whether any discount depends on evidence — Trial court entitled to reach conclusion on evidence

***UCB Corporate Services Ltd* v *Halifax (SW) Ltd (CA),* [2000]** 1 **87**

Solicitors

Assessment of damages

Negligent conveyance advice — Plaintiffs paying unlawful premiums — Law later changed making premiums lawful — Whether plaintiffs entitled to diminution in value basis and recovery of premiums — Held, plaintiffs had suffered no loss by date of trial — Diminution in value basis inappropriate

***Kennedy* v *Emden (CA),* [1997]** 2 **137**

Solicitor failed to report fraudulent creation of second mortgage — Whether damages should be assessed at date of cause of action — Whether damages nominal — Whether proceeds of mortgage indemnity policy should be allowed for in damages assessment — Whether costs and fees of internal lawyers and estate agents recoverable — Held, damages properly assessed at date when property sold — Damages not nominal — Mortgage indemnity policy proceeds disregarded — Internal costs recoverable

***Portman Building Society* v *Bevan Ashford (a firm) (CA),* [2000]** 1 **81**

Conveyancing

Breach of trust — Limitation — S14A of Limitation Act 1980 — Whether solicitor entitled to rely on admitted negligence not alleged by plaintiff as defence to s14A — Whether admitted negligent act same or different cause of action as that relied on by the plaintiff — Held, negligence admitted by defendants separate cause of action, defence of limitation fails

***Birmingham Midshires Building Society* v *Wretham (QB:TCC),* [1999]** 1 **69**

Failure of solicitors to make optional inquiry of local authority relating to rights of way — Whether expert evidence of conveyancing practice should be ordered — Held, expert evidence ordered

***May* v *Woollcombe Beer & Watts (QB),* [1998]** 3 **94**

Foreign property — Whether English solicitor liable for failure to conduct search — Whether liable for handing over purchase money without confirmation of clear search — Held, solicitor liable for failure to ensure search carried out and for obtaining confirmation from Spanish lawyer that purchase money could be handed safely to vendor

***Gregory* v *Shepherds (a firm) (CA),* [2000]** 3 **49**

Damages

Chance — Agricultural land — Negligent conveyancing advice — Late completion — Failure to advise on service of completion notice — Measure of damages — Whether loss of profits on lamb crop — Chance of profits if completion served — Whether loss of a chance of profits — Held, true measure of loss restricted to chance of profit on 1994 lamb crop — Chance depending on consequence of service of completion notice — 40% chance that profit would have been made

***Williams* v *Glyn Owen & Co (CA),* [2004]** 1 **67**

Part II of Landlord and Tenant Act 1954 — Solicitors failing to make application for grant of new tenancy — Claimant tenant seeking damages — Assessment of damages — Annual value of premises — Value of lease contracted-out of Act — Appropriate multiplier — Addition for loss of disturbance compensation — Whether relocation costs payable — Held, £315,000 damages awarded — multiplier of 2.75 applied to annual profit figure — Loss of compensation added — Relocation costs disallowed

***Aran Catering Ltd* v *Stepien Lake Gilbert & Paling (Ch),* [2002]** 1 **6**

Landlord and tenant

Tenant's option to determine lease — Condition for exercise of option — Condition requiring payment of £11,500 prior to termination date — Solicitor retained to serve termination notice — Whether solicitor owing duty to warn tenant to make payment of £11,500 prior to termination date — Whether breach of duty — Held, duty to warn within retainer — Solicitor in breach of duty of care

***Credit Lyonnais* v *Russell Jones & Walker (Ch),* [2002]** 2 **65**

Mortgage loan report

Restaurant premises — Defective title — Plan excluded additional land — Whether lenders contributorily negligent — Held, solicitors entitled to rely on valuers' answer as to land valued

***UCB Bank plc* v *David J Pinder plc (Off Ref),* [1998]** 2 **203**

NEGLIGENCE — *continued*
Solicitors — *continued*

Surveyors
Structural survey — Valuation approach where no comparables of property as it should have been described — Whether purchase price best evidence of market value — Whether deductions for repair cost increases and betterment — Held, repair costs relevant where no comparables — Proper to make adjustments for increase in repair costs and for betterment
***Oswald* v *Countrywide Surveyors Ltd (CA),* [1996]** 2 **104**

Valuers
Valuations for mortgage lenders — Negligence — Assessment of damages — Whether valuers liable for fall in property market — Appeal from [1995] 1 129 — Held, valuers' liability in contract and tort limited to consequences of lenders having less security than they believed — Damages were difference between negligent valuation and correct valuation
***South Australia Asset Management Corporation* v *York Montague Ltd; United Bank of Kuwait plc* v *Prudential Property Services Ltd; Nykredit Mortgage Bank plc* v *Edward Erdman Group Ltd (HL),* [1996]** 2 **93**

Shared ownership scheme
Negligence — Solicitors — Mortgage loan — Duty of care — Summary judgment — Solicitors acting for lender in shared ownership scheme — Whether solicitors failed to obtain written approval to mortgagee protection clauses — Whether solicitors have triable defence — Held, solicitors had triable defence
***Halifax plc* v *Gould & Swayne (CA),* [1998]** 3 **177**

Surveyors
Damages
Aircraft noise — Whether surveyor giving negligent advice about aircraft noise liable for damages — Whether any physical inconvenience or discomfort — Exceptional contracts in respect of which damages for annoyance payable — Whether survey contract an exceptional contract to provide pleasure, relaxation or peace of mind — Held, contract was a contract to provide advice relating to pleasure, relaxation or peace of mind — Aircraft noise capable of physical inconvenience or discomfort — Non-pecuniary damages awarded.
***Farley* v *Skinner (HL),* [2001]** 3 **57**

Evidence
Structural survey — Chartered surveyor — Whether court entitled to find negligence by chartered surveyor on evidence of structural engineer — Whether evidence of structural engineer admissible — Held, court cannot normally find negligence of professional on evidence of witness of different profession — Evidence inadmissible
***Sansom* v *Metcalfe Hambleton & Co (CA),* [1998]** 2 **103**

Liability
Advice relating to letting of premises — Contribution notice — Failure to include "upwards-only" rent review provision in lease — Whether surveyors liable to contribute to damages paid by solicitors on claim by landlord — Held, surveyors liable to make 20% contribution
***Theodore Goddard* v *Fletcher King Services Ltd (QB),* [1997]** 2 **131**

Limitation
Tort — Negligent survey — Whether cause of action accrues at contract date or completion — Held, action accrued at date of contract
***Byrne* v *Hall Pain & Foster (CA),* [1999]** 1 **73**

Mortgage valuation report
Industrial building system — Practice of valuation surveyors — Published warnings of BRE relating to building system — Whether surveyor required to warn purchaser of construction system and potential future service charge increases — Held, surveyor negligent in failing to warn of building system and potential service charge increases
***Izzard* v *Field Palmer (a firm) (CA),* [2000]** 1 **177**

Valuation
Independent expert
Rent review determination — Retail premises — Whether expert negligent in determining level of headline rent — Held, expert not negligent — Claimant must show that defendant's determination was one that no reasonably competent surveyor could have reached — Not established that no competent surveyor would have dealt with evidence in way that defendant did
***Currys Group plc* v *Martin (QB),* [1999]** 3 **165**

Valuers — See also **Expert, appointment of** (Negligent valuation); **Negligence** (Damages: measure and quantification — Valuers) (Duty of care and liability — Valuers) (Limitation Act— Latent damage) (Mortgage loan report — Damages) (Mortgage loan report — Duty of care)

Contributory negligence
Valuation for non-status mortgage — Contributory negligence — Law Reform (Contributory Negligence) Act 1945 — Meaning of "damage" — Whether damage overall transaction loss or part of loss attributable to negligent valuation — Appeal from [1998] 1 108 — Held, contributory negligence percentage deducted from total loss — Contributory negligence percentage not deducted from amount of overvaluation under the principle in *South Australia Asset Management Corporation* v *York Montague Ltd* [1996] 2 EGLR 93
***Platform Home Loans Ltd* v *Oyston Shipways Ltd (HL),* [1999]** 1 **77**

Duty of care
Common law duty of care — Bank holding debenture as security for loan to borrower — Borrower in turn holding charge on properties of third party as security for loans — Receiver for bank appointed on default of borrower — Receiver selling charged properties — Valuer advising receiver and bank — Claim by third party that sale at undervalue — Whether valuer owing common law duty of care to third party — Whether third party having claim for breach of duty in equity against receiver — Whether receiver liable for negligence of valuer — Held, valuer owing no duty of care to third party — Receiver owing duty in equity to third party — Receiver liable to third party for negligence of valuers
***Raja (administratrix of the estate of Raja (deceased))* v *Austin Gray (a firm) (CA),* [2003] 1** 91

Common law duty of care — Bank holding debenture as security for loan to borrower — Borrower in turn holding charge on properties of third party as security for loans — Receiver for bank appointed on default of borrower — Receiver selling charged properties — Valuer advising receiver and bank — Claim by third party that sale at undervalue — Whether valuer owing common law duty of care to third party — Held, reasonable and in public interest for professionals to exercise skills with reasonable competence — Valuer owed common law duty of care to third party
***Raja* v *Austin Gray (QB),* [2002]** 3 **61**

Duty of care in tort
Mortgage valuation — Interest on damages — Accrual of cause of action in tort — Date for commencement of interest on damages — Whether plaintiff suffered damages at date of loan transaction or date of realisation of property — Held, action accrued when plaintiff suffered loss — Loss suffered when the lender can show that he is worse off than he would have been if the security had been worth the sum advised by the valuer; the comparison is between the lender's actual position and what it would have been if the valuation had been correct — No need to wait until lender realises security
***Nykredit Mortgage Bank plc v Edward Erdman Group Ltd (No 2) (HL),* [1998]** 1 **99**

Valuer
Liability
Valuation provided to rugby football club — Club transferring land on basis of valuation — Whether valuer liable for negligent valuation — Measure of damages — Prospect of planning permission — *Red book* practice on instructions — Whether club would have pursued planning possibilities and would have sold at non-negligent valuation — Mitigation — Contributory negligence — Held, non-negligent valuation would have advised on better than 50% chance of planning permission — Damages payable — No breach of duty to mitigate — No contributory negligence
***Montlake* v *Lambert Smith Hampton (QBD),* [2004]** 3 **149**

Limitation
Limitation period — Action founded in tort — Whether loss and accrual of cause of action within limitation period — Burden of proof relating to limitation period — Whether burden of proof solely on claimant — Value of mortgagors' personal covenant — Whether "falling market" graph showed loss within limitation period — Held, claimant must prove loss within limitation period — Where rebuttable presumption raised, then for defendant to rebut presumption — Value of personal covenant correctly found — "Falling market" graph rejected
***DNB Mortgages* v *Bullock & Lees (CA),* [2000]** 1 **92**

Mortgage loan report
Property valued different from property the subject-matter of loan — Whether valuers liable for negligent valuation — Whether negligent valuation causative of loss — Held, if the difference between the property valued and the property used as security is significant, the valuer will not be liable for negligent valuation
***Western Trust & Savings Ltd* v *Strutt & Parker (CA),* [1998]** 3 **89**

Restaurant premises — Profits method valuation — Whether valuers negligent — Whether responsible for plan of land valued — Whether lenders contributorily negligent — Held, valuers liable on net profit to turnover percentage — Lender contributorily negligent for not obtaining accounts and bank statements
***UCB Bank plc* v *David J Pinder plc (Off Ref),* [1998]** 2 **203**

NEGLIGENCE — *continued*
Valuer — *continued*

Mortgage valuation
Negligent survey — Valuer an employee — Whether employee owed duty of care to purchasers — Claimant had half interest in property — Whether claimant only entitled to half of the amount of damages and interest — Whether joint purchaser can be added as second claimant — Held, employee assumed responsibility and owed duty of care — Joint purchaser added as second claimant — Valuer liable to both claimants for full loss
***Merrett* v *Babb (CA),* [2001]** 1 **145**

Limitation Act 1980 — Whether loss accrued more than six years before writ — Whether period of limitation extended by s14A of 1980 Act — Whether plaintiff had material that suggested negligent valuation more than three years before writ — Held, loss under *Nykredit Mortgage Bank plc* v *Edward Erdman Group Ltd (No 2)* [1998] 1 EGLR 99 principles accrued more than six years before writ — But s14A applied to extend time — Valuation estimates of unqualified persons not material excluding s14A
***Mortgage Corporation* v *Lambert & Co (Ch),* [1999]** 3 **59**

Undervaluation — Duties of valuer — Sale of mortgage by mortgagee at discount — Whether mortgagee's loss crystallised upon sale of mortgage — Whether valuer's liability for breach of duty of care limited to loss on sale of property on realisation of security — Whether valuer's liability included loss on sale of mortgage — Held, valuer not liable for loss by mortgagee on sale of mortgage at a discount
***Europe Mortgage Co Ltd* v *GA Property Services Ltd (QB),* [1999]** 3 **53**

Procedure
Valuation report — Interrogatories required defendant valuer to provide details of assumed yield, comparables and basis of annual rent — Whether defendant required to answer interrogatories — Held, interrogatories not premature and information might not be provided from other sources — Defendant valuer required to provide answers
***UCB Bank plc* v *Halifax (SW) Ltd (QB),* [1996]** 2 **102**

NOTICES — See **Agricultural holdings** (Notice to quit — Case D) (Notice to quit — Head tenancy); Housing Act 1988 (Assured shorthold tenancy); **Landlord and tenant** (Assured tenancy — S20 notice) (Assured tenancies — Notice of rent increase) (Housing Act 1988 — Assured shorthold tenancy) (Joint tenancy — Notice to quit) (Landlord and Tenant Act 1954 — Notices) (Landlord and Tenant Act 1954 Part II) (Landlord and Tenant Act 1927) (Lease — Notice) (Long lease — Notice to terminate) (Rent review — Trigger notice) (Landlord and Tenant Act 1954 Part II — Counternotice); **Leasehold enfranchisement** (Leasehold Reform Act 1967 — Enfranchisement notice) (Leasehold Reform, Housing and Urban Development Act 1993 — Initial notice under s42) (Leasehold Reform, Housing and Urban Development Act 1993 — Lease extension) (Leasehold Reform, Housing and Urban Development Act 1993 — Notice claiming new lease)
Break clause — See **Landlord and tenant** (Break clause) (Landlord and Tenant Act 1954 — Request for new tenancy)(Landlord and Tenant Act 1954 Part II — Notice to terminate tenancy) (Notice to terminate — Validity)
Notice to complete — See **Vendor and purchaser** (Contract for sale of land — Notice to complete)
Notice to quit — See **Agricultural holdings** (Notice to quit); **Landlord and tenant** (Break clause) (Landlord and Tenant Act 1954 Part I — Validity of s4 notice) (Notices)
Notice to treat — See **Compensation for acquisition of land** (Notice to treat — Withdrawal)
Purchase notice — See **Landlord and tenant** (Landlord and Tenant Act 1987 — Sale of freehold); **Town and country planning** (Purchase notice)
Service — See also **Agricultural holdings; Joint tenants** (Severance — Notice); **Landlord and tenant** (Landlord and Tenant Act 1987 — Landlord's address for service of notices) (Leasehold enfranchisement)(Notice to quit)
S196(4) of Law of Property Act 1925
Landlord and tenant — Rent review — Trigger notice — "Ordinary course of post" — Whether recipient required for deemed service — Whether *Stephenson & Son* v *Orca Properties Ltd* [1989] 2 EGLR 129 correctly decided — Whether implied term that recipient of notice precluded from avoiding service — Held, no requirement in deeming provision in s196(4) that no service without available recipient — Question of fact when service would be effected in the "ordinary course of post" — No implied term relating to the avoidance of service
***WX Investments Ltd* v *Begg (Ch),* [2002]** 3 **47**

Statutory repair notice — See **Landlord and tenant** (Landlord and Tenant Act 1987 — Sale of freehold)
Termination of tenancy — See **Landlord and tenant** (Landlord and Tenant Act 1954 Part II — Notice to terminate tenancy)
Trigger notice — See **Landlord and tenant** (Rent review clause — Time-limits)
Validity — See **Landlord and tenant** (Break clause) (Break clause — Estoppel) (Leasehold enfranchisement — Leasehold Reform, Housing and Urban Development Act 1993); **Leasehold enfranchisement** (Leasehold enfranchisement — Leasehold Reform, Housing and Urban Development Act 1993)

NUISANCE — See also **Landlord and tenant** (Covenant for quiet enjoyment — Rubbish chute) (Secure tenancy — Suspended possession order); **Public health** (Statutory nuisance)
Compensation — See **Compensation** (Land Compensation Act 1961 — Article 4 direction)
Flooding
Local highway authority
Culvert — Culvert not nuisance when constructed — Culvert becoming nuisance by reason of later developments — Whether local highway authority liable for flood damage — Held, decision in *Radstock Co-operative & Industrial Society Ltd* v *Norton—Radstock Urban District Council* [1967] Ch 1094 followed — Local authority did not construct or adapt culvert — Local authority not liable in nuisance
***Bybrook Barn Garden Centre Ltd* v *Kent County Council (QB),* [2000]** 2 **63**

Landlord and tenant
Quiet enjoyment
Nuisance — Inadequate sound insulation between flats — Whether breach of landlord's covenant for quiet enjoyment — Whether landlord liable in nuisance — Held, no breach of covenant for quiet enjoyment — Landlord not liable in nuisance
***Southwark London Borough Council* v *Mills (HL),* [1999]** 3 **35**

Neighbours — See **Housing** (Secure tenancy — Possession claim)
Quiet enjoyment — See **Landlord and tenant** (Quiet enjoyment — Breach of covenant) (Repudiation — Breach of covenant)
Proprietary interest
Related companies
Claim in nuisance for disturbance, noise and vibration — Property occupied by subsidiary companies contributing to costs through inter-company accounting — Whether related companies have sufficient proprietary interest to pursue nuisance actions — Held, claimants had insufficient proprietary interest to maintain claims in nuisance and negligence
***Butcher Robinson & Staples Ltd* v *London Regional Transport (QB: TCC),* [1999]** 3 **63**

Sewers
Human Rights Act 1998
Nuisance — Flooding — Liability of water undertaker — Drains and sewers — Statutory duty to provide effectual system of drainage — Non-feasance — Negligence — Nuisance — *Rylands* v *Fletcher* — Whether duty owed under *Leakey* v *National Trust* [1980] 1 QB 405 — Whether defence of statutory authority — Whether failure to act incompatible with European Convention on Human Rights — Held, no liability at common law — Liability arising under s6 of Human Rights Act 1998 — Measure of damages on difference in value basis
***Marcic* v *Thames Water Utilities (QB:TCC),* [2001]** 3 **111**

Statutory nuisance — See **Housing** (Environment); **Landlord and tenant** (Repairs — Statutory nuisance)
Subsidence
Tree roots
Subsidence damage from tree roots — Trees in highway — City council maintaining highway under agency agreement — Whether council proper defendant — Whether council should have been given opportunity to abate the nuisance — Whether underpinning works necessary — Held, council having control under agency agreement as proper defendant — Council not showing that abatement opportunity diminishing damage — Underpinning works constituting proper works
***LE Jones (Insurance Brokers) Ltd* v *Portsmouth City Council (CA),* [2003]** 1 **99**

Tree
Damage
Cause of action — Change of ownership after physical damage — Whether continuing nuisance — Whether new owner entitled to costs of remedial works — Held, tree root damage was continuing nuisance — New owner entitled to cost of remedial works
***Delaware Mansions Ltd* v *Westminster City Council (CA),* [1999]** 3 **68**

Liability
Falling tree branch caused personal injury — Whether cause attributable to decay — Whether tree owner liable to person using highway — Held, tree owner failed to inspect tree — Liability under categories (ii) and (iii) of *Noble* v *Harrison* [1926] 2 KB 332
***Chapman* v *Barking and Dagenham London Borough Council (QB),* [1997]** 2 **141**

OPTION — See **Landlord and tenant** (Option); **Vendor and purchaser**

Undue influence

Validity

Grant of option to acquire life tenancy — Whether grantee of option exercising undue influence on grantor — Whether option valid — Whether correct tests of undue influence applied — Whether prima facie proof of undue influence — Held, undue influence proved — Relevance tests were: (i) that complainant reposed trust and confidence in the other party, or the other party acquired ascendancy over complainant; and (ii) that transaction not readily explicable by relationship of the parties — Appellant establishing both elements on a prima facie basis.

***Macklin* v *Dowsett (CA),* [2004] 2 75**

OPTION TO PURCHASE — See **Capital gains tax** (Option to purchase land — Expenditure); **Land Registration Act 1925** (Overriding interests)

PARTNERSHIP

Assets — See **Milk quota** (Partnership asset)

PARTY WALL — See also **London Building Acts (Amendment) Act 1939** (Party wall)

Party Wall etc Act 1996

Damages

Works carried out without notice — Claim for damages — Party wall subjected to forces from Kango hammer — Causation —Whether building owner entitled to take forensic advantage of lack of pre-works survey as to condition — Held, absence of pre-works survey owing to failure to serve notice under 1996 Act — In dispute as to condition of adjoining building, building owner not entitled to take advantage of lack of evidence of condition in claim for damage

***Roadrunner Properties Ltd* v *Dean (CA),* [2004] 1 73**

PIPELINE — See **Rates and rating** (Rateability — Plant and machinery)

PLANNING OBLIGATION — See **Vendor and purchaser** (Sale of land — Law of Property (Miscellaneous Provisions) Act 1989)

"POINTE GOURDE" PRINCIPLE — See **Compensation for acquisition of land**

POSSESSION — See also **Adverse possession; Housing** (Secure tenancy); **Landlord and tenant** (Assured tenancy — County Courts Act 1984) (Assured tenancy — Housing Act 1988) (Possession — European Con vention on Human Rights) (Secure tenancy — European Court of Human Rights) (Landlord and Tenant Act 1987 — S48 notice) (Possession — Suspended order) (Rent Acts); **Mortgage** (Order for possession) (Possession proceedings — Limitation period); **Trespass** (Remedy — Erection of building)

Rights to possession

Contractual rights

Possession — RSC Ord 113 — Trespassers — Protesters against development — Whether licensee having contractual rights to "occupy" land entitled to possession under Ord 113 — Held, licensee with contractual rights of access not entitled to possession

***Countryside Residential (North Thames) Ltd* v *Tugwell (CA),* [2000] 2 59**

RSC Ord 113

Licence

Contractual licence — Whether licensee entitled to possession order — Held, licensee entitled to possession order under Ord 113

***Manchester Airport plc* v *Dutton (CA),* [1999] 1 147**

PRACTICE AND PROCEDURE — See also **Housing** (Secure tenancy — Right to buy); **Public health** (Statutory nuisance — Procedure in Magistrates' Court)

Summary judgment

Possession proceedings

Judge concluding before commencement of trial that no arguable defence — Parties attending for three-day trial — Whether summary judgment under CPR 24.2 at commencement of trial where no case management identified issues — Held, CPR 24.2 should not be used at commencement of trial — There should have been earlier case management conference to identify issues — Summary judgment upheld

***James* v *Evans (CA),* [2000] 3 1**

PRE-EMPTION RIGHT — See also **Deed**

Enforcement

S2 Law of Property (Miscellaneous Provisions) Act 1989

Lease containing pre-emption right in favour of landlords — Whether pre-emption right equitable interest in land — Whether right amounting to conditional contract or option — Whether s2 of 1989 Act satisfied — Held, offer under pre-emption right capable of being withdrawn at any time — Revocable offer not capable of being equitable interest in land — No conditional contract or option arose — S2 of 1989 Act not satisfied

***Bircham & Co Nominees (No 2) Ltd* v *Worrell Holdings Ltd (CA),* [2001] 3 83**

PRESCRIPTION ACT 1832 — See **Easement** (Acquisition)(Right to light)

PROCEDURE — See also **Lands Tribunal** (Costs — Equivalence principle) (Jurisdiction — Abuse of process) (Leasehold valuation tribunal — Appeal to Lands Tribunal) (Procedure)

Civil Procedure Rules

Expert evidence

Expert witness — Employee — Local authority landlord — Part 35 of CPR — Whether employee of local authority entitled to give expert opinion evidence — Held, court entitled to see employee's report before permitting use of employee as expert under Part 35 of CPR — Court needs to be satisfied of issues in question and ability of employee

***Field* v *Leeds City Council (CA),* [2000] 1 54**

Pre-action discovery for documents

Intended proceedings against builder — Allegations of non-compliance with building regulations — Whether documents relating to building regulations should be disclosed — Whether order amending Supreme Court Act 1981 *ultra vires* — Held, order validly made — Requirements of CPR 31.16 satisfied — Pre-action discovery ordered

***Burrells Wharf Freeholds Ltd* v *Galliard Homes Ltd (TCC),* [1999] 2 81**

Expert

Surveyor

Civil Procedure Rules — Failure of expert to comply with court directions — Expert debarred from acting as expert witness — Consequential order that claim against third party failed — Whether appropriate for court to debar expert — Whether claim against third party should continue — Whether expert should be imposed on judge by consent of parties — Held, expert properly debarred — Expert should not give any evidence even of fact — Claim against third party dismissed

***Stevens* v *Gullis (CA),* [1999] 3 71**

Pleadings

Amendment

Mortgage — Possession proceedings — Expiration of limitation period — Application to amend pleadings to add claim for money judgment — Whether mortgagee entitled to add money claim to possession proceedings after expiration of limitation period — Whether new remedy a new claim under s35(5)(*a*) of Limitation Act 1980 — Whether new claim raised new factual issues — Discretion — Held, additionalremedy not a new claim under s35(5)(*a*) — If a new claim, based on substantially the same facts as pleaded — Discretion exercised under RSC Ord 20 r 5(5)

***Lloyds Bank plc* v *Rogers (CA),* [1999] 3 83**

Possession proceedings

Appeal

Suspended possession order — Application to set aside — Appeal — Tenant not appearing at possession order hearing — Tenant seeking to set aside possession order — Whether merits of defence or reason for non-appearance predominant consideration — Held, non-appearance at hearing predominant consideration

***Lambeth London Borough Council* v *Henry (CA),* [2000] 1 33**

Res judicata

Agricultural tenancy

Successive proceedings relating to protection of tenant's rights — Whether earlier injunctive and damages proceedings estopping hearing of claim to determine entitlement to possession of land — Held, earlier proceedings limited in issues decided — Claim to entitlement to lawful possession could be pursued at subsequent proceedings

***JS Bloor (Measham) Ltd* v *Calcott (Ch),* [2002] 1 1**

PROPERTY MISDESCRIPTIONS ACT 1991 — See also **Estate agents**

False and misleading statements

Pictures and show houses

Housebuilder — Houses as built differed materially from pictures and show house — Whether statements in form of pictures and show house mere promises to future or statements of existing facts — Held, pictures and show house were false or misleading statements of existing facts as developer knew that house to be built would materially differ

***Lewin* v *Barratt Homes Ltd (QB),* [2000] 1 77**

PROPRIETARY ESTOPPEL — See also **Agricultural Holdings** (Agricultural tenancy — Proprietary estoppel)

PROPRIETARY ESTOPPEL — *continued*

Agreement
Negotiations "subject to contract"
Negotiations for 10-year term — Intended tenant allowed into occupation — Intended landlord dying before agreement completed — Whether tenancy granted — Whether proprietary estoppel — Whether s2(5) of Law of Property (Miscellaneous Provisions) Act 1989 applied — Whether solicitor had authority to conclude agreement to grant tenancy — Held, no proprietary estoppel within s2(5) of 1989 Act — Solicitor had no authority to conclude agreement
***James* v *Evans (CA),* [2000]** 3 **1**

Constructive trust
"Subject to contract" negotiations
Agreement for lease — Licence to occupy — Termination of licence — Repossession of premises — Proprietary estoppel and constructive trust — "Subject to contract" negotiations — Whether serious issue to be tried — Held, grounds for relief not established — No estoppel or trust where negotiations "subject to contract"
***Edwin Shirley Productions Ltd* v *Workspace Management Ltd (Ch),* [2001]** 2 **16**

Possession proceedings
Expectation of renewing leases to perpetuity
Practice of predecessor in title to renew leases — Whether practice founded proprietary estoppel — Whether requirements of proprietary estoppel satisfied — Whether freeholder required by estoppel to renew leases — Held, practice of renewing leases did not found proprietary estoppel
***Keelwalk Properties Ltd* v *Waller (CA),* [2002]** 3 **79**

PUBLIC HEALTH
Statutory nuisance
Procedure in magistrates' court
Costs — Environmental Protection Act 1990 s82(12) — Abatement notices — Construction of s82 — Whether costs of proceedings recoverable — Whether magistrates have jurisdiction to permit adjournment of allow nuisance to be abated — Held, costs recoverable if statutory nuisance exists on date of complaint — Court cannot adjourn to allow landlord to remedy statutory nuisance
***R* v *Dudley Magistrates' Court, ex parte Hollis (QB),* [1998]** 2 **19**

PUBLIC HEALTH ACT 1936 — See **Lands Tribunal** (Jurisdiction — Interest)

PUBLIC HOUSE — See **Landlord and tenant** (Agreement to grant lease — Misrepresentation); **Rates and rating**

QUASI-CONTRACT
Restitution
Remedy of account and payment
Solicitors — Mortgage advance — Advance made under mistake of fact — Whether solicitors liable to account and payment of sum advanced as loan — Held, solicitors not enriched — Mortgage advance paid out under instructions — Instructions never revoked — Solicitors liable to account — Account not ordered as no payment due
***Portman Building Society* v *Hamlyn Taylor Neck (CA),* [1998]** 2 **113**

RANSOM STRIP — See **Compensation for acquisition of land** (Pointe Gourde scheme) (Valuation)

RATES AND RATING
Assessment
Building in disrepair
Whether hereditament should be valued in disrepair — Whether hypothetical landlord would have repaired building — Held, hereditament in disrepair on material date must be valued in that state taking into account any disrepair which affects the rental value
***Benjamin (VO)* v *Anston Properties Ltd (LT),* [1998]** 2 **147**

Chemical works
Rateable value — Contractor's method — Plant and machinery — Effect of capital grants — Determination of rateable value — Held, application of principles in *Gilmour (VO)* v *Baker-Carr* (1963) 10 RRC 205 LT — Determination of appropriate allowances and adjustments
***Monsanto plc* v *Farris (VO) (LT),* [1999]** 1 **199**

Historic houses
Rateable values — Whether National Trust assumed to be hypothetical tenant — Whether full or modified profits method appropriate — Rentalisation yield — Held, Trust can assume to be hypothetical tenant and its general financial resources can be considered — Modified profits method applied using gross receipts and 3% rentalisation yield
***Hoare* v *National Trust; National Trust* v *Spratling (LT),* [1997]** 2 **229**

Multi-storey car park
Effect of free local car parking policy — Effect of local shop rents, operating costs and car parking receipts — Whether valuation on basis of profits or tone of the list — Held, profits method rejected
***Church Commissioners for England* v *Mummery (VO) (LT),* [1998]** 3 **209**

Offices
Shop premises used by solicitors — Whether demand by prospective shop tenants to be considered — Whether to be valued as shop and premises used as offices — Held, demand from shop tenants relevant — Premises to be valued as shop premises used as offices
***Irving Brown & Daughter* v *Smith (VO) (LT),* [1996]** 2 **183**

Council tax
Self-contained unit
Whether "granny" annex a self-contained unit separately liable for council tax — Tribunals relying on terms of planning permission, community of living and impracticability of separate sale as factors — Held, decisions founded on such factors disclosed errors of law and quashed
***Batty* v *Burfoot; Batty* v *Merriman; Gilbert* v *Childs; Rodd* v *Ritchings (QB),* [1995]** 2 **142**

Dwelling-house
Rating list
S66(2B) of Local Government Act 1988 — Available for letting — Whether domestic property — Whether s66(2B) applied — Whether available for letting throughout the year — Whether property correctly entered in rating list — Held, statutory test not satisfied — Intention of owners to only make available for letting for periods not exceeding 139 days pa
***Godfrey* v *Simm (VO) (LT),* [2000]** 3 **85**

Forfeiture of lease
Liability of tenant
Forfeiture of lease by writ — Tenant later accepting forfeiture and gave up possession — Whether tenant or landlord entitled to possession and liable for rates after forfeiture accepted — Held, after forfeiture accepted landlord and not tenant liable for rates
***Kingston upon Thames Royal London Borough Council* v *Marlow (QB),* [1996]** 1 **101**

Public house
Approved guide for rating valuation of licensed premises
Turnover valuation — Relevance of guide and comparables — Whether actual and prospective profits figures relevant — Lands Tribunal simplified procedure — Held, against evidence of losses nominal rateable value appropriate — Approved guide not applicable without fine tuning
***Sharp* v *Griffiths (VO) (LT),* [1999]** 3 **113**

Rateability
Plant and machinery
Pipelines connecting oil wells with gathering station — Whether pipelines rateable under Calculation for Rating (Plant and Machinery) Regulations 1989 — Whether a "mineral field" — Held, on the facts the pipelines were within "mineral field" and exempt from rateability
***Ottewell (VO)* v *BP Exploration Operating Co Ltd (LT),* [1996]** 1 **203**

Rateable value
Historic houses
Profits method — Whether "overbid" would be made by National Trust — Whether any basis for taking 3% of gross receipts for overbid — Whether Lands Tribunal erred — Held, wrong to disregard evidence that Trust would not make overbid in real world — No justification for 3% of gross receipts
***Hoare* v *National Trust (CA),* [1999]** 1 **155**

Offices
Evidence of rent passing, other rents and assessments — Weight to be given to categories of evidence — Held, no weight to rent of hereditament — Other rental and assessment relevant
***Hospital Plan Insurance Services Ltd* v *Persaud (VO) (LT),* [1998]** 3 **189**

Rebus sic stantibus ("the actual state rule")
Public houses
Large shopping centre with units containing two public houses — Units capable of being stripped to shells and used as retail units — Whether in rating hypothesis units used as public houses should be valued as capable of potential retail use — Held, public houses in units in shopping centre must be valued in actual physical state as occupied for public house purposes ignoring alterations and changes of mode of user
***Scottish & Newcastle Retail Ltd* v *Williams (VO) (LT),* [2000]** 2 **171**

RATES AND RATING — *continued*
Rebus sic stantibus — *continued*
Public houses — continued

Large shopping centre with units containing two public houses — Units in shopping centre capable of being stripped to shell finish and used as retail units — Whether in rating hypothesis units used as public houses should be valued as capable of potential retail use — Held, public houses in units in shopping centre must be valued in actual physical state as occupied for public house purposes ignoring alterations and changes of mode of user
***Scottish & Newcastle Retail Ltd* v *Williams (VO) (CA)*, [2001]** 1 **157**

Unoccupied property
Liability of receivers
Receivers appointed by banks holding charges over properties — Receivers agents of companies owning properties — Whether receivers liable for rates — Held, receivers not liable for rates — Rates not a receivership expense — No fraudulent trading
***Brown* v *City of London Corporation (Ch)*, [1996]** 1 **139**

Transitional certificates
Non-domestic Rating (Transitional Period) Regulations 1990 — Meaning of Regulation 2(3)(a) — Whether applicants entitled to transitional certificates — Whether relevant rateable value actual value in valuation list at 31 March 1990 or value following hypothetical proposal to reflect factors affecting value — Held, relevantrateable value is that which would result from hypothetical proposal from valuaiton officer made prior to 31 March 1990
***R* v *Huelin (VO), ex parte Murphy Ltd; R* v *Hodgetts (VO), ex parte Nationwide Building Society (CA)*, [2000]** 1 **97**

Valuation
State of disrepair
Hereditament in disrepair — Whether premises must be valued in state of disrepair on material date — Held, premises must be valued in actual state of disrepair as affects rental value on material date — Local Government Finance Act 1988, Sched 6
***Benjamin (VO)* v *Anston Properties Ltd (LT)*, [1998]** 2 **147**

RECEIVER — See **Mortgage** (Receiver — Duty of care); **Liability** — See **Rates and rating** (Unoccupied property); **Negligence** (Valuers — Duty of care)
Power of sale
Duties
Mortgages — Receivers selling mortgaged properties — Receivers agents of mortgagors — Alleged breach of duty — Alleged sale at undervalue — Whether receivers under duty to pursue planning applications in respect of properties prior to sale — Whether receivers under duty to prelet vacant commercial property — Held, receivers not under obligation to pursue applications for planning permission or to prelet vacant properties
***Silven Properties* v *Royal Bank of Scotland plc (CA)*, [2003]** 3 **49**

RECEIVERSHIP
Powers
Powers of appointment
Power of sale — Law of Property Act 1925 — Receivers appointed by bank over properties — Whether breach of contract by bank in appointing receivers — Whether bank estopped from appointing receivers — Whether demands for payments in excess of sums due — Whether demand for payment must allow sufficient time to pay — Whether receivers validly appointed — Whether receivers appointed under contractual power or under statutory power of appointment — Whether receivers can be appointed under s101(1)(iii) of Law of Property Act 1925 for non-tenanted property with no income — Duties of receivers — Whether receivers under a duty to add value to property before sale — Whether receivers negligent in relation to sale of development properties — Whether receivers acted in bad faith — Held, legal charge permitted valid appointment immediately following making of deed — Bank did not agree delay and not estopped from appointing receivers — Sufficient time for payment given — Receivers validly appointed — Receivers not under any duty to add value to property — Receivers owed primary duty to mortgagee — Receivers not negligent in sale of development property — Receivers did not act in bad faith
***Silven Properties Ltd* v *Royal Bank of Scotland (Ch)*, [2002]** 3 **115**

RECTIFICATION — See also **Contract** (Rectification); **Mistake**
Lease
Bona fide purchaser
Assignment of term — Whether landlord entitled to rectification against assignee — Whether assignee bona fide purchaser for value without notice — Whether landlord's right of rectification binds assignee under s70(1)(*g*) of Law of Property Act 1925 — Burden of proof for rectification — Whether landlord had "clean hands" — Held, assignee bona fide purchaser for value without notice — Right of rectification bound assignee as a right under s70(1)(*g*) — Burden of proof not discharged for rectification
***Nurdin & Peacock plc* v *DB Ramsden & Co Ltd (Ch)*, [1999]** 1 **119**

REFUSE TIP
Licence
Rent formula
Exclusive right to deposit refuse — Rental payments — Whether payment to be made for capping materials as well as dry waste — Held, payments must include airspace occupied by capping materials
***Hunts Refuse Disposals Ltd* v *Norfolk Environmental Waste Services Ltd (CA)*, [1997]** 1 **16**

RELIEF — See **Landlord and tenant** (Forfeiture — Equitable charge)

RENT — See **Election** (Waiver — Landlord and tenant); **Landlord and tenant** (Rent — Set-off) (Service charges — Interest on late payments)
Arrears — See **Insolvency** (Individual voluntary arrangement — Rent arrears); **Landlord and tenant** (Assignment — Liability of original tenant) (Forfeiture — Possession) (Forfeiture — Relief) (Forfeiture — Surety covenants) (Forfeiture — Validity) (Insolvency — Disclaimer of lease) (Possession — Procedure) (Rent arrears); **Leasehold enfranchisement** (Leasehold Reform Act 1967)
Formula — See **Refuse tip** (Licence)
Rack-rents — See **Landlord and tenant** (Rent review clause — Construction: general)

RENT (AGRICULTURE) ACT 1976 — See **Insolvency** (Surety — Duty to mitigate)

RENT ACTS — See also **Housing** (Secure tenancy); **Landlord and tenant** (Landlord and Tenant Act 1954 Part I — Validity of s4 notice) (Rent Acts) (Rent Acts — Fair rent) (Side agreement or letter — Turnover rent); **Negligence** (Duty of care and liability — Letting agents)
Rent Act 1977
Possession
Statutory tenancy — Case 1 of Sched 15 — Whether proviso for re-entry on bankruptcy of tenant an obligation within case 1 — Whether proviso for re-entry carried through from contractual lease to statutory tenancy — Held, proviso for re-entry on bankruptcy within case 1 — Proviso travels through to statutory tenancy
***Cadogan Estates Ltd* v *McMahon (CA)*, [1999]** 2 **67**

RENT ASSESSMENT COMMITTEE — See **Landlord and tenant** (Assured tenancy — Notice of increase of rent) (Long lease — Notice to terminate) (Rent Acts — Fair rents) (Rent Acts — Fair rent determination) (Rent Acts — Rent assessment) (Tenancy at will — Calculation of rent)

RENT REVIEW — See **Agricultural holdings** (Arbitration); **Arbitration** (Appeal — S68 of Arbitration Act 1996); **Landlord and tenant** (Assignment — Liability of original tenant) (Assured tenancy — Contracting-out) (Insolvency — Vesting order) (Landlord and Tenant Act 1954 Part II — Rent review) (Rent review) (Rent review clause) (Rent review — Construction) (Rent review — Hypothetical term) (Rent review — Trigger notice)
Arbitration — See **Arbitration** (Arbitrator — Misconduct)

RENTCHARGES
Rentcharges Act 1977
Validity of rentcharge
Service charges — Payment of service charges for maintenance of industrial estate — Variable rentcharge — Service charge expenditure including liability for rates — Whether rentcharge voided by s2(4) of 1977 Act — Whether rentcharge voided by absence of express reaonableness limitation — Held, payment of rates on land for benefit of rentcharge payers — Rentcharge not voided by ss2(4) and 2(5) of 1977 Act
***Orchard Trading Estate Management Ltd* v *Johnson Security Ltd (CA)*, [2002]** 2 **1**

REPAIRING COVENANT — See **Landlord and tenant** (Landlord and Tenant Act 1954 Part II — Landlord's opposition to new tenancy) (Repairs — Breach of implied covenant) (Repairs — Covenant)

REPAIRS — See **Landlord and tenant** (Covenant for quiet enjoyment — Repairs), (Landlord and Tenant Act 1987 — Manager) (Repairs) (Repairs — Dilapidations) (Repairs — Implied covenant) (Rent Acts — Fair rent) (Rent review — Disregard of improvements) (Repairs — Steel framework) (Secure tenancy — Repairs) (Service charges — Landlord and Tenant Act 1985) (Subletting — Enforcement of covenants); **Rates and rating** (Assessment — Building in disrepair) (Valuation — State of disrepair)

RESCISSION — See **Contract for sale of land**; **Landlord and tenant** (Assignment)

RES JUDICATA — See **Estoppel** (Cause of action estoppel — Legal charge); **Procedure** (Res judicata — Agricultural tenancy)

RESTRICTIVE COVENANT — See also **Landlord and tenant** (Covenants — Covenant by landlord) (Covenants — Letting scheme) (Leasehold enfranchisement — Covenants)

Annexation

S78 of Law of Property Act 1925

Building and user restrictions — Post-1925 covenants — Land sold off by series of conveyances — Conveyances in two principal forms — Express words of annexation referring to "for the benefit of the property at [named locality] belonging to the Vendors or part thereof for the time being remaining unsold" — Conveyances without express words of annexation identifying property on "Fee Farm Estate" — Whether statutory annexation under s78 — Whether land intended to benefit identifiable — Whether s78 subject to contrary intention — Held, restrictive covenants not annexed — S78 had no application — *Federated Homes Ltd* v *Mill Lodge Properties Ltd* [1980] 1 WLR 594 not affecting requirement that land intended to be benefited must be so defined as to be easily ascertainable — Requirement identified in *Marquess of Zetland* v *Driver* [1939] Ch 1 remaining necessary condition for annexation — S78 subject to contrary intention — Express words of annexation excluding s78 — Conveyances not identifying benefited land

***Crest Nicholson Residential (South) Ltd* v *McAllister (CA),* [2004]** 2 **79**

Application under s84(1) of Law of Property Act 1925

Density restriction

Development in rear gardens — S84(1)(aa) and impeding reasonable user — Whether restriction securing practical benefits of substantial value or advantage to objectors — S84(1)(c) and lack of injury to persons entitled to benefit — Whether objectors' properties would suffer any injury — Planning permission for proposed development — Relevance of planning permission to application under s84 — Plans restriction — Held, applications refused — Restrictions giving objectors substantial practical benefits, including protection of views, spaciousness and freedom from building noise — Planning permission granted by inspector relevant but given little weight — Different exercise of tribunal under s84 — Tribunal concerned with protection of property rights and inspector with planning control

***In Diggens' Application (No 2) (LT),* [2001]** 2 **163**

Application to modify

Grounds (a), (aa) and (c)

Construction of covenant — Whether requirement for approval of plans applied to extension to original building — Whether grounds made out — Determination of compensation payable — Held, covenant applied to extension to original building — Application allowed under ground (aa) of s84(1) of Law of Property Act 1925 — Compensation of £10,000 ordered

***In Jillas' application (LT),* [2000]** 2 **99**

Breach — See also **Damages** (Lord Cairns' Act — Assessment)

Damages

Assessment of damages — Lord Cairns Act — Restriction prohibited additional bedrooms to hotel — Relevant factors in assessment of damages — Held, incremental approach appropriate measure of damages — Hypothetical negotiations — Damages assessed at £375,000

***Amec Developments Ltd* v *Jury's Hotel Management (UK) Ltd (Ch),* [2001]** 1 **81**

Remedies

Restrictions against building and use — Breach — Whether delay or acquiescence — Failure to apply for interlocutory relief — Whether grant of injunctive relief — Whether award of damages —Whether assessed by reference to sum payable for release of restrictive covenant — Held, failure to apply for interlocutory relief bars final injunctive relief — Damages payable for breach — Damages assessed by reference to sum plaintiff would have sought for release of covenant

***Gafford* v *Graham (CA),* [1999]** 3 **75**

Burden

Restriction

Use for purposes of private dwelling-house — Whether restriction preventing erection of more than one dwelling-house — Enforcement — No erection of buildings without prior approval of plans — Plans requiring approval of named company — Company dissolved and incapable of restoration — Whether restriction remaining enforceable — Held, restriction for purposes of "a" dwelling-house preventing erection of more than one dwelling-house — Restriction against building without approval of plans discharged upon dissolution of company

***Crest Nicholson Residential (South) Ltd* v *McAllister (Ch),* [2003]** 1 **165**

Successors in title

Covenant restrictive of user on land retained by vendor — Whether covenant binding on successors in title — Whether covenant enforceable under s79 of Law of Property Act 1925 — Held, covenant personal and not intended to bind successors in title — S79 excluded by contrary intention

***Morrells of Oxford Ltd* v *Oxford United Football Club (Ch),* [2000]** 3 **70**

Covenant restrictive of user on land retained by vendor — Whether covenant binding on successors in title — Whether covenant enforceable under s79 of Law of Property Act 1925 — Held, covenant personal and not intended to bind successors in title — S79 excluded by contrary intention

***Morrells of Oxford Ltd* v *Oxford United Football Club (CA),* [2001]** 1 **76**

Construction

Restriction on use

Covenant restricting use of buildings to any purpose other than as a private dwelling-house — Whether indefinite article limiting number or manner of use of buildings — Whether more than one dwelling-house could be erected and used as such on burdened land — Held, indefinite article not connoting singularity — Literal meaning in favour of several buildings — Factual matrix not disturbing literal meaning — Decided authority distinguished

***Martin* v *David Wilson Homes Ltd (CA),* [2004]** 3 **77**

Use as private residence only

Covenant restricting use of land for private residence only — Whether part of land can be used for roadway to access adjoining development site for 10 houses — Whether restriction concerning land as well as dwelling — Whether restriction binding persons other than successor residential owner of burdened land — Held, land as well as dwelling subject to restriction — Restriction binding persons other than successor residential owner of burdened land

***Jarvis Homes Ltd* v *Marshall (CA),* [2004]** 3 **81**

Discharge or modification

Discretion of Lands Tribunal

Unopposed application under s84 of Law of Property Act 1925 — Whether tribunal exercised discretion properly in modifying rather than discharging covenants — Held, tribunal retains discretion in unopposed applications — On the facts modification rather than discharge appropriate

***In University of Westminster's Application (LT),* [1997]** 1 **191**

Obsolete covenant

Application under Law of Property Act 1925, s84(1) (*a*), (*aa*), (*b*) and (*c*) — Whether restriction obsolete — Whether objectors enjoyed practical benefits of substantial value or advantage — Whether tribunal's discretion can be exercised in absence of planning permission — Failure to provide agreed statement of facts — Held, application allowed under paras (*a*), (*aa*) and (*c*)

***Re Nichols' Application (LT),* [1997]** 1 **144**

Building scheme — Mutual restrictions on erecting buildings — Application to modify or discharge under paras (*a*), (*aa*) and/or (*c*) of s84(1) of Law of Property Act 1925 — Held, paras (*aa*) and (*c*) not satisfied — Purpose of restrictions now obsolete — Restrictions modified under para (*a*) — Compensation payable to those objectors only who remained bound by restrictions — Applicants pay objectors' costs

***Re Kennet Properties Ltd's Application (LT),* [1996]** 2 **163**

Planning agreement

Law of Property Act 1925 s84(1) — Town and Country Planning Act 1990 s106 — Whether restrictions in 1990 planning agreement obsolete — Held, where no change in planning policies restriction unlikely to be obsolete

***Caton's application, In re (LT),* [1999]** 3 **121**

Right to buy restriction

Right to buy sale — Restriction against trade or business — Whether objecting council entitled to practical benefits of substantial value or advantage — Whether application under s84(1) (*aa*) of Law of Property Act 1925 can succeed where objector custodian of public interest — Whether compensation payable — Held, application allowed — Restriction did not secure any benefit or advantage to objecting council

***Re Willis' Application (LT),* [1997]** 2 **185**

S157 of Housing Act 1985

Covenant imposed in "right to buy" conveyance restricted disposal to locally qualified purchaser — Whether covenant within tribunal's jurisdiction — Whether ground (*aa*) of s84(1) of Law of Property Act 1925 satisfied — Held, no discharge — Covenant still of benefit to council

***Re Milius' Application (LT),* [1996]** 1 **209**

Enforcement

Division of burdened land

Whether benefit passed on division of plot subject to restriction against building — Held, restriction intended to benefit land outside original plot not part of plot following subdivision

***Briggs* v *McCusker (Ch),* [1996]** 2 **197**

RESTRICTIVE COVENANT — *continued*
Enforcement — *continued*

Statutory powers
Use restricted to particular type of hospital — Approval required to elevations of proposed buildings — Whether National Health Trust hospital bound by restrictive covenants — Whether hospital entitled to exercise its statutory powers — Held, hospital entitled to exercise statutory powers and use hospital in breach of restriction on use — Statutory powers will not inevitably authorise breach of obligation to obtain approval of building elevations
***Cadogan* v *Royal Brompton Hospital National Health Trust (Ch),* [1996]** 2 **115**

Meaning and construction
Residential development
Whether restriction against use otherwise than as a private residence prevents construction of a residence — Whether restriction became absolute on death of person whose approval was required for any buildings — Held, restriction was use as a residence did not prevent building a residence — Restriction became absolute on death of person able to grant consent
Briggs* v *McCusker (Ch), [1996] 2 197

Use
Recreational purposes — Whether football club stadium use breach of covenant against use other than for recreational purposes — Held, recreational purposes given narrow meaning — Commercial football stadium use not recreational purpose
***Thames Water Utilities Ltd* v *Oxford City Council (Ch),* [1999]** 1 **167**

Modification
Application to suspend proceedings
S84(2) of Law of Property Act 1925 — Whether proceedings should be suspended to allow application to High Court — Whether matters raised within s84(2) — Whether application should be granted on paper as of right — Whether successful objectors should bear costs — Held, proceedings suspended — Matters raised within s84(2) — Objectors disallowed costs — Application unnecessary and adding to costs — Motives of delay
***In The Girls' Day School Trust (1872) Application (LT),* [2002]** 2 **89**

S84 Law of Property Act 1925
Restriction against alterations imposed in 1992 — Grounds (*a*) and (*c*) — Whether restriction obsolete — Whether discharging or modifying restriction would injure objector — Whether compensation should be adjusted for rise in property values — Held, restriction not obsolete — Restriction modified — Compensation of £2,450 after adjustment for rise in property values
***In Davies' application (LT),* [2001]** 1 **111**

Overriding private rights — See **Town and country planning** (Overriding private rights)
Right to buy
Housing Act 1985
Restriction to single dwelling–house use — Payment demanded for release of covenant — Whether covenant imposed for improper purpose — Whether covenant lawfully imposed — Held, covenant unlawfully imposed — Covenant *ultra vires*
***R* v *Braintree District Council, ex p Halls (CA),* [2000]** 3 **19**

Private dwelling-house restriction
Holiday lets — Whether use for holiday purposes occupation for dwelling-house purpose — Held, use of properties for holiday lets not private dwelling-house use
***Caradon District Council* v *Paton (CA),* [2000]** 3 **57**

S84(1) Law of Property Act 1925
Amendment of application
Application to discharge or modify restriction against erection of more than one house — Whether applicants entitled to amend application to add further restriction to be modified — Lands Tribunal Rules 1996 — Held, the rules do not permit amendment of substantive application to add additional restriction to be modified. Only the grounds in paras (*a*), (*aa*), (*b*) and (*c*) of s84(1) of the 1925 Act can be amended under r 36 of the rules
***In Diggens' Application (LT),* [2000]** 3 **87**

S237 Town and Country Planning Act 1990
Use
Breach of restrictive covenant — Proposed commercial football stadium — Whether use of stadium and ancillary buildings and land in breach of covenant against use other than for recreational purposes authorised — Whether restrictive covenant against use overridden by s237 Town and Country Planning Act 1990 — Held, s237 did not authorise use of stadium by club
***Thames Water Utilities Ltd* v *Oxford City Council (Ch),* [1999]** 1 **167**

RIGHT OF LIGHT — See also **Easements**
Express grant
Interference
Effect of express grant in lease — Whether servient owner entitled to obstruct light by new building contemplated prior to grant of lease — Whether dominant owner entitled to level of light subject only to interference within test in *Colls* v *Home & Colonial Stores* [1904] AC 179 — Construction of lease — Held, tenant entitled to benefit of express grant without interference subject to test in *Colls* — Construing lease as a whole, plaintiff not entitled to constructbuilding interfering with right to light
***Frogmore Developments Ltd* v *Shirayama Shokusan Co Ltd (Ch),* [2000]** 1 **121**

RIGHT OF PRE-EMPTION — See **Landlord and tenant** (Assignment — Right of pre-emption)

RIGHT OF SUPPORT
Natural right
Landslip
Whether landowner entitled to damages from owner of lower land — Held, on the facts support not withdrawn.
***Holbeck Hall Hotel Ltd* v *Scarborough Borough Council; Scarborough Borough Council* v *Geotechnical Engineering (Northern) Ltd (Off Ref),* [1997]** 2 **213**

RIGHT OF WAY — See **Easement** (Right of way) (Creation — Prescription) (Acquisition by prescription — Village green); **Human rights**; **Valuation** (Property — Stokes principle)

RIGHT TO BUY — See also **Housing** (Right to buy — Defective house type) (Secure tenancy — Right to buy); **Restrictive covenant** (Discharge or modification — Right to buy restriction) (Discharge or modification — S157 of Housing Act 1985)
Housing Act 1985
Restrictive covenant
Restriction to single dwelling-house use — Payment demanded for release of covenant — Whether covenant imposed for improper purpose — Whether covenant lawfully imposed — Held, covenant unlawfully imposed — Covenant ultra vires
***R* v *Braintree District Council, ex p Halls (CA),* [2000]** 3 **19**

SECURE TENANCY — See **Compensation** (Land Compensation Act 1973 Part I — Valuation); Landlord and tenant (Secure tenancy — Suspended possession order)

SERVICE CHARGES — See **Landlord and tenant** (Landlord and Tenant Act 1987 — Manager); Landlord and tenant (Service charges — Statutory trust); **Leasehold valuation tribunal**
Reasonable costs
Porter and cleaning services
Small block of flats — S19 of Landlord and Tenant Act 1985 — Whether expenditure on full-time porter reasonably incurred — Leasehold valuation tribunal — Appeal to Lands Tribunal — Whether rehearing or review — Burden of proof — Whether claimant before LVT or appellant to Lands Tribunal having burden of proof — Held, expenditure on full-time porter unreasonably incurred — Appeal from LVT treated as a rehearing — Burden of proof on appellant
***Veena SA* v *Cheong (LT),* [2003]** 1 **175**

SET-OFF — See **Landlord and tenant** (Equitable set-off — Assignment of right to recover rent) (Landlord and Tenant Act 1987 — Manager) (Secure tenancy) (Service charges — Set-off)

SEWERS AND DRAINS — See also **Easements**
Public sewers
Meaning of "sewer"
Public Health Act 1875 — Pipe serving dwelling and highway gully — Whether highway "premises" — Whether pipe a sewer — Whether sewerage undertaker liable for sewage discharge — Held, purpose of highway gully to protect house — Pipe a drain and not public sewer — Sewerage undertaker not liable for sewage spill — Highway capable of being "premises" under s4 of 1875 Act — Highway not "premises" on facts
***City of Bradford Metropolitan District Council* v *Yorkshire Water Services Ltd (CA),* [2002]** 1 **85**

SHELTERED HOUSING — See **Compensation** (Land Compensation Act 1973 Part I — Valuation)

SHOPPING CENTRE
Pedestrian malls
Excluding public
Whether malls subject to walkway agreement — Whether public have equitable licence or public rights to use malls — Held, public had no enforceable rights to enter and any licence could be revoked
***CIN Properties Ltd* v *Rawlins (CA),* [1995]** 2 **130**

SOLICITORS — See **Agency** (Solicitors — Tenancy); **Conveyancing**; **Quasi-contract** (Restitution — Remedy of account and payment)
Negligence — See **Negligence** (Duty of care and liability) (Limitation Act) (Solicitors) (Solicitors — Assessment of damages) (Solicitors — Damages) (Solicitors — Landlord and tenant) (Solicitors — Conveyancing)

SPECIFIC PERFORMANCE — See **Landlord and tenant** (Repairs — Covenant)

SQUATTER — See **Adverse possession** (*Animus possidendi* — Limitation Act 1980) (Lease — Registered land)

STAMP DUTY
Lease premium
Consideration determinable after execution
Grant of lease for premium — Premium defined as product of stated figure and price of 13.75% Treasury Loan Stock 25 days following execution — Whether instrument chargeable — Assessment — Held, calculation could be worked out on day of execution by reference to closing price of loan stock previous day — Premium subject to *ad valorem* stamp duty
***LM Tenancies 1 plc* v *Inland Revenue Commissioners (Ch)*, [1996]** 2 **119**

"STOKES" PRINCIPLE — See **Valuation** (Property — Stokes principle)

SUBPOENAS — See **Arbitration** (Subpoenas *duces tecum*)

SUCCESSION — See **Agricultural holdings** (Succession — Principal source of livelihood); **Covenants** (Freehold — Positive covenants); **Restrictive covenant** (Burden — Successors in title)

SUPREME COURT ACT 1981 — See **Landlord and tenant** (Forfeiture)

SURETY — See **Landlord and tenant** (Assignment) (Termination of lease)
Covenant — See **Landlord and tenant** (Assignment — Liability of surety) (Forfeiture — Surety covenants) (Surety covenant)
Insolvency — See **Insolvency** (Surety) (Voluntary arrangement — Co-debtors)
Lease — See **Landlord and tenant** (Surrender of lease — Implied surrender)

SURVEYORS — See **Compensation for sea defence works** (Loss of profits claim); **Negligence** (Architects — Tortious liability) (Surveyors — Damages) (Surveyors — Mortgage valuation loan); **Procedure** (Expert)
Fees — See **Landlord and tenant** (Leasehold enfranchisement — Surveyor's fees)
London Building Acts (Amendment) Act 1939
Award of third surveyor — See **London Building Acts (Amendment) Act 1939** *(Party wall)*
Negligence — See **Experts** (Negligent valuation); **Negligence** (Damages: measure and quantification) (Limitation Act) (Surveyors) (Valuation — Independent expert)

TAXATION
Capital gains tax — See **Capital gains tax** (Retirement relief)
Council tax — See **Rates and rating** (Council tax)
Inheritance tax
Agricultural land
Agricultural property relief — Whether land consisting of buildings "agricultural land or pasture" under Inheritance Tax Act 1984 — S5 of Interpretation Act 1978 — Held, contrary intention clear from 1984 Act precluding wide definition of land in 1978 Act — Land not agricultural land
***Starke* v *Commissioners of Inland Revenue (CA)*, [1996]** 1 **157**

Valuation
Agricultural land
Agricultural tenancy held partnership — Deceased held 50% share in partnership — Whether value of partnership share 50% of value of tenancy — Whether reluctance of actual freehold owners to participate in paying 50% of vacant possession premium relevant — Held, although notional transaction in open market involved seller and purchasers who were abstractions, actual freeholder's reluctance to participate in paying vacant possession premium cannot be ignored
***Walton's Exors* v *Commissioners of Inland Revenue (CA)*, [1996]** 1 **159**

TELECOMMUNICATIONS ACT 1984
Telecommunications Code
Objection to apparatus
Three-month time limit — Alleged illegible para 18 notice — Para 17(2) notice of objection — Whether valid para 18 notice — Whether three-month time limit for para 17(2) objection running from date of valid para 18 notice or completion of apparatus —Whether para 18 notices required to be seen from land to which public has access — Whether completion under para 17 meaning physical or operational completion — Held, time limit running from date of physical completion — Para 18 notice need not be legible from land to which public have right of access
***Lloyd Jones* v *T-Mobile (UK) Ltd (CA)*, [2003]** 3 **55**

TENANCY AT WILL — See **Landlord and tenant** (Landlord and Tenant Act 1954 Part II — S43(3)) (Surrender of lease — Surrender by operation of law) (Tenancy at will — Use and occupation)

TIME OF THE ESSENCE — See **Landlord and tenant** (Rent review)

"TOMLIN" ORDER — See **Valuation** (Property — Stokes principle)

TOWN AND COUNTRY PLANNING — See also **Easement** (Acquisition — Prescription)
Appeal
Human rights
Appeal against enforcement notice dismissed by inspector and High Court — Whether hearings in breach of art 6(1) of the European Convention of Human Rights — Held, hearings by inspectors do not comply with convention as to independence because Secretary of State can revoke inspector's appointment to hear appeal
***Bryan* v *United Kingdom (ECHR)*, [1996]** 2 **123**

Blight notice
Counternotice
Town and Country Planning Act 1990 — Paras 3 and 4 of Sched 13 — Whether property "allocated" or "indicated" for transport purposes — Held, although property blighted by authority's consultation exercise, no legal blight — Land not indicated or allocated on development plan
***Jones* v *Greater Manchester Passenger Transport Executive (LT)*, [1998]** 3 **121**

Serious effect on amenity or convenience
Part of front garden proposed to be acquired for road scheme — Whether part could be acquired without seriously affecting the amenity or convenience of subject dwelling-house — Held, loss of adjoining tree screen and construction works amounted to serious effect on amenity and convenience respectively
***Smith* v *Kent County Council (LT)*, [1995]** 2 **196**

Validity
Claimant selling property after service of blight notice — Whether deemed withdrawal — Whether property blighted — Whether claimant entitled to compensation — Held, deemed withdrawal of blight notice — Property not blighted at material date of counternotice — No entitlement to compensation
***Carrel* v *London Underground Ltd (LT)*,[1996]** 1 **179**

Validity of counternotice
Blight served in respect of cottage and adjoining highway subsoil — Whether counternotice valid — Whether highway subsoil part of hereditament — Held, blight notice valid — Highway subsoil part of hereditament
***Norman* v *Department of Transport (LT)*, [1996]** 1 **190**

Counternotice relying on Town and Country Planning Act 1990 s151(4) (a) — Land ceasing to be blighted after counternotice — Whether counternotice should be upheld — Held, where objection in counternotice based on ground (a) of s151(4) of the 1990 Act, date of counternotice material date — Blight notice upheld
***Sinclair* v *Secretary of State for Transport (LT)*, [1997]** 2 **173**

Local plan allocated area for redevelopment in connection with polytechnic — Polytechnic later becoming university — Whether local plan allocated land for the functions of the local authority — Held, land not so allocated — *Bolton Corporation* v *Owen* followed
***Elcock* v *Newham London Borough Council (LT)*, [1996]** 2 **137**

Deceit
Development
Misrepresentation by local authority — Development of sites HC and G — Local planning authority in favour of both developments — Authority owning interest in site G and representing existence of tenant mix agreement to minimise comptetition to site HC — Developer induced by representation to develop site HC — Secret agreements relaxing tenant mix agreement — Whether authority liable for deceit
***Slough Estates plc* v *Welwyn Hatfield District Council (QB)*, [1996]** 2 **219**

Demolition
Listed building consent
Whether removal of chimney breasts and consequential alterations demolition or alterations — Whether compensation payable for refusal of consent — Held, building includes part of a building and works demolition of part — No compensation payable
***Shimizu (UK) Ltd* v *Westminster City Council (CA)*, [1995]** 1 **167**

Enforcement notice

Human Rights Treaty

Planning legislation — Enforcement appeal procedure — Whether enforcement notice violation of treaty — Whether right to appeal inspector's decision to High Court sufficient compliance with treaty

***Bryan* v *United Kingdom (ECHR),* [1996]** 2 **123**

Stop notice

Compensation — Claimant for compensation holding licence to use shipping berth and transit shed — S186 of Town and Country Planning Act 1990 — Qualifying requirements for compensation of interest in or occupation of land — Whether interest under licence sufficient interest in land — Whether claimant in occupation of land — Whether letter sufficient claim for purposes of s186 and reg 12 of the Town and Country Planning (General) Regulations 1992 — Held, interest under a licence not an interest in land for purposes of s186 of 1990 Act — Claimant not in occupation — Letter of claim valid claim for compensation

***International Traders Ferry Ltd* v *Adur District Council (CA),* [2004]** 2 **89**

Material consideration

Planning obligation

Developers seeking planning permission for rival superstore sites — One developer offering funding for link road — Whether offer material consideration — Whether decision of Secretary of State to refuse planning permission to developer making funding offer valid — Held, funding offer a material consideration — Secretary of State considered it and entitled to give no weight to offer

***Tesco Stores Ltd* v *Secretary of State for the Environment (HL),* [1995]** 2 **147**

Overriding private rights

Restrictive covenants

Restrictive covenants overridden following acquisition and development of land for planning purposes — Site being redeveloped for second time — Whether s237 of Town and Country Planning Act 1990 overrides restrictive covenant on second development — Held, restrictive covenants overriden by second development

***R* v *City of London Council, ex parte Mystery of the Barbers of London (QB),* [1996]** 2 **128**

Planning agreement — See **Restrictive covenant** (Discharge or modification)

Planning conditions

Unauthorised development

Planning permission granted for erection of building subject to agricultural occupancy condition — Building erected in breach of planning control — Whether building subject to planning condition — Held, *Kerrier District Council* v *Secretary of State for the Environment* wrongly decided — Condition not enforceable in relation to unauthorised building

***Handoll* v *Warner Goodman & Streat (CA),* [1995]** 1 **173**

Purchase notice

Validity

Whether purchase notice served in respect of part only of land the subject of planning refusal valid under s137 of Town and Country Planning Act 1990 — Held, notice invalid — Notice must be served in respect of the same land the subject of the planning refusal

***Cook* v *Winchester City Council (LT),* [1995]** 1 **179**

Restrictive covenant

S237 Town and Country Planning Act 1990

Breach of restrictive covenant — Proposed commercial football stadium — Whether use of stadium and ancillary buildings and land in breach of covenant against use other than for recreational purposes authorised —Whether restrictive covenant against use overridden by s237 Town and Country Planning Act 1990 — Held, s237 did not authorise use of stadium by club

***Thames Water Utilities Ltd* v *Oxford City Council (Ch),* [1999]** 1 **167**

S106 agreement

Enforcement

Planning obligation — S106 agreement — Obligation to sell land — S2 of Law of Property (Miscellaneous Provisions) Act 1989 — Whether agreement void for failing to contain signature — Whether implication of term — Whether rectification of agreement — Held, s106 agreement containing obligation to sell land within s2 of the 1989 Act — Agreement unenforceable without signature of purchaser

***Jelson Ltd* v *Derby City Council (QB),* [1999]** 3 **91**

Tree preservation order

Compensation

Refusal of consent to lop tree — Refusal delayed building works — Whether fees of arboriculturalist, costs of appeal and increased interest charges allowable — Held, such costs were recoverable as reasonably foreseeable

***Factorset Ltd* v *Selby District Council (LT),* [1995]** 2 **190**

Confirmation

Tree preservation order — Town and Country Planning Act 1990 — Whether consent of Forestry Commission required under s200 — Whether local planning authority wrong to confirm TPO against recommendation of an officer and without considering national guidelines — Held, failure to comply with s200 or to consider national guideline — Application to quash dismissed as TPO would have been confirmed in any event

***Armitage* v *South Oxfordshire District Council (QB),* [1998]** 2 **121**

TOWN AND VILLAGE GREENS

Class (c) green

Registration

Commons Registration Act 1965 — Commons Registration (New Land) Regulations 1969 — Whether 20-year qualifying period any period of 20 years — Whether any application after 30 January 2001 engages only amended definition of class (c) green — Whether continuity of user in relation to class (c) green within amended definition required to date of application or of registration — Whether applications can be amended — Whether less land than applied for capable of registration — Effect of s12 of Inclosure Act 1857 and s29 of Commons Act 1876 — Effect of inaccessible areas on land subject to application for registration — Effect of tracks on land capable of being public rights of way — Held, any 20-year period of user sufficient for class (c) greens — Application after 30 January 2001 engages only amended definition where 20-year period of user not satisfied by that date — Applications not amendable but registration authority entitled to have regard to all circumstances subject to materiality — Guidance given on many issues concerning registration

***Oxfordshire County Council* v *Oxford City Council (LT),* [2004]** 1 **105**

Commons Registration Act 1965

Registration

Lawful sports and pastimes — 20 years' user — Evidence of user — Whether use of footpaths wrongly taken into account — Agricultural use — Whether recreational use must interfere with agricultural user or vice versa — User as of right — Whether public statements of local inhabitants inconsistent with recreational use admissible — Identification of locality — Held, evidence of use of footpaths not relevant to use of land — Recreational use must interfere with agricultural use — Public statements of inhabitants inconsistent with claim to recreational use as of right — Ecclesiastical parish consituting locality

***R (on the application of Laing Homes Ltd)* v *Buckinghamshire County Council (Admin),* [2003]** 3 **69**

Registration

Commons Commissioners

European Convention on Human Rights — S13 of Commons Registration Act 1965 — Whether applications under s13 relating back to entry on the registers of greens made prior to 3 January 1970 — Whether Commons Commissioners having jurisdiction to determine applications to register new commons — Whether determination of s13 applications by registration authorities provisional only — Whether determination of s13 applications by registration authorities violating Art 6 of Convention — Held, s13 registration process not relating back to greens registered prior to 3 January 1970 — Commons Commissioners having no jurisdiction under s13 — Determination of s13 applications not violating Art 6

***R (on the application of Whitmey)* v *Commons Commissioners (CA),* [2004]** 3 **1**

Sports and pastimes as of right

Commons Registration Act 1965 — Overgrown site owned by developer registered as green on basis of 20-years user as of right — Procedure for determination of factual issues — Locality delineated on plan — Remedy — Whether user as of right — Whether fairness requiring non-statutory inquiry — Whether any area arbitrarily delineated on application plan a locality — Meaning of "locality" and "neighbourhood" — Whether remedy judicial review or application under s14 of 1965 Act — Whether user "continued" as of right after objection to earlier application for registration — Held, user of overgrown site not capable of user as of right for claimed pastimes — Fairness requiring non-statutory inquiry or hearing — Red line on map insufficient to identify "locality" — Challenge to registration can be brought by judicial review or under section 14 of 1965 Act — User not continued "as of right" after objection to first application for registration — Locality must be administrative unit

***R (on the application of Cheltenham Builders Ltd)* v *South Gloucestershire District Council (HL),* [2004]** 1 **85**

TOWN AND VILLAGE GREENS — continued

S22 of Commons Registration Act 1965
Registration
Sports and pastimes indulged in for 20 years — User tolerated and encouraged by owner — Whether user as of right — Whether user with implied permission — Whether implied permission always negating user as or right — Held, implied permission normally negating as of right— Permission of public authority insufficient to negate as of right — User as of right
***R (on the application of Beresford)* v *City of Sunderland (Ch),* [2004] 1 94**

TREES — See **Nuisance** (Subsidence — Tree roots)

TREE PRESERVATION ORDER — See **Town and country planning** (Tree preservation order)

TREE ROOT DAMAGE — See **Compensation** (Refusal of consent to fell tree)
Adopted highway
Liability of highway authority
Post-adoption tree causing tree root damage to adjoining building — Whether highway authority liable in nuisance and negligence — Held, highway authority liable for pre- and post-adoption trees provided reasonable foreseeability of damage
***Hurst* v *Hampshire County Council (CA),* [1997] 2 165**

TRESPASS — See also **Agricultral holdings** (Agricultural tenancy — Proprietary estoppel); Commons (Rights of Common — Extent of grazing rights); **Easement** (Rights of way — Remedy for trespass); **Landlord and tenant** (Secure tenancy — Suspended possession order); **Licence** (Eviction — Damages); **Possession** (Rights to possession — Contractual rights); **Water**
Damages — See **Damages** (Lord Cairns' Act)
Injunction
Acquiescence
Defence — Estoppel — Defendant's objects on claimant's land — Objects not objected to for three years — Whether defence of acquiescence or estoppel based on mere delay in objecting — Whether defence based on reliance and detriment — Held, defence of acquiescence or estoppel based on reliance and detriment not mere delay
***Jones* v *Stones (CA),* [1999] 3 81**

Possession order
Land occupied by trespassers — Owner fearing trespassers would move onto other areas — Whether possession order can include areas of land not occupied by trespassers — Whether jurisdiction to make possession order — Whether real danger of trespass by defendants to other areas — Held, jurisdiction to include areas in possession order additional to area occupied — Must be real danger of trespass to other areas
***Drury* v *Secretary of State for the Environment, Food and Rural Affairs (CA),* [2004] 3 85**

Remedy
Erection of building
Building erected partly on plaintiff's land — Whether plaintiff entitled to damages or mandatory relief — Held, court has no real discretion — Plaintiff has discretion whether to accept trespass as accretion to his land or to obtain possession order or injunction
***Harrow London Borough Council* v *Donohue (CA),* [1995] 1 257**

Shopping centre
Pedestrian malls — See **Shopping centre** *(Pedestrian malls)*

TRUSTS — See also **Agricultural holdings** (Notice to quit); **Co-ownership**; **Insolvency** (Bankruptcy — Foreign property); **Landlord and tenant** (Service charges — Statutory trust); **Beneficial interest** — See **Trusts of Land and Appointment of Trustees Act 1996** (Mortgage — Charge of beneficial interest)
Joint ownership
Trusts of Land and Apppointment of Trustees Act 1996
Family home — Trusts of Land and Appointment of Trustees Act 1996 — Whether order for sale — Whether previous law on s30 of Law of Property Act 1925 relevant — Held, order for sale contemplated if compromise loan transaction not achieved — Previous law on s30 not necessarily relevant
***Mortgage Corporation* v *Shaire (Ch),* [2000] 3 131**

Trustees
Liabilities
Environmental Protection Act 1990 — Whether trustee entitled to lien over trust fund for future and contingent liabilities under 1990 Act — Whether trustee entitled to invest and vary investments — Appropriate directions — Held, trustee has lien over trust fund in respect of future and contingent liabilities under 1990 Act — Trustee entitled to invest trust fund with regard to his interest as trustee
***X* v *A (Ch),* [2000] 1 19**

TRUSTS OF LAND AND APPOINTMENT OF TRUSTEES ACT 1996 — See also **Co-ownership; Landlord and tenant** (Joint tenancy)
Mortgage
Charge of beneficial interest
Application for orders of possession and sale — Whether principles under s30 of Law of Property Act 1925 relevant — Whether orders should be made — Held, s30 principles relevant — In absence of exceptional circumstances chargee entitled to orders
***TSB Bank plc* v *Marshall (CC),* [1998] 3 100**

UNFAIR CONTRACT TERMS ACT 1977 — See **Estate agents** (Negligence — Duty of care and liability)

USE AND OCCUPATION
Compensation
Tenancy at will
Valuation — Failure of negotiations to grant lease of commercial premises — Value of use and occupation — Whether rent agreed for proposed lease best evidence of value of use and occupation — Whether value tenant at will actually obtained relevant — Held, unless basis of agreement substantially different from tenancy actually held, rent agreed for proposed lease good evidence of value of use and occupation
***Lewisham London Borough Council* v *Masterson (CA),* [2000] 1 134**

USER COVENANT — See **Landlord and tenant** (Covenants) (User covenant — Consent for change of use) (User covenant — Valuation)

VALUATION — See **Agricultural holdings** (Agricultural tenancy — Valuation); **Compensation for acquisition of land** (Open market value) (Value of land); **Compulsory Purchase** (Compensation — Post-valuation date damage); **Experts** (Third-party determination — Non-speaking valuation); **Landlord and tenant** (Assignment — Landlord and Tenant Act 1988) (Dilapidations — Damages) (Dilapidations — Valuation) (Leasehold enfranchisement — Acquisition of freehold) (Leasehold enfranchisement — Leasehold Reform, Housing and Urban Development Act 1993) (Rent review — Valuation hypothesis) (User covenant); **Leasehold enfranchisement** (Leasehold Reform Act 1967 — Determination of price) (Leasehold Reform Act 1967 — Enfranchisement price) (Leasehold Reform Act 1967 — S9(1A) valuation) (Leasehold Reform, Housing and Urban Development Act 1993 — Collective enfranchisement) (Leasehold Reform, Housing and Urban Development Act 1993 — Extended lease); **Mortgage** (Sale — Power of sale); **Negligence** (Mortgage loan report — Valuation) (Surveyors — Mortgage valuation report) (Valuers — Duty of care) (Valuers — Limitations) (Valuer — Mortgage valuation); **Rates and rating** (Rebus sic stantibus ("the actual state rule")
Compensation — See **Compensation** (Land Compensation Act 1973 Part I — Valuation evidence); **Compensation for acquisition of land** (Valuation)
Damages — See **Landlord and tenant** (Housing Act 1988 — Damages for eviction)
Marriage value
Lotting
Farmhouse and adjoining land — Valuation under Inheritance (Provision for Family and Dependants) Act 1975 — Whether three parcels of adjoining land having marriage value — Held, asset valuation reflecting marriage value of surrounding land to principal farmhouse
***Quilter* v *Quilter (Ch),* [2004] 1 79**

Negligence — See **Expert, unlawful appointment of** (Negligent valuation); **Negligence** (Damages: measure and quantification) (Duty of care and liability — Disclaimer) (Limitation Act — Non-transaction mortgage loan) (Mortgage loan report — Contributory negligence) (Valuers)

Private property company
Discount to net asset value
Valuation under Inheritance (Provision for Family and Dependants) Act 1975 — Private property companies — Discount to net asset value — Whether discount 15% or 30% — Held, value of net assets of private property companies discounted by 15%
***Quilter* v *Quilter (Ch),* [2004] 1 79**

Property
Stokes principle
Dispute over right of way compromised by Tomlin order incorporating agreement to purchase access by reference to *Stokes* principle — Whether costs of development to be deducted from site value — Held, no costs to be deducted from valuation already on net basis
***Challock Parish Council* v *Shirley (Ch),* [1995] 2 137**

Rates and rating — See **Rates and rating** (Valuation)
Rent review clause — See **Leasehold enfranchisement** (Leasehold Reform Act 1967 — Determination of enfranchisement price)
Taxation purposes — See **Taxation** (Valuation — Agricultural land)

VALUE ADDED TAX
Alterations
Listed buildings
Value Added Tax Act 1983, sched 5, group 8A — Grade II house damaged by fire — Whether an alteration or reconstruction — Held, tribunal failed to consider whether works were reconstruction
***Commissioners of Customs & Excise* v *Moorish (QB),* [1998]** 3 **117**

VENDOR AND PURCHASER — See also **Mistake** (Rectification — Form of relief)
Condition of sale
Implied easements
Law Society's General Conditions — Effect of condition 5(3) on sale of part of vendor's land — Quasi–easements without grant or prescription — Whether vendor entitled to easements over land conveyed — Held, intentions of parties relevant — Not intention of parties that easements should be created on sale of part of vendor's land — Neither rule in *Wheeldon* v *Burrows* nor s62 of the Law of Property Act 1925 had application to reserve easements in favour of vendor's retained land
***Selby District Council* v *Samuel Smith Old Brewery (Tadcaster) Ltd (CA),* [2001]** 1 **71**

Contract for sale of land
Notice to complete
Leasehold interest — Purported forfeiture — Title guarantee — Form of registered title — Statutory power to sell — Whether notice to complete valid — Whether notice valid where forfeiture action founded on ineffective s146 notice — Held, notice to complete valid — Seller not unable to make title at time when seller made defendant to forfeiture action founded on ineffective s146 notice — Other matters relating to title guarantee, form of registered title and statutory power of sale contained no impediment to sale
***Lambeth London Borough Council* v *Vincent (Ch),* [2000]** 2 **73**

Conveyance
Hedge and ditch presumption
Boundaries — Hedge and ditch presumption — Conveyance by reference to Ordnance Survey plan — Whether presumption rebutted — Held, presumption rebutted by vendor's admission that land did not include ditch.
***Alan Wibberley Building Ltd* v *Insley (CA),* [1998]** 2 **107**

Deposit
Purchaser's lien
Purchasers making deposits in respect of proposed flats in refurbishment scheme — Vendor developer insolvent — Whether purchasers entitled to lien on property to protect deposits in priority to successor in title to vendor — Held, irrelevant contracts conditional — Priority given to contract made prior to application by vendor for registration of its title — No constructive trust
***Chattey* v *Farndale Holdings Inc (CA),* [1997]** 1 **153**

Estate agents
Duty to protect vendor's interest — See **Estate agents** *(Duties)*
Lock-out agreement
Breach
Exclusivity of negotiations secured by purchaser — Alleged breach by vendors — Interlocutory relief — Whether injunction appropriate remedy — Held, injunction not appropriate remedy as damages adequate
***Tye* v *House (Ch),* [1997]** 2 **171**

Option
Exercise
Development land — Option excercisable following application for development of property — 45-acre site — Application and permission for building on 2.5 acres plus landscaping — Whether application for development of property — Whether "property" including "any part or parts" — Whether "development" having meaning under planning legislation — Whether engineering operations for landscaping constituting "development" under agreement — Held, "property" meaning "property as a whole" and not "property and any part or parts" — Planning application not for development of property — 'Development' interpreted against context of agreement — "Development" not having wider meaning under planning legislation — Engineering operation for landscaping not constituting "development" under agreement — Option not validly exercised
***Hallam Land Management Ltd* v *UK Coal Mining Ltd (CA),* [2002]** 2 **80**

Time of the essence
Option exercisable on specified events — Whether time of the essence for exercise of option — Held, time of the essence applies as a universal rule to an option to purchase
***Di Luca* v *Juraise (Springs) Ltd (CA),* [1998]** 2 **125**

Valuation
Non-speaking valuation — Determination of option price — Whether expert's determination void — RICS valuation practice statements — Whether expert carried out task directed by option price formulation — Held, inferences cannot be drawn as to reasons behind a non-speaking valuation — Valuer made no mistake — Mistake in formulation of price in option agreement
***Morgan Sindall plc* v *Sawston Farms (Cambs) Ltd (CA),* [1999]** 1 **90**

Sale of land
Law of Property (Miscellaneous Provisions) Act 1989
Letter signed by vendor and accompanying plan signed by vendor and purchaser — Whether letter and plan one document — Whether document signed by purchaser — Whether typing of purchaser's name as addressee on letter a signature — Held, two documents the letter referring to the plan — 1989 Act not satisfied as letter not signed — Typing of addressee's name not signature
***First Post Homes Ltd* v *Johnson (CA),* [1996]** 1 **175**

Planning obligation — S106 agreement — Obligation to sell land — S2 of Law of Property (Miscellaneous Provisions) Act 1989 — Whether agreement void for failing to contain signature — Whether implication of term — Whether rectification of agreement — Held, s106 agreement containing obligation to sell land within s2 of the 1989 Act — Agreement unenforceable without signature of purchaser
***Jelson Ltd* v *Derby City Council (QB),* [1999]** 3 **91**

Property
Flagstones — Vendor removing flagstones prior to exchange of contracts — Vendor giving fraudulent answer about removal of flagstones — Whether flagstones part of "the property" — Whether vendor breaching contract — Whether purchaser entitled to order requiring restoration of flagstones — Held, flagstones part of property — Vendor in breach of contract — Vendor ordered to restore flagstones
***Taylor* v *Hamer (CA),* [2003]** 1 103

VILLAGE GREEN — See Town and village greens
Commons Registration Act 1965
User as of right
Sports and pastimes — Whether activities exercised as of right — Whether subjective test of user as of right — Whether *R* v *Suffolk County Council, ex p Steed* [1995] 2 EGLR 232 correctly decided — Held, no subjective test in establishing user as of right — Belief in existence of right for local inhabitants irrelevant — *Steed* wrongly decided
***R* v *Oxfordshire County Council, ex p Sunningwell Parish Council (HL),* [1999]** 2 **94**

WAIVER — See **Compensation** (Limitation period); **Election** (Waiver — Landlord and tenant); **Landlord and tenant** (Forfeiture); **Leasehold enfranchisement** (Leasehold Reform, Housing and Urban Development Act 1993 — New extended lease application)

WASTE — See **Compensation for acquisition of land** (Valuation — Waste tip); **Environmental Protection Act 1990** (Waste-management licence — Insolvency); **Insolvency** (Disclaimer of licence — Waste-management licence); **Landlord and tenant** (Repairs — Covenant) (Waste, liability for — Tenant for years)

WATER
Trespass
Damages
Water undertaker laying water main in land without serving notice under s159 of Water Industry Act 1991 — Claim for damages — Whether damages loss of development value or restitutionary damages based on value of benefit to water undertaker — Whether damages based on compensation payable under s180 of and Schedule 12 to Water Industry Act 1991 — Held, damages based on loss of opportunity to negotiate had statutory notices been served — Damages based on value of water main to water undertaker during period of trespass not payable
***Severn Trent Water Ltd* v *Barnes (CA),* [2004]** 2 **95**

WALKWAY AGREEMENTS — See **Shopping centre** (Pedestrian malls)

WATER
Compensation for water main construction — See **Compensation for water main construction**
Pipeline
Compensation
Depreciation in value — Pipeline constructed through garden of private house — Whether 50% of value of easement strip plus depreciation to value of house appropriate basis — Date from which statutory interest payable — Items of disturbance claim — Held, 2.5% depreciation in value of whole property appropriate — Compensation for disturbance items — Statutory interest payable from date of formal claim
***Bestley* v *North West Water Ltd (LT),* [1998]** 1 **187**

TABLE OF STATUTES

References are to year, number and volume

eg **[1995]** 2 **127** *et seq* refers to vol 2 of 1995 at p127 and following

Access to Justice Act 1999
- s54(4) . . . **[2003]** 3 **129, 132, 133**

Acquisition of Land Act 1919
- r4 . . . **[2003]** 2 **6**

Acquisition of Land (Authorisation Procedure) Act 1946
- Sched 2
  - para 3 . . . **[2003]** 1 **118**
  - para 3(1) . . . **[2003]** 1 **122**

Administration of Justice Act 1970
- s36 . . . **[1995]** 1 **127** *et seq*, **[1997]** 1 **143**

Administration of Justice Act 1973
- s8 . . . **[1997]** 1 **143**

Agricultural Holdings Act 1923
- s14(2)(*b*) . . . **[2004]** 2 **182**
- s21 . . . **[2004]** 2 **182**
- s24(1) . . . **[2004]** 2 **182**
- s27 . . . **[2004]** 2 **182**

Agricultural Holdings Act 1986 . . . **[1997]** 2 **10** *et seq*, **[1998]** 3 **3** *et seq*, **[2002]** 1 **2** *et seq*
- s2 . . . **[2002]** 1 **2, 5**
- s2(1) . . . **[1998]** 3 **1** *et seq*
- s3 . . . **[1997]** 2 **6** *et seq*
- s5 . . . **[1997]** 2 **6** *et seq*
- s12 . . . **[1996]** 1 **1** *et seq*, **[2000]** 1 **2, 4, 5**
- s25 . . . **[2002]** 1 **5**
- s26 . . . **[2000]** 1 **3, 4, 9**
- s26(1) . . . **[1997]** 1 **5**
- s36(3)(a) . . . **[1995]** 2 **1** *et seq*
- s39 . . . **[1995]** 2 **1**
- Sch 2
  - para 1(3)(a) . . . **[1996]** 1 **5**
  - para 1(3)(b) . . . **[1996]** 1 **6**
- Sch 3
  - Case B . . . **[1995]** 1 **10** *et seq*, **[1996]** 1 **7, [2000]** 2 **1** *et seq*
  - Case D . . . **[2000]** 1 **2, 3, 5, 6**
  - Case E . . . **[1995]** 1 **6** *et seq*
- Sch 11 . . . **[2000]** 1 **2, 3, 4, 6, 7**
  - para 27 . . . **[2000]** 1 **1, 4, 8**

Agricultural Holdings (Scotland) Act 1991
- s22(2)(d) . . . **[2001] 1** 6, *et seq*

Allotments Act 1925
- s8 . . . **[1996]** 1 **147** *et seq*, **2 117**

Arbitration Act 1950
- s12(4) . . . **[1995]** 1 **103**
- s20 . . . **[1998]** 2 **14** *et seq*
- s22 . . . **[1995]** 1 **246** *et seq*, **2 5**
- s22(1) . . . **[1999]** 2 **1, 4**
- s23 . . . **[1995]** 1 **15, [1996]** 2 **23, 26**
- s23(2) . . . **[1999]** 2 **1,4**

Arbitration Act 1979
- s1 . . . **[1996]** 2 **86**
- s1(5) . . . **[1999]** 2 **1, 4**

Arbitration Act 1996
- s1 . . . **[2003]** 2 **150**
- s1(a) . . . **[2003]** 2 **156**
- s9 . . . **[1999]** 3 **7, 10**
- s12 . . . **[1998]** 1 **91, 94, [2004]** 1 **44, 47**
- s12(3) . . . **[2004]** 1 **46**
- s12(3)(a) and (b) . . . **[1998]** 1 **91, 94**
- s33 . . . **[2003]** 1 **1, 3, [2003]** 2 **18, 149, 150**
- s34 . . . **[2003]** 1 **3**
- s45 . . . **[2003]** 3 **25**
- s49 . . . **[1999]** 3 **103, 107**
- s68 . . . **[2002]** 2 **97** *et seq*, **[2003]** 1 **1, 3 [2003]** 2 **16, 17, 18, 20, 21, 149, 150, [2004]** 2 **43, 44, 45, 47**
- s68(2) . . . **[2003]** 1 **1, [2003]** 2 **18**
- s68(2)(c) . . . **[2003]** 1 **3, 4**
- s68(2)(d) . . . **[2003]** 1 **1, 6, [2004]** 2 **44, 47**
- s68(3) . . . **[2003]** 2 **18**
- s69 . . . **[2002]** 2 **97** *et seq*, **[2003]** 2 **18, 21, [2004]** 2 **43, 44, 45, 47**
- s69(1) . . . **[2003]** 2 **18, [2004]** 2 **46**
- s69(2) . . . **[2003]** 2 **18**
- s69(3) . . . **[2003]** 2 **18, 21**
- s69(3)(a) . . . **[2004]** 2 **46**
- s69(3)(b) . . . **[2004]** 2 **44, 47**

Bankruptcy Act 1869
- s125 . . . **[1998]** 3 **76**

Capital Gains Act 1979
- s19(1) . . . **[2003]** 2 **78**
- s20(1) . . . **[2003]** 2 **78**
- s27 . . . **[2003]** 2 **73, 75**
- s27(1) . . . **[2003]** 2 **73, 74, 76** *et seq*
- s46(1) . . . **[2003]** 2 **74, 76, 79**
- s32(1)(*a*) . . . **[2000]** 2 **69, 71**
- s32(*b*) . . . **[2000]** 2 **69, 71**

Chancel Repairs Act 1932
- s2(3) . . . **[2000]** 2 **150**

Chancery Amendment Act 1858 (Lord Cairns' Act) . . . **[2001]** 1 **81,** *et seq*
- s2 . . . **[1995]** 1 **148** *et seq*

Charging Orders Act 1979
- s1(5) . . . **[2002]** 1 **84**
- s2 . . . **[1999]** 2 **52**
- s3(4) . . . **[2001]** 2 **26**

Civil Evidence Act 1968
- s3(1)(a) . . . **[1995]** 1 **107**

Civil Evidence Act 1972
- s3 . . . **[1998]** 2 **103, 104, 105**

Civil Evidence Act 1995
- s1 . . . **[2003]** 1 **178**
- s2(4) . . . **[2003]** 1 **178**
- s11 . . . **[2003]** 1 **178**

Civil Procedure Act 1997
- s 4(2) . . . **[1999]** 2 **81, 82**
- s 8(1) . . . **[1999]** 2 **82**
- s 8(2) . . . **[1999]** 2 **82**

Commons Act 1876
- s29 . . . **[2003]** 1 **24, 28, [2003]** 3 **72, 75, 143, 145, [2004]** 1 **105, 112**

Commons Registration Act 1965
- s1(1) . . . **[2004]** 1 **107, 112**
- s1(2) . . . **[2004]** 1 **107, 108, 112**
- s1(2)(*a*) . . . **[1995]** 2 **233, [1997]** 1 **133, 134, 135, [2003]** 3 **70 71, [2004]** 3 **7**
- s1(3) . . . **[2003]** 3 **70**

s1(3)(*b*) .... **[2004]** 1 **107, 111**
s2 .... **[2003]** 3 **70, [2004]** 1 **108,**
s2(1)(*a*) .... **[2004]** 1 **97**
s3 .... **[2003]** 3 **70, [2004]** 1 **108**
s3(1)(*b*) .... **[2004]** 1 **97**
s5(7) .... **[2004]** 3 **1, 3**
s8 .... **[2004]** 1 **107**
s10 .... **[2003]** 3 **70, [2004]** 1 **108,** **[2004]** 3 **70**
s13 ... **[1997]** 1 **131,** *et seq,* **[2004]** 1 **94, 96, 105, 108, 110, 111, 112, 113, 116,** **[2004]** 3 **1** *et seq*
s13(3) .... **[2003]** 3 **69, 70**
s13(*b*) .... **[2004]** 1 **113**
s14 .... **[2004]** 1 **85, 86, 88, 90, 91, 108, 118, [2003]** 3 **71**
s14(*b*) .... **[2004]** 3 **1, 5, 7**
s15(3) .... **[2004]** 1 **107, 111, [2003]** 3 **71**
s16(2) .... **[2004]** 1 **113, 114**
s19 .... **[2003]** 3 **71**
s22 .... **[1995]** 2 **233, [1997]** 1 **132,** *et seq,* **[2004]** 1 **90, 94, 95, 96, 104, 107, 112, 113, 114, 118, [2003]** 3 **69, 70, 76**
s22(1) .... **[2003]** 3 **69, 72, 73, 75, 76, 77 80, 82, [2003]** 3 **69, 71, 76, [2004]** 1 **86, 96, 99, 100, 101, 102, [2003]** 3 **69, 71, 73, 75, 76, 77, 78, 80, 81, 82**
s22(1)(*c*) .... **[1999]** 2 **95, 97, [2004]** 1 **98**
s22(1A) .... **[2004]** 1 **90, 91, 93, 94, [2003]** 3 **71**
s22(1A)(*b*) .... **[2004]** 1 **86, 109, 115**
s29 .... **[2003]** 3 **74, 75, 76**
Companies Act 1985
s36A(5) .... **[2004]** 1 **32**
s36A(6) .... **[2004]** 1 **32**
s458 .... **[1996]** 1 **139**
s654 .... **[1998]** 2 **63, 66**
s657 .... **[1998]** 2 **66**
s736(1) .... **[1997]** 2 **46, 47**
s736A(5) .... **[1997]** 2 **48**
Compulsory Purchase Act 1965
s1(3) .... **[2003]** 3 **2,**
s3 .... **[2003]** 3 **1** *et seq*
s5(2) .... **[2004]** 2 **175, 178, 179, 181**
s5(6) .... **[2004]** 2 **175, 179, 181, 183**
s6 .... **[1998]** 2 **233, [2004]** 2 **176**
s7 .... **[1995]** 1 **20**
s9 .... **[2003]** 3 **2**
s10 .... **[1996]** 2 **115, [1998]** 2 **222, 229, [2000]** 2 **5** *et seq* **[2000]** 3 **133** *et seq* **[1999]** 2 **109** *et seq*
s11 .... **[1997]** 2 **21,** *et seq,* **[1998]** 3 **18** *et seq,* **[2003]** 3 **1** *et seq* **[2004]** 2 **178**
s11(1) .... **[1998]** 1 **1** *et seq,* **2 15, [2003]** 3 **1** *et seq* **[2004]** 2 **2,**
s14 .... **[1998]** 2 **233** *et seq*
s15 .... **[1998]** 2 **233, 234, 235**
s16 .... **[1998]** 2 **233, 235**
s17(1)(*b*) .... **[1998]** 2 **233, 235**
s20 .... **[1999]** 3 **183, [2004]** 2 **175, 177** *et seq*
s20(1) .... **[2004]** 2 **181**
Compulsory Purchase (Vesting Declarations) Act 1981
s9(1) .... **[2004]** 2 **178**
s10(3) .... **[1995]** 1 **229, [1996]** 2 **143** *et seq,* **[1998]** 3 **11, 12**
s13 .... **[1997]** 1 **131** *et seq*
s22 .... **[1997]** 1 **132** *et seq*
Consumer Credit Act 1974
s127(3) .... **[2004]** 2 **59, 60**
Counter-Inflation Act 1973
s11(1) .... **[2000]** 1 **169, 170**
Countryside and Rights of Way Act 2000
s67 .... **[2003]** 1 **26**
s68 .... **[2003]** 1 **18, 23, 24, 25, 28, [2004]** 2 **16, 17**
s68(1)(b) .... **[2003]** 1 **28**
s98 .... **[2003]** 3 **71**
County Courts Act 1984
s5(1) .... **[1999]** 2 **31, 32, 33**
s7(3) .... **[1999]** 2 **31, 32, 33**
s45(4) .... **[1999]** 2 **31, 32, 33**
s86 .... **[2002]** 1 **83**
s138 .... **[1997]** 2 **96** *et seq,* **[1998]** 2 **47** *et seq,* **[1999]** 2 **30** *et seq*
s138(2) .... **[1995]** 2 **23,** *et seq*
s138(3) .... **[1997]** 2 **97,** *et seq*
s138(9) .... **[1996]** 1 **91**
s138(9A) .... **[1995]** 2 **24**
s138(9B) .... **[1995]** 2 **24**
s138(9C) .... **[1999]** 3 **29** *et seq*
s139 .... **[2000]** 1 **23, 27, 28**
s139(2) .... **[2001]** 2 **30, 3 39**
Courts and Legal Services Act 1990
s17 .... **[2003]** 3 **135, [2004]** 3 **135**
s17(3) .... **[2003]** 3 135
s27 .... **[2003]** 3 **135, 137**
s27(2) .... **[2003]** 3 **135**
s27(4) .... **[2003]** 3 **135**
s28 .... **[2003]** 3 **135**
s119 .... **[2003]** 3 **135**
Crime (Sentences) Act 1997
s29 .... **[2004]** 2 **136**
Criminal Justice Act 1991
s18(3) .... **[2000]** 3 **18**
Criminal Procedure Act 1865
s4 .... **[1995]** 1 **107**
s5 .... **[1995]** 1 **107**
Defective Premises Act 1972
s4 .... **[1998]** 3 **38, 39, [2001]** 1 **53, 55, 56, 57, [2002]** 1 **112, 113, [2002]** 3 **112. 113**
s4(3) .... **[2001]** 1 **53, 55, 56**
Disability Discrimination Act 1995
s22 .... **[2003]** 2 **14, 15**
s22(1)(a) .... **[2003]** 2 **15**
s24 .... **[2003]** 2 **15, 16**
s25 .... **[2003]** 2 **15**
s57 .... **[2003]** 2 **15, 16**
Distress for Rent Act 1737
s10 .... **[1998]** 3 **147**
s18 .... **[1999]** 2 **23** *et seq,* **[1998]** 3 **60. 61**
s19 .... **[1999]** 3 **148**
Electricity Act 1989
s4 .... **[2003]** 1 **111**
s6 .... **[2003]** 1 **111**
s9 .... **[2003]** 1 **112**
s37 .... **[2003]** 1 **112**
Sch 3 .... **[2003]** 1 **112**
Sch 4 .... **[2003]** 1 111
para 6 .... **[2003]** 1 **111** *et seq*
para 6(1) .... **[2003]** 1 **112, 113**
para 6(2) .... **[2003]** 1 **112, 113**
para 6(3) .... **[2003]** 1 **111** *et seq*
para 6(4) .... **[2003]** 1 **112**
para 6(5) .... **[2003]** 1 **112**
para 6(8) .... **[2003]** 1 **112**
para 7 .... **[1998]** 2 **217** *et seq,* **[2003]** 1 **112**
para 7(1) .... **[1999]** 1 **177, 178, [2003]** 1 **112**
para 7(2) .... **[2003]** 1 **112**
para 7(4) .... **[2003]** 1 **112**
Environmental Protection Act 1990
s35(11) .... **[1999]** 3 **23, 24, 25, 26**
s40(1) .... **[1999]** 3 **23, 25**
s78A .... **[2000]** 1 **19**
s78F .... **[2000]** 1 **20**
s78K .... **[2000]** 1 **20**
s79 .... **[1998]** 2 **20**
s79(1)(a) .... **[1998]** 3 **25** *et seq*
s80 .... **[1999]** 3 **13**
s80(2)(b) .... **[1999]** 3 **13, 15**
s80(4) .... **[1999]** 3 **15**
s82 .... **[1998]** 2 **19** *et seq*
s82(12) .... **[1998]** 2 **19** *et seq*
Equal Pay Act 1970
s1(2) **[2004]** 2 **138**
Estate Agents Act 1979
s3 .... **[1998]** 1 **9** *et seq*

s3(1) . . . . **[2001]** 1 **21, 23, 26**
s3(1)(*a*)(i) . . . . **[1996]** 1 **229**
s3(2)(*a*)(i) . . . . **[1998]** 1 **9** *et seq*
s3(1)(d) . . . . **[2001]** 1 **21**, *et seq*
s4 . . . . **[2001]** 1 **21, 27**
s18 . . . . **[1998]** 1 **6**
Finance Act 1972
s45 . . . . **[2004]** 2 **64**
Finance Act 1975
s38(1) . . . . **[1996]** 1 **160** *et seq*
Sch 4
para 4 14(2) . . . . **[1996]** 1 **161**
Finance Act 1985
s69 . . . . **[1999]** 2 **191** *et seq*
s69(2)(*a*) . . . . **[1996]** 1 **164** *et seq*, **[1998]** 2 **117, 118, 119**
s69(2)(*b*) . . . . **[1999]** 2 **192, 193, 194**
s69(4)(*a*) . . . . **[1999]** 2 **194**
s70 . . . . **[1999]** 2 **191, 193**
Sch 20 . . . . **[1998]** 2 **117, 118**
para 4(1) . . . . **[1999]** 2 **191**
para 4(2) . . . . **[1999]** 2 **191**
para 10(1)(*c*) . . . . **[1999]** 2 **192, 193, 194**
para 13(1) . . . . **[1999]** 2 **191**
General Rate Act 1967
s20 . . . . **[2000]** 2 **185**
Highways Act 1971
s18 . . . . **[1995]** 2 **133**
Highways Act 1980
s28 . . . . **[2000]** 2 **84** *et seq*
s28(2) . . . . **[2000]** 2 **85**
s31 . . . . **[2004]** 1 **118**
s31(1) . . . . **[2004]** 1 **98, 99**
s35 . . . . **[1995]** 2 **133**
s38 . . . . **[1996]** 1 **49** *et seq*
s96(6) . . . . **[1997]** 2 **166** *et seq*
s246(2A) . . . . **[1995]** 2 **213** *et seq*, **[1996]** 1 **52**
s261(1)(*a*) . . . . **[1995] 169**
s263 . . . . **[2000]** 2 **17**
s265 . . . . **[2000]** 2 **17, 18**
Housing Act 1980
s36 . . . . **[2002]** 3 **11**
Sch 22
para 2 . . . . **[1999]** 2 **121,123, 131, 132, [1999]** 3 **230, 232**
para 5 . . . . **[1995]** 1 **243, [1999]** 2 **117, 118, 120**
Housing Act 1985
s9(1) . . . . **[2004]** 2 **11**
s15(1) . . . . **[2004]** 2 **11**
s17 . . . . **[2004]** 2 **9, 11, 12**
s17(1)(*b*) . . . . **[2004]** 2 **9, 12**
s18 . . . . **[2004]** 2 **11**
s30 . . . . **[2003]** 3 **124**
s32 . . . . **[2000]** 2 **73, 76, 77, 78, [2000]** 2 **73, 76, 77, 78**
s71 . . . . **[2003]** 3 **114**
Part IV . . . . **[1997]** 1 **33**, *et seq*
s79(1) . . . . **[1995]** 2 **10, [1998]** 1 **22** *et seq*
s79(3) . . . . **[1997]** 1 **33**
s81 . . . . **[2003]** 3 **114, 115**
s82 . . . . **[2003]** 3 **115**
s82(1) . . . . **[1997]** 1 **33** *et seq*
s82(2) . . . . **[1997]** 1 **33** *et seq*
s84 . . . . **[1998]** 1 **20**
s85 . . . . **[1997]** 1 **33** *et seq*, **[2000]** 1 **29, 31, 32, [2000]** 2 **35, 37**
s85(2) . . . . **[2000]** 1 **33, 34, [2001]** 3 **16, 18**
s85(2)(*b*) . . . . . **[2000]** 1 **29** *et seq*, 2 **35, 37, [2002]** 3 **105, 106, 108, 109**
s85(3)(*a*) . . . . **[1997]** 1 **33** *et seq*
s85(4) . . . . **[2000]** 1 **30, 32**
s91(1) . . . . **[2000]** 1 **50** *et seq*
s92 . . . . **[1998]** 1 **42**
s121 . . . . **[1998]** 3 **31, 33**
s127 . . . . **[2000]** 3 **19, 22**
s138 . . . . **[1998]** 3 **32**
s155 . . . . **[2000]** 3 **22**
s157 . . . . **[1996]** 1 **209** *et seq*, **[2000]** 3 **22**
s167 . . . . **[2000]** 3 **19**
s168 . . . . **[2000]** 3 **19**
s187 . . . . **[1996]** 2 **89**
s189 . . . . **[2003]** 3 12
Part XI . . . . **[1999]** 3 **17**
s352 . . . . **[1999]** 3 **17**
s563(1) . . . . **[2000]** 2 **26, 27, 28**
Sch 1
para 2(1) . . . . **[1999]** 1 **33, 34**
Sch 2 . . . . **[1999]** 3 **17, [2004]** 3 118
Part I
ground 2 . . . . **[1998]** 1 **15** *et seq*
Sch 6
para 5 . . . . **[2000]** 3 **22**
para 6 . . . . **[2000]** 3 **19, 22**
Housing Act 1988
s1 . . . . **[2001]** 3 **93**, *et seq*
s1(1) . . . . **[2000]** 1 **157** *et seq*
s5 . . . . **[2001]** 2 **38, 41, [2003]** 3 **23**
s5(1) . . . . **[1999]** 2 **139** *et seq*, **[2003]** 3 **23**
s7 . . . . **[2001]** 2 **38, 41, [2002]** 3 **44**
s8 . . . . **[1999]** 2 **139, 140, 141, [2000]** 2 **39, 40**
s8(1)(b) . . . . **[1999]** 2 **139, 141, 142**
s9(1) . . . . **[2003]** 2 **22, 23**
s9(6) . . . . **[2003]** 2 **22**
s13(4) . . . . **[2003]** 1 **58, 59**
s14(1) . . . . **[1997]** 2 **135** *et seq*
s20 . . . . **[1997]** 1 **26, 27, [1998]** 2 **25** *et seq*, **[1999]** 2 **63** *et seq*, **[2000]** 2 **40** *et seq*, **[2002]** 1 **10** *et seq*, 3 **28** *et seq*, **[2003]** 3 **24, 25,**
s20(1)(c) . . . . **[1998]** 2 **25** *et seq*
s20(4) . . . . **[1996]** 2 **67** *et seq*
s21 . . . . **[1999]** 2 **14, 15, 63** *et seq*, **[2003]** 3 **23**
s21(1) . . . . **[2003]** 3 **23**
s21(4) . . . . **[1996]** 2 **67** *et seq*, **[2003]** 3 **22, 23**
s21(4)(a) . . . . **[2003]** 3 **22, 25**
s27 . . . . **[1998]** 1 **30** *et seq*
s27(1) . . . . **[1999]** 1 **28**
s27(7)(a) . . . . **[1999]** 1 **28** *et seq*
s27(8) . . . . **[1999]** 1 **27, 28, 30**
s27(8)(a) . . . . **[1999]** 1 **28** *et seq*
s28 . . . . **[1998]** 1 **30** *et seq*
s28(1) . . . . **[1999]** 1 **27** *et seq*, 3 **158, 160**
s28(3)(a) . . . . **[1999]** 1 **31**
s39(3) . . . . **[1995]** 2 **75** *et seq*
s45(1) . . . . **[2002]** 1 **10**
Sch 1
para 6 . . . . **[1995]** 2 **75**
Sch 2
Part I
ground 1 . . . . **[1997]** 1 **26, 27**
ground 7 . . . . **[2000]** 2 **39, 40**
ground 8 . . . . **[2002]** 3 **42** *et seq*, **[2003]** 2 **21** *et seq*
Part II
ground 10 . . . . **[2000]** 2 **40, [2002]** 3 **42** *et seq*
ground 14 . . . . **[1999]** 2 **139** *et seq*
Housing Act 1996
Part V . . . . **[2003]** 3 **118**
s193(5) . . . . **[2004]** 1 **36**
s193(7F) . . . . **[2004]** 1 **36**
Housing Finance Act 1972
s90 . . . . **[2002]** 1 **35**
Housing Grants, Construction and Regeneration Act
s108 . . . . **[1999]** 3 **8**
Human Rights Act 1998
s3 . . . . **[2001]** 2 **73**, *et seq*, 3 **52, [2003]** 2 **16, [2003]** 3 **86, 90, 114 [2004]** 1 **3, [2004]** 1 **50, 54, 55, 56, [2004]** 2 **57, 58, 132, 134, 135, 136, 137, 138, 139, 140, 141, 143, 149, [2004]** 3 **85, 90**
s3(1) . . . . **[2004]** 1 **1, 3, [2004]** 2 **136, 137, 138, 143, 144, 145, 146, 147, 149,**
s3(2) . . . . **[2004]** 2 **136**
s3(2)(b) . . . . **[2004]** 2 **143**
s3(2)(c) . . . . **[2004]** 2 **143**
s4 . . . . **[2004]** 2 **136, 137, 138, 139, 143, [2003]** 3 **69, 70**
s4(2) . . . . **[2003]** 3 **83**
s4(6) . . . . **[2004]** 2 **139**
s6 . . . . **[2004]** 2 **143**

s6(1) . . . [2001] 1 33, 34, [2004] 2 143, [2004] 3 114
s6(1) . . . [2001] 3 118
s6(2) . . . [2001] 1 33, 34, 44, 45, 46, [2004] 2 143
s6(2)(a) . . . [2004] 2 143
s6(2)(b) . . . [2004] 2 143
s6(3) . . . [2004] 2 143
Inclosure Act 1857
s12 . . . [2003] 1 24, [2003] 3 72, 75, 76, 143, 145, [2004] 1 112,
Ss13(3) . . . [2003] 3 69, 70
Inheritance Tax Act 1984
s115(2) . . . [1996] 1 157 *et seq*
s115(4) . . . [1996] 1 157 *et seq*
Insolvency Act 1986 . . . [1995] 2 40 *et seq*
s4(3) . . . [1996] 2 40
s5(2) . . . [1995] 2 19
s10 . . . [2000] 1 113 *et seq*
s10(1)(*b*) . . . [1999] 2 39, 41, 42
s11 . . . [2000] 1 113, 116, 117
s11(3)(*c*) . . . [1997] 2 55
s11(3)(*d*) . . . [2000] 1 115 *et seq*
s178 . . . [1997] 2 33, 34, 36, [1999] 3 21 *et seq*
s178(4) . . . [1998] 2 66
s178(6) . . . [1996] 2 49 *et seq*, [1997] 2 36, 40, [1999] 1 2, 3, 4
s181(3) . . . [1998] 2 58, 61, 62
s248 . . . [1999] 2 39, 40, 41, 42
s252 . . . [1995] 2 33
s252(2) . . . [1995] 2 33 *et seq*
s252(4)(*b*) . . . [1996] 1 94
s253 . . . [1996] 2 33
s260 . . . [1995] 2 41
s260(2) . . . [1997] 1 43, [1998] 3 75
s260(2)(*b*) . . . [1996] 2 35
s262 . . . [1996] 2 35
s281 . . . [1998] 3 75
s283 . . . [1999] 1 36, 40, 41
s283(1)(*a*) . . . [1999] 1 36, 40
s283(3A) . . . [1999] 1 40, 41
s285(3) . . . [1997] 2 54, 55
s285(4) . . . [1997] 2 54
s306(1) . . . [2000] 2 29
s306(2) . . . [2000] 2 29, 30
s315 . . . [1995] 2 43
s320(2) . . . [1996] 1 151 *et seq*
s383(2) . . . [1997] 2 55
s423 . . . [1995] 1 1 *et seq*
s423(1)(*c*) . . . [1995] 1 1 *et seq*
s436 . . . [1995] 1 1, [1999] 3 21 *et seq*, [2000] 2 29
Interpretation Act 1978
s5 . . . [1995] 1 157, [2003] 1 111, 112, [2004] 1 2
s7 . . . [2003] 3 85, 86, 87, 89, 90, [2004] 1 50 *et seq*,
s26 . . . [2003] 3 86, 87
Judicature Act 1873
s25 . . . [2003] 3 3, 16
s25(6) . . . [2003] 3 3, 16
Land Charges Act 1925
s4(6) . . . [1997] 1 162
s10(1) . . . [1999] 1 62, 63, [2004] 2 84
s50 . . . [2004] 2 84
s50(1) . . . [2004] 2 84
Land Charges Act 1972
s2(1) . . . [1999] 2 22
s2(5) . . . [2004] 2 84
s2(7) . . . [1999] 2 22
s3(5) . . . [2004] 2 84
s11 . . . [2004] 2 84
Class F . . . [1999] 2 22
Land Compensation Act 1961
s1 . . . [1995] 1 267 *et seq*, [1999] 3 2, [2004] 2 176
s2 . . . [2004] 2 111
s4 . . . [1997] 1 167, 168, 179
s4(1) . . . [2003] 1 10, 11, 12
s4(1)(*a*) . . . [2000] 2 81, 84, [2003] 1 10
s5 . . . [2004] 2 177, 178, 179, 181
s5(2) . . . [1995] 1 20, [1996] 1 29 *et seq* [1998] 1 2, [1999] 2 9, 10, 11, [2000] 2 81 *et seq* [2003] 1 124, 126, 127,128, [2003] 2 1, 2, [2004] 2 1 *et seq*, [2004] 2 107
s5(3) . . . [2001] 1 185, [2002] 2 110, 113 [2004] 2 103, 106, 107
s5(4) . . . [1996] 1 244, [1997] 2 181 *et seq* [2003] 2 1, 5, 6
s5(5) . . . [1998] 1 2, [2000] 2 81 *et seq*, 165 *et seq*, [2001] 3 33, *et seq*
s5(6) . . . [1998] 1 137, 2 233, [1999] 2 8, 9, [2001] 1 101, [2001] 3 2, 5, 6, 212, 215, 222, 223, [2000] 2 81, 165, 166, 167, 168, [2004] 2 1 *et seq*
s6 . . . [2002] 2 114, [2004] 2 108, 109, 120
s6(1)(a) . . . [2004] 2 108
s6(1)(b) . . . [2004] 2 108
s9 . . . [2001] 2 145, *et seq*
s11 . . . [1995] 1 272
s14 . . . [1999] 3 3
s14(2) . . . [2001] 1 168, [2003] 2 2
s14(3A) . . . [2003] 2 2
s14(4) . . . [2001] 1 168
s15 . . . [1999] 3 3, [2003] 2 2
s15(1) . . . [2003] 2 1, 2
s15(2) . . . [2003] 2 2
s15(3)(*a*) . . . [2000] 3 153, 158
s15(5) . . . [2003] 2 2
s17 . . . [1995] 2 175, [1996] 2 10 *et seq*, [1998] 2 7, 219 *et seq*, [2001] 1 169
s17(4) . . . [1996] 1 29 *et seq*, [1998] 2 9 *et seq*, 3 13 *et seq*, [2000] 1 13 *et seq*
s17(4)(*b*) . . . [2000] 1 13, 14, 15
s18 . . . [1995] 2 175 *et seq*
s22(2) . . . [1998] 2 9
s22(2)(*a*) . . . [1998] 3 13 *et seq*, [2000] 1 13 *et seq*
s29(1) . . . [2002] 2 161
s31(1) . . . [1996] 2 15 *et seq*
s31(2) . . . [1996] 2 15 *et seq*
s31(3) . . . [1999] 2 196, 197, 201, 202
s32(1) . . . [1998] 2 15
s37 . . . [2004] 2 178, 179, 182, 184
s37(3)(*a*) . . . [2004] 2 184
s38 . . . [2004] 2 182, 184
s38(2) . . . [2004] 2 180, 184
s39(1) . . . [2003] 2 6, 178
s39(2) . . . [2003] 2 2
Land Compensation Act 1973
Part I . . . [1997] 1 245 *et seq*, 2 177 *et seq*, [1999] 2 203, 204, 205
s1 . . . [1995] 1 191 *et seq*, [1996] 1 185 *et seq*, 2 148
s1(6) . . . [1997] 2 178, 179
s3(1)(*f*) . . . [1995] 2 167
s3(2) . . . [1999] 2 205
s6(1) . . . [1995] 1 193 *et seq*
s8 . . . [1995] 2 157
s9(5) . . . [1997] 1 245, 246
s16 . . . [1995] 2 168
s19(2A) . . . [1999] 2 205
s52 . . . [1998] 2 219, 229, [1999] 2 10, [2003] 3 3
s52(1) . . . [1996] 2 15 *et seq*
s52(5) . . . [1996] 2 15 *et seq*
s58 . . . [1995] 2 197
Land Registration Act 1925
s22 . . . [1997] 1 40
s34 . . . [2003] 1 87
s35(1) . . . [2003] 1 87
s70(1)(*a*) . . . [1996] 1 49 *et seq*
s70(2) . . . [1997] 1 154
s110(1) . . . [2003] 3 42
s110(3) . . . [2003] 3 42
Landlord and Tenant Act 1927
s3 . . . [2004] 2 50, 51
s3(1) . . . [2004] 2 50, 51, 52, 53
s3(4) . . . [2004] 2 51, 52
s18 . . . [2001] 2 129 *et seq*

s18(1) [1995] 1 55 *et seq*, [2000] 1 129, 138, 139, 143, 148, [2000] 2 115, 116, [2003] 3 91, 92, 92, 97, 102, 105
s19(1) [2000] 2 50, 51, 53
s19(3) [1995] 1 68 *et seq*
s23 [2002] 2 29 *et seq*, [2003] 3 85 *et seq*, [2004] 1 50 *et seq*,
s23(1) [1998] 1 63, [1999] 2 45, [2004] 3 86

Landlord and Tenant Act 1954
Part I [1999] 1 36 *et seq*, [2004] 2 31, 32, 36, 37,
s3 [2001] 2 52
s3(3) [1997] 2 62 *et seq*
s4 [1997] 2 62 *et seq*, [2004] 2 36
s9 [2004] 2 36
s12 [2004] 2 36
s13 [2004] 2 36
s17 [2003] 2 31, 32, 36, 37
s18(1) [2004] 3 91, 92, 97, 105
Part II [2004] 2 58, 59, 60, 61, 62
s23 [1995] 2 61 *et seq*, [1996] 1 109 *et seq*, [2003] 1 142, [2003] 3 91, [2004] 2 61 *et seq*,
s23(1) [1999] 1 139 *et seq*, [2003] 1 142, [2003] 3 88, [2004] 2 63, 64,
s23(2) [2003] 1 142, [2004] 2 63, 64
s24 [1999] 1 44, 45, [2000] 1 40, [2004] 2 55, 59, 176, 179, 184
s24(1) [1997] 1 74 *et seq*, [2000] 1 40, [2004] 2 56, 60, [2003] 3 28,
s24A [1997] 2 211, [1998] 1 69, 70
s24(3)(*a*) [1997] 1 74, 75
s25 [1997] 1 88 *et seq*, 2 107, 109, [1998] 1 65 *et seq*, [1999] 3 46, 48, [2000] 1 68, 70, 71, [2001] 3 41, *et seq*, [2003] 3 28, 30, [2004] 2 55, 56, 58,
s25(1) [2004] 2 55
s25(3) [2003] 3 28
s25(4) [2003] 3 28
s25(5) [2003] 3 28 *et seq*, [2004] 2 55, 56, 60
s26 [1998] 1 63, 64, 2 73 *et seq*, [2000] 1 138 *et seq*, [2001] 2 59, [2004] 2 60, 62
s26(1) [2004] 2 55
s26(2) [1996] 1 113, [1998] 2 73, 75, [2004] 2 55
s26(3) [1998] 1 63, 64, [2000] 1 145
s26(6) [2003] 3 29
s27(2) [1995] 2 61 *et seq*, [1996] 1 106, [1997] 1 73 *et seq*, [2004] 2 58
s28 [2004] 2 176, 179, 184
s29 [2003] 3 29
s29(2) [2003] 3 28 *et seq*, [2004] 2 55, 56, 60,
s29(3) [2003] 3 29, [2004] 2 62
s30(1)(*a*) [1996] 1 109, [2002] 1 45, [2004] 1 123
s30(1)(*b*) [2002] 1 45, 49
s30(1)(*f*) [1996] 1 109, [1997] 2 110 *et seq*, [1999] 3 46, 47, [2002] 2 147 *et seq*, [2002] 2 147 *et seq*, [2003] 2 42 *et seq*, [2003] 2 87, 88, 90 *et seq*, [2004] 1 136, 137, 141
s30(1)(*g*) [1998] 2 75 *et seq*, [1999] 3 46, 47, [2000] 1 57, 58, 59, 66 *et seq*, 2 45 *et seq*, [2002] 1 50 *et seq*, [2003] 2 37, 38, 39
s30(2) [2000] 1 58, 59, 60
s30(3) [2002] 1 50 *et seq*
s31 [2000] 1 167 *et seq*
s31A [2002] 2 147 *et seq*
s34 [1997] 2 114, [2000] 1 60, 61
s34(1) [2002] 2 48 *et seq*
s34(1)(*d*) [1995] 1 240
s34(2) [2000] 1 60, 61, 62
s37 [2000] 1 138, 146, 148, [2001] 2 61, [2003] 1 141, 142, [2004] 2 56
s37(1)(*e*) [2003] 1 140
s37(1)(*f*) [2003] 1 140
s37(1)(*g*) [2003] 1 140
s37(3) [1999] 3 46
s37(7) [1999] 3 46, 48
s38 [2003] 1 141, 142, [2004] 2 36, 37, [2004] 2 65
s38(1) [1999] 1 45, 46, [2004] 2 36
s38(2) [1998] 3 157 *et seq*, [2003] 1 142
s38(3) [2003] 1 141
s38(4) [1997] 2 105, 106, [2000] 1 64, 65, 66, [2004] 2 67
s38(4)(*a*) [2000] 1 63, 64, 65, [2004] 2 67
s43(3) [1999] 2 58, 59
s43(3)(*b*) [1999] 2 58, 59
s44(1) [1998] 1 70, 72
s64 [1997] 1 233 *et seq*, [2004] 2 61
s64(1)(*c*) [2002] 2 43 *et seq*
s66 [1998] 1 66, 67, 68
s66(4) [2003] 3 85, 86

Landlord and Tenant Act 1961
s32 [2002] 1 39, 40
s11 [1998] 3 38, 39, [2001] 1 53, 54, 55, 56, [2002] 1 25, 26, [2002] 3 33, 34, 35, [2004] 1 62, 63
s11(1)(*a*) [2002] 1 101, 110, 111
s11(1)(*b*) [2002] 1 38 *et seq*
s11(1)(*b*) [2004] 1 63
s11(1A) [2004] 1 63, 64
s11(1A)(*a*) [2004] 1 64
s11(1A)(*b*)(i) [2004] 1 64
s11(3A) [2004] 1 64
s18 [1997] 1 49, 51, [2003] 1 176, 181, [2003] 3 11
s18 (1) [2003] 3 11
s18(2) [2000] 2 106
s19 [1997] 1 49, 51, [1999] 2 53, 54, 57, [2000] 2 105, 106, 109, [2000] 3 89, [2003] 1 176, 178, 179, 181
s19(1) [2003] 1 176, [2003] 3 10, 11
s19(1)(*a*) [1998] 2 155, 156, [2000] 2 105, 106, 110, 111, [2003] 1 176, 177, 182
s19(2) [2003] 3 10, 11
s19(2A) [1999] 2 131, 132, [2000] 3 44 *et seq*, [2001] 2 173, *et seq*, [2003] 1 176
s19(2A)(*a*) [2003] 1 176, 181, 182
s20 [1999] 2 53 *et seq*, [2002] 1 35 *et seq*
s20(4) [1998] 2 155, 156
s20(9) [1998] 2 155, 156, [1999] 2 55, 56
s20B [2003] 3 9 *et seq*
s20B(1) [2003] 3 10, 11
s20B(2) [2003] 3 11
s20C [1995] 2 47 *et seq*, [2001] 2 50, [2003] 1 175, 176, 179, 181, 184, 185, 186
s30 [1999] 2 132
s30A [1997] 1 49, 51
s31 [2000] 1 167 *et seq*, [2001] 1 131, 138, 142
s31A(6) [1999] 2 131, 132
s31A(7)(*a*) [2003] 1 177
Sch
para 8 [1997] 1 49, 52

Landlord and Tenant Act 1987 [1998] 1 131
Part I [2003] 1 56, [2004] 2 121
s1 [2004] 2 123
s1(2) [2004] 2 125
s2 [2004] 2 123
s3 [2004] 2 123
s4 [2004] 2 121, 123
s4(1) [2004] 2 127
s4(4) [2004] 2 127
s5 [1996] 2 25 *et seq*, [2004] 2 121, 123, 125, 126, 127
s5(3) [2004] 2 121, 125, 126, 127
s5(5) [2004] 2 125
s5B [2004] 2 121, 123
s6 [1997] 1 93 *et seq*, [2004] 2 124
s7 [2004] 2 124
s8E [2004] 2 124
s10A [2004] 2 124
s11 [1997] 2 44 *et seq*, [1998] 1 34 *et seq*, [2004] 2 124
s11(1)(c) [1997] 2 49
s11(3) [1998] 1 38, 39
s11A [2004] 1 24, 27
s12 [1995] 1 200, [1996] 1 59, [1996] 2 43 *et seq*, [1997] 1 181, 2 45 *et seq*, 193 *et seq*, [1998] 2 130 *et seq*, [2004] 2 124, 127
s12(1) [1997] 2 193
s12(3)(a) [1997] 2 207
s12(3)(a)(i), [1996] 2 43
s12(4) [1996] 1 59
s12(4)(b) [1995] 1 202 *et seq*
s12(6) [1997] 1 219, [1998] 2 129 *et seq*

s12A . . . . . [2004] 2 **124**
s12B. . . . . [2004] 2 **124**
s13(1)(a) . . . . . [1997] 2 **207**
s13(1)(b) . . . . . [1997] 2 **207**
s16 . . . . . [1996] 1 **59**
s16(1)(b)(ii) . . . . . [1996] 1 **59** *et seq*
s18 . . . . . [1996] 2 **25** *et seq*, [2003] 3 **11**
s18(1) . . . . . [2003] 3 **11**
s18A . . . . . [2004] 2 **124**
s19(1) . . . . . [2003] 3 **10, 11**
s19(2) . . . . . [2003] 3 **9, 11**
s19(2)(A) . . . . . [2001] 3 **28**, [2002] 2 **130**
Part II . . . . . [2003] 1 **52, 54, 56, 57**
s20B . . . . . [2003] 3 **9, 10, 11**
s20B(1) . . . . . [2003] 3 **10, 11**
s20B(2) . . . . . [2003] 3 **1**
Part III . . . . . [2003] 1 **56**
s24 . . . . . [2003] 1 **56, 57**
s24(1) . . . . . [2003] 1 **57**
s24(2) . . . . . [2003] 1 **56, 57** [2003] 2 **28** *et seq*
s24(2)(b) . . . . . [2003] 1 **57**
s24(5) . . . . . [2003] 1 **57**
s24(5)(a) . . . . . [2003] 1 **57**
s24(9) . . . . . [2003] 2 **28** *et seq*
s24(9A) . . . . . [2003] 2 **29, 30**
s31C . . . . . [2002] 2 **130**
Part IV . . . . . [2003] 1 **56**
Part V . . . . . [2003] 1 **56**
s42 . . . . . [2000] 3 **47**, [2003] 1 **42, 45, 52, 54, 55**
s42(3) . . . . . [2003] 1 **45**
s48 . . . . . [1995] 1 **72** *et seq*, [1999] 2 **13, 14, 15**
s54 . . . . . [2004] 1 **24, 27**
s60 . . . . . [2004] 2 **124**
Landlord and Tenant Act 1988
s1 . . . . . [1996] 2 **31**, [2001] 3 **25**, [2002] 3 **21**, [2003] 1 **133**
s1(3) . . . . . [1995] 1 **43** *et seq*, [1996] 2 **205**, [1997] 1 **23**, [1998] 2 **38** *et seq*, **172, 187, 193**, [2002] 1 **24**, [2003] 1 **133, 136, 137, 139, 140**, [2004] 1 **122**
s1(3)(a) . . . . . [2003] 1 **56**, [2003] 1 **133, 136**
s1(3)(b) . . . . . [2003] 1 **133, 136, 137, 140**
s1(6) . . . . . [1997] 1 **23**, [2002] 1 **24**, [2004] 1 **122**
s2 . . . . . [2003] 1 **56**, [2003] 1 **134**
s4 . . . . . [1995] 1 **43**, [2003] 1 **134, 140**, [2004] 1 **122**
Landlord and Tenant (Covenants) Act 1995 . . . . . [2002] 2 **125**
s2 . . . . . [2003] 1 **56**, [2004] 1 **21**
s3 . . . . . [2003] 1 **56**, [2004] 1 **21, 22, 23**
s3(5) . . . . . [2000] 1 **148** *et seq*
s3(6) . . . . . [2003] 1 **56** [2004] 1 **22, 23**
s4 . . . . . [2003] 1 **56**, [2004] 1 **21**
s8 . . . . . [2001] 2 **11, 12, 14**
s15 . . . . . [2003] 1 **56** [2004] 1 **21, 22, 23**
s16(3)(b) . . . . . [2000] 2 **50, 51**
s17 . . . . . [1998] 1 **27** *et seq* [1999] 2 **45** *et seq*
s27(5) . . . . . [1999] 2 **45, 46**, [2003] 3 **87**
s28(1) . . . . . [2004] 1 **21**
Land Registration Act 1925
s19(2) . . . . . [2000] 1 **148, 152**
s20(1) . . . . . [2000] 1 **148, 153, 154**
s50(1) . . . . . [2000] 1 **148, 153, 154, 155**
s70(1)*(a)* . . . . . [2000] 2 **18, 19**
s70(1)*(g)* . . . . . [1999] 1 **86** *et seq*, **121, 124, 125**, [2000] 2 **19**, [2002] 3 **76** *et seq*
s75(1) . . . . . [1998] 3 **55**
Lands Clauses Consolidation Act 1845
s68 . . . . . [1998] 3 **140**
s76 . . . . . [2003] 1 **118**
s77 . . . . . [2003] 1 **117, 118, 120, 121, 123**
s121 . . . . . [2004] 2 **179, 181**
s121(1) . . . . . [2004] 2 **181**
s121(3) . . . . . [2004] 2 **181**
s123 . . . . . [2003] 1 **119**
Lands Clauses Consolidation (Scotland) Act 1845
s19 . . . . . [1995] 1 **232** *et seq*
Lands Tribunal Act 1949
s1(3) . . . . . [2003] 3 **3**
s1(5) . . . . . [2003] 3 **1, 3** [2004] 2 **176**
s3(4) . . . . . [1999] 2 **210, 212**, [2003] 1 **186**, [2003] 2 **29**
s3(5) . . . . . [1998] 2 **15**, [2003] 1 **178**, [2003] 1 **10, 14**
Law of Distress (Amendment) Act 1908
s1 . . . . . [1995] 2 **50**
s4(1) . . . . . [1995] 2 **50**
Law of Property Act 1925
s26(3) . . . . . [2001] 3 **13**
s30 . . . . . [2000] 3 **131**
s36(2) . . . . . [1998] 3 **114**
s40(1) . . . . . [2000] 3 **158**
s52 . . . . . [1999] 2 **36**
s53(1)*(a)* . . . . . [1999] 2 **34, 37**
s54(2) . . . . . [1997] 1 **85** *et seq*, [1999] 2 **34, 35, 36, 37**
s56 . . . . . [1997] 2 **80** *et seq*
s61 . . . . . [2003] 1 **165**
s62 . . . . . [2000] 3 **74** *et seq*, [2001] 1 **71, 119**
s62(2) . . . . . [2003] 1 **29**
s73(1) . . . . . [2004] 1 **31**
s74 . . . . . [2004] 1 **32, 33**
s74(1) . . . . . [2004] 1 **30, 31**
s74(3) . . . . . [2004] 1 **31**
s77 . . . . . [2003] 3 **13**,
s78 . . . . . [1997] 2 **81, 82**, [2004] 2 **79, 82** *et seq*, [2004] 2 **79, 82, 85, 86**
s79 . . . . . [1995] 1 **49** *et seq*, [2000] 3 **70** *et seq* [2001] 1 **76**, *et seq*, [2004] 2 **85**
s79(1) . . . . . [2004] 2 **85, 86**
s79(2) . . . . . [2004] 2 **85**
s84(1) . . . . . [1996] 1 **209** *et seq*, 2 **163** *et seq*, [2000] 2 **99** *et seq*, 3 **87**, [2001] 2 **163**, *et seq*, [2003] 1 **172**, [2003] 2 **157, 164**
s84(1)*(a)* . . . . . [1996] 2 **163** *et seq*, [1997] 1 **145** *et seq*, **192**, ... [2000] 2 **99** *et seq*, [2001] 1 **111**, [2003] 2 **157, 158, 159, 164**
s84(1)*(aa)* . . . . . [1997] 1 **145, 148, 149, 192**, 2 **186** *et seq*, [2000] 2 **99** *et seq*, [2003] 2 **157, 158, 159, 164**
s84(1)*(b)* . . . . . [1997] 1 **145, 149, 192**, [2001] 2 **168**
s84(1)*(c)* . . . . . [1996] 2 **163** *et seq*, [1997] 1 **145, 149, 192** [2000] 2 **99, 100, 103**, [2001] 1 **111**, 2 **171**, [2003] 2 **157, 158, 159, 164**
s84(1A)*(a)* . . . . . [1997] 2 **186** *et seq*
s84(1C) . . . . . [1997] 1 **193**
s84(2) . . . . . [2002] 2 **89** *et seq*
s86(2) . . . . . [2003] 1 **73**
s87(1) . . . . . [2003] 1 **73**
s90 . . . . . [2001] 3 **39**
s99 . . . . . [1995] 1 **1**, [2000] 1 **101**
s99(2) . . . . . [1999] 2 **93, 94**
s99(13) . . . . . [2000] 1 **101**
s101(1)(iii) . . . . . [2002] 3 **76** *et seq*
s109(8) . . . . . [1996] 1 **139**
s136 . . . . . [2003] 3 **16**
s136(1) . . . . . [2003] 3 **13**
s139 . . . . . [1995] 1 **7** *et seq*, [2004] 2 **179**
s141 . . . . . [1997] 1 **182, 183**, [2003] 3 **12** *et seq*, [2004] 1 **49**,
s141(2) . . . . . [2003] 3 **15**
s141(3) . . . . . [1998] 1 **39** *et seq*, [2003] 3 **15**
s142 . . . . . [1995] 1 **49** *et seq*, [1995] 2 **14**, [2003] 3 **13** *et seq*,
s146 . . . . . [1995] 1 **63**, [1996] 2 **65** *et seq*, [1998] 2 **69** *et seq*, [2000] 1 **20** *et seq*, 2 **73, 75, 76**, [2003] 1 **70** *et seq*
s146(1) . . . . . [2003] 1 **70, 72, 73**
s146(2) . . . . . [1995] 2 **23**
s146(3) . . . . . [2003] 1 **70**
s146(4) . . . . . [1995] 2 **24**, [1998] 1 **44** *et seq*, [1999] 2 **51**, 3 **29** *et seq*, [2000] 1 **27, 28**, [2003] 1 **70, 72, 73**
s146(5) [2003] 1 **70**
s146(5)*(b)* . . . . . [1995] 2 **24** *et seq*, [1996] 2 **40** *et seq*, [2003] 1 **70, 73**
s149(6) . . . . . [2000] 2 **91, 92, 93**
s149(6)*(c)* . . . . . [2000] 2 **91, 92**
s193 . . . . . [2003] 1 **17, 18, 21**, [2003] 3 **6**, [2004] 2 **15, 17, 18**,
s193(1) . . . . . [2004] 2 **17**
s193(2) . . . . . [2003] 1 **18**, [2004] 2 **17, 18**

s193(4) . . . . **[2003]** 1 **17, 18, 24, [2003]** 3 **6,**
[2004] 2 **15, 16, 17, 18, 20** *et seq,*
s194 . . . . **[2003]** 3 **143, 144**
s194(1) . . . . **[2003]** 3 **144**
s194(2) . . . . **[2003]** 3 **143, 144**
s196(3) . . . . **[1998]** 3 **112** *et seq,* **[2003]** 3 **90**
s196(4) . . . . **[2002]** 3 **47** *et seq,* **[2003]** 3 **90**
s196(5) . . . . **[1996]** 1 **57** *et seq*
s198(1) . . . . **[1999]** 2 **52**
s203(2)(*a*) . . . . **[1996]** 2 **40** *et seq*

Law of Property (Miscellaneous Provisions) Act 1989
s2 . . . **[1995]** 2 **113** *et seq,* **[1997]** 2 **194,** *et seq,* **[1999]** 2 **37, 181** *et seq,*
[1999] 3 **91, 95, 96, [2001]** 2 **82,** *et seq,*
[2004] 2 **31, 33, 36**
s2(1) . . . . **[2000]** 3 **158**
s2(3) . . . . **[1996]** 1 **175** *et seq*
s2(5) . . . . **[1999]** 2 **183** *et seq*

Law of Property (Miscellaneous Provisions) Act 1994
s2(1) . . . . **[2000]** 2 **19, 20, 22**
s2(5) . . . . **[2000]** 3 **1, 2, 3**
s3 . . . . **[2003]** 3 **14**
s3(1) . . . . **[2000]** 2 **19, 20, 22**

Law Reform (Contributory Negligence) Act 1945
s1 . . . . **[1998]** 1 **107, 108, 109, [2000]** 1 **23**
s1(1) . . . . **[1999]** 1 **77** *et seq*
s1(3) . . . . **[2000]** 1 **26, 27**

Leasehold Property (Repairs) Act 1938
s1 . . . . **[1996]** 1 **78, [1997]** 2 **194** *et seq,* **[2003]** 1 **70, 73**
s1(4) . . . . **[2003]** 1 **70**
s1(5) . . . . **[1998]** 2 **36, 37, [2003]** 2 **30, 31, 32, 33, 35**
s1(5)(*a*) . . . . **[2003]** 2 **30, 32, 34, 35**
s1(5)(*d*) . . . . **[2003]** 2 **34, 35**
s1(5)(*e*) . . . . **[2003]** 2 **30, 32, 34, 35**

Leasehold Reform Act 1967 . . . . **[1995]** 2 **111**
s1 . . . . **[2003]** 1 **156, [2003]** 3 **36, [2004]** 2 **72,**
s1(1) . . . . **[1995]** 2 **69**
s1A . . . . **[2004]** 2 **72**
s1A(2) . . . . **[2004]** 2 **71**
s1(2) . . . . **[2003]** 1 **159**
s2 . . . . **[2002]** 2 **61, [2003]** 1 **156, 158, 159, 160, [2003]** 3 **35**
s2(1) . . . . **[1995]** 1 **89** *et seq,* **[1997],** 1 **96** *et seq,*
[2003] 1 **151, 152, 153, 154, 155, 156, 160, 162, 163,**
[2003] 3 **35, 38**
s2(1)(*a*) . . . . **[2003]** 1 **152, 153, 154, 155, 158, 159, [2003]** 3 **39,**
s2(1)(*b*) . . . . **[2003]** 1 **152, 153, 154, 155, 159, 162,**
[2003] 3 **35** *et seq*
s2(2) . . . . **[1995]** 1 **89** *et seq,* **[2001]** 3 **47,** *et seq,*
[2003] 1 **151, 152, 153, 154, 155, 158, 160** *et seq*
s2(3) . . . . **[1997]** 1 **96, 97, [1998]** 2 **92** *et seq,* **[2003]** 1 **155, 162,**
[2003] 3 **39, [2004]** 2 **71,**
s2(5) . . . . **[2003]** 1 **142, 152, 155, 158, 159, 160, 161, 162**
s3 . . . . **[1998]** 2 **92, [2004]** 2 **71**
s3(1) . . . . **[2002]** 2 **53** *et seq,* **[2004]** 2 **71**
s3(3) . . . . **[1999]** 2 **214, 216, [2004]** 2 **69, 71**
s3(6) . . . . **[2004]** 2 **71, 74**
s4 . . . . **[2004]** 2 **70**
s4(1) . . . . **[2004]** 2 **70, 71**
s4A . . . . **[2004]** 2 **70, 71, 72**
s4A(1) . . . . **[2004]** 2 **69, 70, 71, 74**
s4A(2) . . . . **[2004]** 2 **71, 74**
s5(*b*) . . . . **[2003]** 1 **161**
s8 . . . . **[1996]** 2 **80, [2004]** 2 **73**
s8(1) . . . . **[2004]** 2 **72**
s9 . . . . **[1995]** 1 **243, [2003]** 2 **50, [2004]** 2 **72**
s9(1) . . . . **[1996]** 1 **199, [1999]** 1 **181, 182, 185, [2000]** 3 **79**
s9(1A) . . . . **[1995]** 1 **213, [1996]** 2 **179, [1999]** 2 **213, 216,** 3 **103**
[2001] 1 **59, 60, 61,** 3 **150, [2003]** 2 **49, 50**
s9(1A)(*a*) . . . . **[2003]** 2 **49**
s9(1A)(*d*) . . . . **[1999]** 2 **213, 215, 216, [2001]** 1 **59, 60, 61, 62,**
[2002] 2 **55** *et seq,* **[2003]** 2 **49, 50, 51, 52**
s9(1C) . . . . **[1996]** 1 **217** *et seq* **[1999]** 2 **123, [2001]** 3 **149,** *et seq*
s9(4) . . . . **[1999]** 2 **117** *et seq,* **[1999]** 3 **103, 106**
s9(4)(*e*) . . . . **[1995]** 1 **243**
s10(4) . . . . **[1999]** 1 **61, 62, 63**
s10(4)(*b*) . . . . **[1999]** 1 **62, 63**
s20(2)(*c*) . . . . **[2001]** 3 **56**
s21 . . . . **[2000]** 2 **91**
s23(1)(*d*) . . . . **[2004]** 2 **72**
s37(1)(*d*) . . . . **[2003]** 2 **50, [2004]** 2 **72**
s101 . . . . **[2003]** 1 **162**
Sch 3
Part II
para 6 . . . . **[2004]** 2 **69, 72, 73**
para 6(1). . . . . **[2004]** 2 **72, 73**
para 6(1)(*b*) . . . . **[2004]** 2 **73**
para 6(1)(*d*) . . . . **[2002]** 1 **56** *et seq*
para 6(3) . . . . **[2002]** 1 **55** *et seq,* **[2004]** 2 **69, 72, 73**
para 7(1) . . . . **[2004]** 2 **73**

Leasehold Reform, Housing and Urban Development Act 1993
s1 . . . . **[2003]** 3 **41, 151, 152**
s1(1) . . . . **[2003]** 3 **40, 40, 44, 145, 149, [2004]** 1 **146,**
s1(2) . . . . **[2003]** 3 **151, [2004]** 1 **146**
s1(2)(a) . . . . **[2003]** 3 **151, 152**
s1(3) . . . . **[2003]** 3 **149, 150, 151, [2004]** 1 **146**
s1(3)(a) . . . . **[2003]** 3 **149, 150**
s1(3)(b) . . . . **[2003]** 3 **149, 150, 151, 152, 153**
s1(4) . . . . **[2003]** 3 **147, 152, 153, [2004]** 1 **148,**
s1(4)(a)(i) . . . . **[2003]** 3 **147** *et seq*
s1(7). . . . . **[2003]** 3 **151, [2004]** 1 **146**
s1(8) . . . . **[2003]** 3 **40, 150**
s1(12) . . . . **[2003]** 3 **151**
s2(2) . . . . **[2004]** 1 **146**
s2(3) . . . . **[2004]** 1 **146**
s3 . . . . **[2003]** 3 **41, 149, 151, [2004]** 1 **146**
s5 . . . . **[2003]** 3 **41**
s6 . . . . **[2003]** 3 **41**
s7(6)(b) . . . . **[1996]** 2 **75** *et seq*
s13 . . . . **[1996]** 2 **79, [1998]** 2 **81, 82, 96, 97, [2003]** 3 **40, 42, 45**
s13(1) . . . . **[2003]** 3 **40, 151, [2004]** 1 **146**
s13(1)(a) . . . . **[2004]** 1 **146**
s13(2)(b) . . . . **[2003]** 3 **40, 41**
s13(3) . . . . **[2003]** 3 **151, [2004]** 1 **146**
s13(9) . . . . **[2003]** 3 **40**
s13(12) . . . . **[2004]** 1 **146**
s14 . . . . **[2003]** 3 **40**
s20 . . . . **[2003]** 3 **40, 41, 45**
s20(1) . . . . **[2003]** 3 **40, 41, 43** *et seq*
s20(2) . . . . **[2003]** 3 **40, 41, 45**
s20(3) . . . . **[2003]** 3 **40, 41, 45**
s21 . . . . **[2000]** 2 **55, 56, [2003]** 3 **40, 42, 45**
s21(1) . . . . **[2003]** 3 **43, [2004]** 1 **146**
s21(2) . . . . **[2003]** 3 **45**
s21(2)(b) . . . . **[2003]** 3 **40, 43, 44, 45** L
s21(3) . . . . **[2003]** 3 **151, [2004]** 1 **147**
s21(4) . . . . **[2004]** 1 **148**
s22 . . . . **[1996]** 2 **78** *et seq,* **[2003]** 3 **40, 42, 43, 44**
s22(1) . . . . **[2004]** 3 **43**
s24 . . . . **[1995]** 1 **216**
s24(1) . . . . **[2003]** 3 **151, 152, [2004]** 1 **147**
s24(3) . . . . **[2003]** 2 **54, [2003]** 3 **152, [2004]** 1 **147**
s24(4) . . . . **[2003]** 2 **54, [2003]** 3 **152,**
[2004] 1 **147**
s24(4)(a) . . . . **[2003]** 2 **54, 55, [2003]** 3 **46, 47**
s24(4)(b) . . . . **[2003]** 3 **46** *et seq*
s24(4)(b)(i) . . . . **[2003]** 3 **46** *et seq* M
s24(6)(a) . . . . **[2003]** 3 **46, 47**
s24(7) . . . . **[2003]** 3 **47**
s24(8) . . . . **[2003]** 3 **151, 152, [2004]** 1 **147**
s25(1) . . . . **[2000]** 2 **55** *et seq*
s25(3) . . . . **[2000]** 2 **55** *et seq*
s32 . . . . **[1995]** 1 **216, [1996]** 1 **213**
s32(2) . . . . **[2003]** 2 **55, [2003]** 3 **47**
s33 . . . . **[1995]** 1 **219, [2004]** 1 **152**
s34 . . . . **[2004]** 1 **148**
s34(1) . . . . **[2004]** 1 **147**
s34(2) . . . . **[2004]** 1 **147**
s34(9) . . . . **[2004]** 1 **145, 147**

s40 .... **[2003]** 1 **83**
s40(2) .... **[2003]** 1 **75, 80**
s42 .... **[1998]** 1 **122, 123, 124,** 2 **90** *et seq,* **[1999]** 1 **59,** 2 **69,** 3 **49, 50, 109, 110, 111, [2002]** 3 **53** *et seq,* **[2003]** 1 **75, 78** *et seq*
s42(2) .... **[1999]** 3 **51, 230**
s42(3) .... **[1997]** 2 **102, 103, [1999]** 3 **50, [2003]** 1 **75**
s42(3)(b) .... **[1999]** 3 **51**
s42(3)(c) .... **[1997]** 2 **104, [1999]** 1 **60, 61, [2003]** 1 **75, 77**
s42(5) .... **[1998]** 2 **90, 91, 92, [1999]** 3 **50, 51**
s43(3) .... **[1999]** 2 **70, 71, 73**
s45 .... **[2003]** 1 **78, 80, [2003]** 3 **43**
s45(1) .... **[2002]** 1 **63, 64, [2003]** 1 **78, 83**
s45(2) .... **[2001]** 1 **63**
s45(2)(a) .... **[2001]** 1 **63, 64, [2002]** 1 **63, 64** *et seq*
s45(2)(b) .... **[2001]** 1 **63, 64**
s45(3) .... **[2001]** 1 **63, 64, [2002]** 1 **62, 64**
s45(3)(b) .... **[1999]** 1 **60**
s45(3)(c) .... **[1999]** 1 **60, 61**
s46 .... **[1997]** 2 **101** *et seq,* **[2002]** 1 **63, 64**
s47 .... **[2002]** 1 **63, 64**
s48 .... **[2002]** 1 **63, 64, [2003]** 1 **80**
s49 .... **[2003]** 1 **78, 80**
s49(1) .... **[1999]** 1 **60, [2002]** 1 **62**
s54 .... **[1999]** 3 **109, 110, 111**
s56 .... **[1999]** 1 **189**
s62(2) .... **[1996]** 2 **75** *et seq*
s70 .... **[1995]** 2 **202**
s70(2) .... **[1995]** 2 **202**
s90 .... **[1998]** 2 **90, 91, [2004]** 1 **147**
s91 .... **[2004]** 1 **147**
s91(1) .... **[2004]** 1 **147**
s91(2) .... **[2004]** 1 **147**
s99(5) .... **[2002]** 1 **56** *et seq*
s97 .... **[1999]** 3 **109, 110, 111**
s99(5) .... **[1998]** 2 **90** *et seq*
s101(3) .... **[1998]** 2 **96**
Sch 3
Part I
para 3(2) .... **[1998]** 2 **81, 82, 83**
para 6(1)(d) .... **[2002]** 1 **56** *et seq*
para 6(3) .... **[2002]** 1 **55** *et seq*
Part III
para 15 .... **[1996]** 2 **79**
Sch 5 .... **[2003]** 2 54, 55, **[2004]** 1 147
para 3(1)(b) .... **[2003]** 3 **46** *et seq*
Sch 6
Part I
para 1 .... **[2004]** 1 **145**
para 1(b) .... **[2004]** 1 **145**
para 1(1) .... **[2003]** 3 **48, [2004]** 1 **146**
para 1(1)(a) .... **[2004]** 1 **148**
Part II
para 3(1A) .... **[2004]** 1 **150**
para 4(2) .... **[2004]** 1 **151**
Sch 6 .... **[1995]** 1 **219, [1995]** 2 **207, [1996]** 1 **213, [1997]** 1 **187, 209, [1997]** 2 **198** *et seq,* **[2001]** 1 **105**
Part I
para 1 .... **[2002]** 3 **60**
Part II
para 3 .... **[1998]** 3 **125**
para 3(1) .... **[1997]** 1 **186, 187, 212** *et seq*
para 3(1)(b) .... **[1998]** 1 **125, 126**
para 4 .... **[1999]** 1 **101, 102**
para 4(2) .... **[1998]** 1 **121**
Sch 7 .... **[2004]** 1 **145, 147, 148**
para 2(ii)(a) .... **[2004]** 1 **145**
para 2(ii)(b) .... **[2004]** 1 **145**
Sch 11
Part 1 .... **[2002]** 2 **58**
Sch 12
Part II
para 9(1) .... **[1999]** 1 **60, [1999]** 3 **50, 51**
Sch 13 .... **[1995]** 1 **223, [1998]** 2 **139, [1999]** 1 **189** *et seq,* **[2000]** 2 **93, 94, [2002]** 2 **87**
Part I
para 1 .... **[1997]** 1 **198, 199, [2004]** 1 **145**
Part II .... **[2003]** 1 **75**
para 3(2)(a) .... **[1999]** 1 **196**
para 4(1) .... **[1999]** 1 **99**
para 4(1)(a) .... **[1995]** 1 **224**
Limitation Act 1939
s7(5) .... **[1999]** 2 **100, 101**
Limitation Act 1980
s2 .... **[1995]** 1 **126**
s5 .... **[2003]** 1 **86**
s8 .... **[2000]** 3 **47, [2003]** 1 **86**
s9 .... **[1997]** 2 **21** *et seq,* **[1998]** 3 **18** *et seq,* **[1999]** 2 **196, 197, [2000]** 2 **84** *et seq,* 3 **100, [2003]** 3 **1**
s9(1) .... **[1998]** 1 **1, 3, [1999]** 2 **205**
s14A .... **[1995]** 1 **123** *et seq,* **143, [1996]** 1 **123** *et seq,* **133, [1996]** 2 **106, [1999]** 1 **71, 72, 73, [1999]** 3 **59, 61**
s14A(7) .... **[1996]** 1 **133**
s14A(10) .... **[1999]** 3 **61**
s15 .... **[2001]** 2 **70,** *et seq,* **[2003]** 1 **118, 119, 120, 122**
s15(1) .... **[1997]** 1 **80, [1999]** 2 **86, 101, [2000]** 2 **138** *et seq*
s15(6) .... **[1997]** 1 **80, [1999]** 2 **101**
s17 .... **[2003]** 1 **119, 120, 122**
s19 .... **[1996]** 1 **103**
s20 .... **[2003]** 1 **85, 86**
s20(1) .... **[2003]** 1 **89**
s20(5) .... **[2003]** 1 **89**
s29(5) .... **[2003]** 3 **3**
s29(5) .... **[2004]** 3 **3**
s32(1)(b) .... **[2001]** 1 **67, 68, 69, 70**
s35(2) .... **[1999]** 3 **85, 86**
s35(3) .... **[1999]** 3 **84, 85**
s35(5)(*a*) .... **[1999]** 3 **84** *et seq*
s36(2) .... **[1998]** 3 **43, 45, 46**
s39 .... **[2000]** 2 **84, 85, 86, 87**
Sch 1
Part I
para 1 .... **[1999]** 2 **86**
para 5 .... **[2002]** 3 **70**
para 5(1) .... **[1997]** 1 **79** *et seq*
para 8 .... **[2002]** 3 **70, [2003]** 1 **122**
para 8(1) .... **[1995]** 2 **124, [1997]** 1 **80**
para 9 .... **[1999]** 2 **101**
Local Government Act 1933
s164 .... **[2004]** 1 **12**
s172(3) .... **[1998]** 3 **62, [2004]** 1 **12**
s305 .... **[1998]** 3 **63, [2004]** 1 **12**
Local Government Act 1972
s111 .... **[1999]** 1 **114**
s122 .... **[2004]** 1 **97, 98, 104**
s123 .... **[1995]** 2 **127, [2002]** 1 **81** *et seq,* **[2004]** 1 **97, 98, 100, 104**
s123(2B)(*b*) .... **[2004]** 1 **98**
Local Government Finance Act 1988
s3 .... **[1995]** 2 **143**
s43(1) .... **[2000]** 2 **177**
s45 .... **[1996]** 1 **101**
s65 .... **[1996]** 1 **139**
s66(2B) .... **[2000]** 3 **85**
s123 .... **[2001]** 1 **15,** *et seq*
s128(2) .... **[2001]** 1 **16,** *et seq*
Sch 6
para 2 .... **[2000]** 2 **177, 187, [2001]** 1 **159,** *et seq*
para 2(1) .... **[1997]** 2 **238** *et seq,* **[1998]** 2 **147** *et seq,* **[1999]** 1 **155, 162, 165**
para 2(3)(b) .... **[1996]** 2 **187**
para 3(2) .... **[2003]** 2 **135, 136**
para 3(2)(b) .... **[2003]** 2 **136**
para (4)(b) .... **[1996]** 2 **187**
Local Government and Housing Act 1989
s186(3) .... **[2001]** 2 **53, 56**
Local Land Charges Act 1975
s10 .... **[2002]** 3 **2**

London Buildings Acts (Amendment) Act 1939 .... [1997] 1 137 et seq
s55(a) to (e) ..... [1995] 1 157
s55(e) ..... [2001] 2 90
s55(n) ..... [1995] 1 156 et seq
s55(n)(ii) ..... [1995] 1 156
s55(m) ..... [1995] 1 156
s55(o) ..... [1995] 1 156
s62 ..... [2001] 2 11, 13
Matrimonial Causes Act 1973
s11(c) ..... [2004] 2 136
Matrimonial Homes Act 1983
s1(1)(a) ..... [1999] 2 21, 22
1(2)(a) ..... [1999] 2 22
2(6) ..... [1999] 2 22
Metropolitan Paving Act 1817
s90 ..... [2004] 2 183, 184
Misrepresentation Act 1967
s2 ..... [1997] 2 42
s2(1) ..... [2002] 2 141, [2004] 1 137
Mobile Homes Act 1983 ..... [2002] 3 65 et seq
National Health Service Act 1977
s54 ..... [2001] 3 78
National Trust Act 1907 ..... [1997] 2 140
New Forest Act 1877
s25(4) ..... [2004] 1 3
s29 ..... [2004] 1 3
New Towns Act 1981
s21(1) ..... [2004] 1 97, 100
s21(1)(a) ..... [2004] 1 97
s36(1) ..... [2004] 1 97
s36(2)(a) ..... [2004] 1 97
s41 ..... [2004] 1 97
s80(1) ..... [2004] 1 97
Open Spaces Act 1906
s10 [2004] 1 98, 104
Party Wall etc Act 1996
s2(2) ..... [2004] 1 74
s2(2)(f) ..... [2004] 1 74
s3(1) ..... [2004] 1 74
s3(2)(a) ..... [2004] 1 74
s8(5) ..... [2004] 1 74
Perpetuities and Accumulations Act 1964
s9(2) ..... [2003] 2 63 et seq
Planning and Compensation Act 1991
s80(1) ..... [1998] 1 210
Planning (Listed Buildings and Conservation Areas) Act 1990
s8 ..... [1995] 1 170, 173
s27 ..... [1995] 1 167 et seq, [1996] 1 167
Prescription Act 1832
s2 ..... [2004] 1 98, 113, [2004] 2 15, 17, 19, 21
s3 ..... [1995] 1 159, [2002] 1 99, 100
s4 ..... [1995] 1 159, [2004] 1 113, [2004] 2 19
s5 ..... [2004] 1 100
Property Misdescriptions Act 1991
s1(1) ..... [2000] 1 77, 79
s1(5) ..... [2000] 1 78
s1(5)(b) ..... [1996] 2 21
s2(1) ..... [1996] 2 21 et seq
Protection from Eviction Act 1977
s5 ..... [1997] 2 10 et seq
Public Health Act 1875 ..... [2002] 1 85 et seq
s4 ..... [2002] 1 86 et seq
Public Health Act 1936
s15 ..... [1995] 1 266 et seq
s278 ..... [1995] 1 267 et seq
Railway Clauses Consolidation (Scotland) Act 1845
s6 ..... [1995] 1 235
Rent Act 1977
s3(1) ..... [2001] 1 48, 50, 51
s12(1)(b) ..... [1995] 1 93
s24(3) ..... [1998] 1 73, 74, 78
s67 ..... [2001] 3 31
s67(2) ..... [1998] 1 80
s67(3)(a) ..... [2001] 2 63, et seq
s70 ..... [1995] 2 80, [1997] 1 237, [1998] 1 80, 87, 88, 3 86, [2001] 2 65, 3 31, [2003] 3 17, 18, [2004] 3 17, 18
s70(1) ..... [1998] 2 99, [2003] 3 18 et seq
s70(2) ..... [1998] 1 87, 88, 2 99 et seq, [2003] 3 21, 33
s70(3)(a) ..... [2003] 3 17, 19 20
s70(3)(b) ..... [2003] 3 19
s74(1) ..... [2000] 1 73
s98(1) ..... [2001] 1 48, 50, 51, [2003] 3 12
s100(2) ..... [2000] 3 35
Part IX ..... [1997] 2 115
s137(3) ..... [1998] 1 73 et seq
Sch 1
Part I
para 2 ..... [2004] 2 132, 133, 134, 135, 136, 137, 138, 141, 142, 147, 148
para 2(1) ..... [2004] 2 140, 141, 142, 146
para 2(2) ..... [2004] 2 132, 133, 134, 138, 139, 140, 141, 142, 146, 147, 148
para 3 ..... [2004] 2 132, 133, 138, 146
para 3(1) ..... [2004] 2 133
Part II
para 13 ..... [1995] 2 75 et seq
Sch 15
Part I
Case 1 ..... [1999] 2 67, 68, [2001] 1 48, 50, 51, 52, [2003] 3 12
Case 11 ..... [1995] 2 99 et seq
Rentcharges Act 1977
s2 ..... [2002] 2 1 et seq
Rights of Light Act 1959
s2(3) ..... [1995] 1 161
s3(2)(b) ..... [1995] 1 161
s3(3) ..... [1995] 1 161
Rights of Way Act 1932
s1 ..... [2004] 1 113
s1(6) ..... [2004] 1 113
Road Traffic Act 1930
s14 ..... [2003] 3 6, 8 [2004] 2 21
s14(1) ..... [2004] 2 18
s34 ..... [2003] 3 5
Road Traffic Act 1988
s34 ..... [2003] 1 25, 26, 29, [2003] 3 6, [2004] 2 16
s34(1) ..... [2003] 1 24, 27, [2004] 2 18, 20, 21
s34(1)(a) ..... [2003] 1 24, 25, 26, 27, 30, 31
s34(1)(b) ..... [2003] 1 24, 25, 27, 30, 31
s34(3) ..... [2003] 1 27
s34(5) ..... [2003] 1 27
s192 ..... [2003] 1 26, [2003] 3 6
Sale of Goods Act 1979
s57 ..... [2000] 3 7, 8
s57(2) ..... [2000] 3 8
s57(3) ..... [2000] 3 8
s57(4) ..... [2000] 3 7, 8
Stamp Act 1891
s13 ..... [1996] 2 119
s13(3) ..... [1996] 2 119
Statute of Marlborough 1267 ..... [1999] 3 145
Supply of Goods Act 1982
s13 ..... [2002] 3 111
Supreme Court Act 1981
s35A ..... [1998] 1 99, 103, [1998] 2 14, 16
s38 ..... [1995] 2 23, [2000] 1 35
s138 ..... [1998] 3 151
Taxation and Chargeable Gains Act 1992
s28(1) ..... [2003] 2 73
s28(2) ..... [2003] 2 73
s60(1) ..... [2003] 2 74
Telecommunications Act 1984
Sch 2 ..... [2003] 3 55, 56
para 17 ..... [2003] 3 58
para 17(1) ..... [2003] 3 56, 58, 59
para 17(2) ..... [2003] 3 55 et seq

para 17(5) .... [2004] 3 **56**
para 18 .... **[2003]** 3 **55** *et seq*
para 18(1) .... **[2004]** 3 **56, 57**
para 18(2) .... **[2003]** 3 **55** *et seq*
para 18(3) .... **[2004]** 3 **56**
Town and Country Planning Act 1947
s12(5)(c) .... **[2004]** 2 **19**
Town and Country Planning Act 1971
s52 .... **[1998]** 2 **159** *et seq*
Town and Country Planning Act 1990
s55 .... **[2003]** 2 **6**
s55(2)(a)(ii) .... **[1996]** 1 **167**
s70 .... **[1995]** 2 **149,** *et seq*
s77 .... **[2001]** 1 **33,** *et seq*
s78 .... **[2001]** 1 **33,** *et seq*
s79 .... **[2001]** 1 **33,** *et seq*
s106 .... **[1995]** 2 **147,** *et seq,* **[1999]** 3 **91** *et seq,* **121, 122**
s107 .... **[1997]** 1 **167, 2 181** *et seq*
s108 .... **[1997]** 1 **167,** *et seq*
s137 .... **[1995]** 1 **180** *et seq*
s137(1)(a) .... **[1995]** 1 **180** *et seq*
s137(2)(a) .... **[1995]** 1 **180** *et seq*
s137(3) .... **[1995]** 1 **180** *et seq*
s149(2) .... **[1996]** 1 **180**
s151(4)(a) .... **[1997]** 2 **173** *et seq,* **[1998]** 3 **121**
s153(6) .... **[1995]** 2 **197**
s166(2)(b) .... **[1995]** 2 **196**
s171(1) .... **[1996]** 1 **190**
s174(2) .... **[2004]** 2 **90**
s183(6) .... **[2004]** 2 **92**
s186 .... **[2004]** 2 **89, 92**
s186(1) .... **[2004]** 2 **90**
s186(2) .... **[2004]** 2 **89, 90, 92, 93**
s186(3) .... **[2004]** 2 **90**
s193(3)(b) .... **[1996]** 1 **244** *et seq*
s198(1) .... **[1998]** 2 **121**
s200 .... **[1998]** 2 **121** *et seq*
s203 .... **[1996]** 2 **136**
s237 .... **[1999]** 1 **169, 170, 175, 176**
s237(1) .... **[1996]** 2 **128** *et seq*
Sch 13 .... **[1998]** 3 **121** *et seq*
para 1 .... **[1996]** 2 **137**
para 2 .... **[1996]** 2 **137** *et seq*
para 6 .... **[1996]** 1 **180**
para 21 .... **[1996]** 2 **137**
para 22 .... **[1996]** 1 **190,** 2 **137**
Town and Country Planning (Scotland) Act 1972
Sch 24
para 36 .... **[1995]** 1 **229** *et seq*
Trade Descriptions Act 1968
s13 .... **[2000]** 3 **17**
Transport Act 1968
s9A(6) .... **[1997]** 2 **30**
Trusts of Land and Appointment of Trustees Act 1996
s11 .... **[2001]** 3 **11**
s13 .... **[2001]** 3 **81**
s14 .... **[2000]** 3 **131**
s15 .... **[2000]** 3 **131**
s15(1) .... **[1998]** 3 **102**
s15(3) .... **[2000]** 3 **137**
Unfair Contract Terms Act 1977
s2(2) .... **[1997]** 1 **122**
s11 .... **[1996]** 1 **46**
s13(3) .... **[1997]** 1 **122**
Sch 1
para 1(*b*) .... **[1998]** 1 **53, 55, 56**
Unfair Terms in Consumer Contracts Regulations 1999
reg 5(1) .... **[2004]** 2 **25, 27** *et seq*
reg 6(2) .... **[2004]** 2 **25, 27** *et seq*
Value Added Tax Act 1983
Sch 5
Group 8A .... **[1998]** 3 **117** *et seq*
Water Act 1989
Sch 19
para 4 .... **[1998]** 1 **207**
para 6(1) .... **[1998]** 1 **207, 210**
para 6(2) .... **[1998]** 1 **207, 210**
Water Industry Act 1991
s18(8) .... **[2001]** 3 **117, 119**
s94 .... **[2001]** 3 **117**
s98 .... **[2001]** 2 **157,** *et seq*
s106 .... **[2001]** 3 **116**
s159 .... **[2001]** 2 **157,** *et seq,* **[2004]** 2 **95**
s159(1) .... **[2004]** 2 **95**
s159(4) .... **[2004]** 2 **96**
s180 .... **[2004]** 2 **95, 96**
Sch 12 .... **[2001]** 2 **157,** *et seq,* **[2004]** 2 **95, 96**
para 2(1) .... **[1997]** 1 **176, 177**
para 2(2) .... **[1997]** 1 **176, 177**
Youth Justice and Criminal Evidence Act 1999
s41 .... **[2004]** 2 **136, 137**
s41(1) .... **[2004]** 2 **137**

TABLE OF STATUTORY INSTRUMENTS

References are to year, number and volume

eg **[1995]** 2 **127** *et seq* refers to vol 2 of 1995 at p127 and following

Access to Justice Act 1999 (Destination of Appeals) Order 2000 (SI 2000/No 1071)
- Art 1 . . . . . **[2003]** 3 **57**
- Art 1(2)(c) . . . . . **[2003]** 3 **57**
- Art 1(3) . . . . . **[2003]** 3 **57**
- Art 4 . . . . . **[2003]** 3 **57**

Agricultural Holdings (Arbitration on Notices) Order 1987 (SI 1987/No 710)
- para 3 . . . . . **[2000]** 1 **2, 3**
- para 4 . . . . . **[2000]** 1 **2, 3**

Assured Tenancies and Agricultural Occupancies (Forms) Regulations 1988 (SI 1988/No 2203) . . . . . **[2002]** 1 **10**
- reg 2 . . . . . **[1998]** 2 **26, [2002]** 1 **10, 13**
- reg 3(7) . . . . . **[2002]** 1 **10**
- Sch 1
- form 7 . . . . . **[1998]** 2 **26, [2000]** 2 **41**

Civil Procedure Rules 1998 (SI 1998/No 3132)
- Pt 3.4 . . . . . **[2003]** 1 **124**
- Part 20 . . . . . **[2000]** 1 **35**
- Part 24 . . . . . **[2000]** 3 **1, 5, [2003]** 1 **124**
- Part 24(2)(a)(ii) . . . . . **[2000]** 3 **1, 5**
- Part 31.16(3) . . . . . **[1999]** 2 **81, 83**
- Part 31.16(3)(a), (b) and (c) . . . . . **[1999]** 2 **81, 84**
- Part 31.16(3)(d) . . . . . **[1999]** 2 **81, 83, 84**
- Part 35 . . . . . **[2000]** 1 **55**
- Part 39.6 . . . . . **[2003]** 3 **136**
- Part 52 . . . . . **[2003]** 3 **132**
- Part 52.3(2) . . . . . **[2003]** 1 **125, [2003]** 3 **55**
- Part 52.4(2) . . . . . **[2003]** 3 **55**
- Sch 2
- rC49.6A . . . . . **[2000]** 2 **40, 41, 42, 43**

Civil Procedure (Modification of Enactments) Order 1998 (SI 1998/No 2940)
- art 5 . . . . . **[1999]** 2 **81, 82**

Commons Registration (New Land) Regulations 1969 (SI 1969/No 1843)
- reg 3 . . . . . **[2003]** 3 **71, [2004]** 1 **111**
- reg 3(1) . . . . . **[2004]** 1 **108, 113**
- reg 3(4) . . . . . **[2004]** 1 **108**
- reg 3(7) . . . . . **[2004]** 1 **108**
- reg 3(7)(*a*) . . . . . **[2003]** 3 **71**
- reg 4(1) . . . . . **[2004]** 1 **108**
- reg 5 . . . . . **[2003]** 3 **71**
- reg 5(1) . . . . . **[2004]** 1 **108**
- reg 5(4) . . . . . **[2004]** 1 **108**
- reg 5(4)(*a*) . . . . . **[2003]** 3 **71**
- reg 5(4)(*b*) . . . . . **[2003]** 3 **71**
- reg 5(4)(*c*) . . . . . **[2003]** 3 **71**
- reg 5(7) . . . . . **[2004]** 1 **108**
- reg 6 . . . . . **[2004]** 1 **108, 116**
- reg 7 . . . . . **[2003]** 3 **71, [2004]** 1 **108**
- reg 7(1) . . . . . **[2004]** 1 **116**
- reg 8 . . . . . **[2004]** 1 **108**
- reg 8(1) . . . . . **[2004]** 1 **116**
- reg 8(3) . . . . . **[2004]** 1 **116**

Council Tax (Chargeable Dwellings) Order 1992
- Art 2 . . . . . **[1995]** 2 **142**
- Art 3 . . . . . **[1995]** 2 **142**

County Court Rules 1981 (SI 1981/No 1687)
- Ord 6 r3(2) . . . . . **[1999]** 3 **29** *et seq*
- Ord 13, r5(1)(c) . . . . . **[1998]** 3 **43, 44, 46**
- Ord 16(2) . . . . . **[1997]** 2 **105**
- Ord 37, r2 . . . . . **[2000]** 1 **33**
- Ord 38, r1(2) . . . . . **[1997]** 1 **15**

Dairy Produce Quota Regulations 1984 (SI 1984/No 1087)
- reg 7 . . . . . **[1997]** 2 **16** *et seq*

Dairy Produce Quota Regulations 1993
- reg 32(4) . . . . . **[1996]** 1 **16**

Dairy Produce Quota Regulations 1997 (SI 1997/No 733) . . . . . **[2001]** 3 **101,** *et seq*

Estate Agents (Provision of Information) Regulations 1991
- (SI 1991/No 859) . . . . . **[2001]** 1 **21**
- r5 . . . . . **[1998]** 1 **7, 8**

Estate Agents (Undesirable Practices)(No 2) Order 1991 (SI 1991/No 1032)
- Art 2(c) . . . . . **[2001]** 1 **21, 25, 27**
- Sch 3
- para 1(a) . . . . . **[2001]** 1 **25, 26, 27**
- para 2 . . . . . **[2001]** 1 **25, 26, 27**

Insolvency Rules 1986
- r5.17 . . . . . **[1996]** 2 **35**
- r6.25 . . . . . **[1999]** 3 **28**
- r11.3 . . . . . **[1999]** 1 **2**
- r11.13 . . . . . **[1997]** 2 **37, 40**

Landlord and Tenant Act 1954 Part II (Notices) Regulations 1983 (SI 1983/No 133)
- regs 2 and 3 . . . . . **[1998]** 1 **66**

Sch 2
- form 1 . . . . . **[1997]** 1 **88, 89**
- form 13 . . . . . **[1997]** 1 **88, 89**

Land Registration Rules 1925
- r40 . . . . . **[1997]** 1 **154** *et seq*
- r77A . . . . . **[2000]** 2 **21**
- r278 . . . . . **[1999]** 2 **90, 92, [2000]** 2 **18**

Lands Tribunal Rules 1949
- R50(4) . . . . . **[2003]** 2 **2**

Lands Tribunal Rules 1975
- r49 . . . . . **[1995]** 1 **179**

Lands Tribunal Rules 1996 (SI 1996/No 1022)
- r2 . . . . . **[2001]** 2 **99**
- r6(1)(a) . . . . . **[1999]** 3 **230, 232**
- r13 . . . . . **[2000]** 3 **88**
- r14 . . . . . **[2000]** 3 **88**
- r32 . . . . . **[1998]** 2 **219**
- r36 . . . . . **[2000]** 3 **88**
- r38 . . . . . **[2000]** 3 **88**
- r52 . . . . . **[2001]** 2 **86, 99**
- r52(1) . . . . . **[2003]** 1 **10**

Leasehold Reform (Collective Enfranchisement and Lease Renewal) Regulations 1993 (SI 1993/No 2407)
- reg 2 . . . . . **[2003]** 3 **42**
- Sch 1
- para 2 . . . . . **[2003]** 3 **42, 44**
- para 3 . . . . . **[2003]** 3 **42**

Leasehold Reform (Notices) Regulations 1997
- form 1 . . . . . **[2002]** 1 **58** *et seq*

Non-Domestic Rating (Transitional Period) Regulations 1990 (SI 1990/No 608)
- reg 2(3)(a) . . . . . **[2000]** 1 **97, 98, 99, 100**

Notice to Quit etc (Prescribed Information) Regulations 1988
- reg 2 . . . . . **[1997]** 2 **10**
- Sch . . . . . **[1997]** 2 **10** *et seq*

Public Path Orders and Extinguishment of Public Rights of Way Orders Regulations 1983 (SI 1983/No 23)
- reg 16 . . . . . **[2000]** 2 **84, 85, 86**

Rent Acts (Maximum Fair Rent) Order 1999
- (SI 1999/No 6) . . . . . **[2000]** 1 **167** *et seq*, **[2001]** 1 **129,** *et seq*

Art 2(7) . . . . [2001] 3 **30**
Rent Assessment Committees (England and Wales)(Rent Tribunal) Regulations 1980 (SI 1980/No 1700)
para 3 . . . . [2003] 1 **59**
Rules of the Supreme Court
Ord 5 r4(1) . . . .[1996] 2 **40** *et seq*
Ord 18 r8(2) . . . . [1999] 3 **89**
Ord 18 r12(4) . . . . [1999] 3 **89**
Ord 19 r5(3) . . . . [1998] 1 **51**
Ord 20 r5 . . . .[1999] 3 **84** *et seq*
Ord 40 . . . .[1995] 2 **134**, [1996] 2 **99**
Ord 40 rl(3) . . . . [1995] 2 **135**
Ord 40 r10 . . . . [1995] 2 **135**
Ord 58 r4(b) . . . . [1995] 2 **107**
Ord 59 r5 . . . .[1997] 1 **233** *et seq*
Ord 61 . . . .[1999] 2 **210, 211, 212**
Ord 113 . . . . [1997] 2 **65**, [2000] 2 **59** *et seq*
Ord 113 rl . . . .[1999] 1 **147** *et seq*
Statutory Nuisances (Appeals) Regulations 1995 . . . . [1999] 3 **13**
Town and Country Planning (Use Classes) Order 1963
Class III . . . . [2000] 1 **45**, [2002] 1 **16** *et seq*
Class IV . . . . [2000] 1 **45**, [2002] 1 **16** *et seq*
Class X . . . . [2000] 1 **45**, [2002] 1 **16** *et seq*
Town and Country Planning (General) Regulations 1992 SI 1992/No 1492
reg 12 . . . . [2004] 2 **89, 90**
Valuation for Rating (Plant and Machinery) Regulations 1989
reg (2)(a) . . . . [1996] 1 **204**
Class 5 . . . . [1996] 1 **204**
Class 5(c) . . . . [1996] 1 **204**
Vehicular Access across Common and Other Land (England) Regulations 2002 (SI 2002 No 1711)
reg 11 . . . . [2003] 1 **18**
reg 11(2) . . . . [2003] 3 **8**

TABLE OF EUROPEAN LEGISLATION

Brussels Convention
Art 16.1 .... **[2000]** 2 **30**
Convention for the Protection of Human Rights and Fundamental Freedoms
Art 1 .... **[2000]** 2 **149, 152, 153, [2004]** 1 **50, 54, 56, 58, 61, [2004]** 3 **69, 70, 82**
Art 6 ... **[2000]** 2 **149, 152, 153,** [2003] 3 **116, 122, [2003]** 3 **112, 116, [2004]** 1 **1, 3, [2004]** 1 **50, 51, 54, [2004]** 1 **56, 57, 60, [2004]** 2 **55, 58, 61, 144,**
Art 6(1) ..... **[1996]** 2 **123** *et seq*, **[2001]** 1 **33,** *et seq*, **[2003]** 1 **58, 59, [2003]** 3 **116, [2004]** 1 **50, 56, [2004]** 1 **57, [2004]** 2 **55, 60, 61**
Art 8 ...... **[2003]** 2 **121, 128, [2003]** 3 **109** et seq, **[2004]** 1 **56** *et seq*, **[2004]** 2 **9, 13, 132, 133, 134, 135, 147, 148, [2004]** 3 **109** *et seq*
Art 8(1) .... **[2003]** 2 **127, [2003]** 3 **109, 110, 111, 113, 114, 115, 116, 117, 118, 119, 120, 121**
Art 8(2) .... **[2003]** 2 **127, 128, [2003]** 3 **109** *et seq* **[2004]** 1 **57, 58, 59, 61,**
Art 9 .... **[2000]** 2 **149, 152, 153**
Art 12 .... **[2004]** 2 **148**
Art 14 .... **[2000]** 2 **149, 152, 153, [2003]** 3 **116, 117, 122, 124, 126 [2004]** 1 **56, 57, [2004]** 2 **132, 133, 134, 135, 146, 147**
Council Regulation 3950/92
Art 7 .... **[1997]** 2 **16** *et seq*
EEC Council Regulation 857/84
Art 3a, para 1 .... **[1996]** 1 **16**
Art 7(1) .... **[1996]** 2 **7**
EEC Council Regulation 1962/17 .... **[1999]** 1 **51, 55**
Art 7(1) .... **[1999]** 1 **56, 57**
EEC (Commission) Regulations 1984/83 .... **[1999]** 2 **146, 147, 161**
Art 6 .... **[1999]** 1 **46, 47, 48**
EEC Commission Regulation 3567/92
Art 13 .... **[1996]** 2 **1** *et seq*
Art 15 .... **[1996]** 2 **1** *et seq*, **[2003]** 3 **70, 82, 111, 116, 119, 124**
EEC Regulation 3886/92
Art 39 .... **[1996]** 2 **1** *et seq*
Art 55 .... **[1996]** 2 **1** *et seq*, **[2003]** 3 **117**
First Protocol to the European Convention on Human Rights
Art 1 .... **[1996]** 2 **123, [2003]** 1 **72, [2003]** 2 **121, 127, 128, [2004]** 1 **11, 13, 14, [2003]** 3 **111, 116, 117, 119, 123, 124, [2004]** 2 **9, 13, [2004]** 2 **58, 59, 60, 93, 94,**
Eleventh Protocol to the European Convention on Human Rights
Art 5
para 2 .... **[2004]** 1 **11**
Treaty of Rome
Art 85 .... **[1999]** 1 **44** *et seq*, 2 **147** *et seq*
Art 85(1) .... **[1998]** 1 **53** *et seq*, **[1999]** 1 **51** *et seq*, 2 **147** *et seq*
Art 85(2) .... **[1999]** 1 **54** *et seq*, 2 **147** *et seq*
Art 85(3) .... **[1999]** 2 **146** *et seq*
Art 177 .... **[1996]** 2 **1**
Universal Declaration of Human Rights ....
Art 12 .... **[2003]** 3 **123**
Art 17(2) .... **[2003]** 3 **123**

TABLE OF HONG KONG LEGISLATION

Crown Lands Resumption Ordinance 1900
s3 .... **[1995]** 1 **20**
s10(2)(d) .... **[1995]** 1 **20** *et seq*

TABLE OF FOREIGN LEGISLATION

Compulsory Purchase of Land (Procedure)(Jersey) Law 1961
Art 4(3) .... **[1998]** 1 **137**
Art 9(1) .... **[1998]** 1 **137, 138**
Art 14(2) .... **[1998]** 1 **137, 138**
Constitution of Belize **[2004]** 2 **145**

TABLE OF CIRCULARS

Crichel Down Rules (October 1992)
r9 .... **[2000]** 3 **11**
r14(2) .... **[2000]** 3 **63**
r14(7) .... **[2000]** 3 **11**
DOE Circular 36/78 (Trees and Forestry) .... **[1998]** 2 **121**